Springer Theses

Recognizing Outstanding Ph.D. Research

Aims and Scope

The series "Springer Theses" brings together a selection of the very best Ph.D. theses from around the world and across the physical sciences. Nominated and endorsed by two recognized specialists, each published volume has been selected for its scientific excellence and the high impact of its contents for the pertinent field of research. For greater accessibility to non-specialists, the published versions include an extended introduction, as well as a foreword by the student's supervisor explaining the special relevance of the work for the field. As a whole, the series will provide a valuable resource both for newcomers to the research fields described, and for other scientists seeking detailed background information on special questions. Finally, it provides an accredited documentation of the valuable contributions made by today's younger generation of scientists.

Theses are accepted into the series by invited nomination only and must fulfill all of the following criteria

- They must be written in good English.
- The topic should fall within the confines of Chemistry, Physics, Earth Sciences, Engineering and related interdisciplinary fields such as Materials, Nanoscience, Chemical Engineering, Complex Systems and Biophysics.
- The work reported in the thesis must represent a significant scientific advance.
- If the thesis includes previously published material, permission to reproduce this must be gained from the respective copyright holder.
- They must have been examined and passed during the 12 months prior to nomination.
- Each thesis should include a foreword by the supervisor outlining the significance of its content.
- The theses should have a clearly defined structure including an introduction accessible to scientists not expert in that particular field.

More information about this series at http://www.springer.com/series/8790

Martin J.A. Schütz

Quantum Dots for Quantum Information Processing: Controlling and Exploiting the Quantum Dot Environment

Doctoral Thesis accepted by
Ludwig-Maximilian University, München, Germany

 Springer

Author
Dr. Martin J.A. Schütz
Max-Planck-Institut für Quantenoptik
Garching
Germany

Supervisor
Prof. Ignacio Cirac
Max-Planck-Institut für Quantenoptik
Garching
Germany

ISSN 2190-5053   ISSN 2190-5061   (electronic)
Springer Theses
ISBN 978-3-319-83972-1   ISBN 978-3-319-48559-1   (eBook)
DOI 10.1007/978-3-319-48559-1

This Springer imprint is published by Springer Nature
The registered company is Springer International Publishing AG
The registered company address is: Gewerbestrasse 11, 6330 Cham, Switzerland

Supervisor's Foreword

Since the discovery of quantum physics more than hundred years ago, our vision of nature has experienced a tremendous change. In the microscopic world, where most of its exotic phenomena manifest themselves, the properties of particles become fuzzy, and the observer is endorsed with an active role in how they change. Thanks to the technological progress experienced during the last 30 years, these phenomena, which used to attract more the attention of philosophers than physicists, may also lead to new paradigms in the way we process and transmit information. In fact, by now we know that if we are able to control and manipulate microscopic systems, so that we can exploit the laws of quantum physics, we will be able to build quantum computers and communication devices, leading to a revolution in our information society.

Most of the pioneering experiments in this front of research were done in the realm of quantum optics. That is the theory that studies fundamental phenomena in the interaction of light with matter, typically atoms. In fact, atoms can reliably store qubits, the unit where (quantum) information is stored, whereas photons can be used to transmit those qubits from one place to another. During the last few years, a large amount of knowledge has been gained in how to manipulate atoms, transpass the information to photons, transmit them through optical fibers, and put it back into different atoms. The field of quantum optics is, in that sense, well established, and a great deal of control has been achieved during the last decade.

In recent years, solid-state systems have appeared as alternative to build quantum computation and communication devices. In particular, electrons in quantum dots embedded in semiconducting materials can store qubits, which can be manipulated using external fields. Furthermore, quantum information can be stored in the nuclear spins of the atoms at the quantum dot. These situations are very reminiscent of what happens in quantum optical systems, if one identifies the electrons or nuclear spins with atoms. Thus, one may expect to observe similar phenomena with these solid-state systems as with atoms, or even use the tools and ideas developed in the field of quantum optics in order to learn how to manipulate or transport the information stored in quantum dots. This is precisely what has been achieved in this thesis.

In the first part of Martin's thesis, the interaction of the nuclear spins of a quantum dot with the electrons that go through it has been exploited in order to predict a very intriguing behavior. This system is very reminiscent of a set of atoms (whose role is played by the nuclear spins) interacting with propagating photons (here the electrons). In the quantum optics setup, one expects to observe phenomena associated with superradiance. Similarly, due to the collective nature of the coupling between the nuclear spins and the central electron spin in the quantum dot, the nuclear system may experience a strong correlation buildup, resulting in a sudden intensity burst in the electron current emitted from the quantum dot, giving rise to a new phenomenon in the solid state, electronic superradiance.

In the second part of Martin's thesis, the possibility of entangling two electrons in neighboring quantum dots has been analyzed. In particular, by sending electric current through both dots, one can obtain such a goal in the steady state. Once again, this situation is very similar to that of two atomic systems interchanging photons propagating from one to the other. In that context, it was predicted and experimentally observed that for some value of the external parameters, atomic entangled states could be produced. Based on that analogy, in this thesis an efficient way for achieving the same task with nuclear spins adjacent to quantum dots was put forward.

In the third part, Martin continues to reveal and systematically explore previously undiscovered connection points between the fields of quantum optics and modern solid-state semiconductor spin systems. In striking analogy to cavity QED, he proposes and analyzes phonon modes associated with surface acoustic waves (SAWs) in piezo-active materials as a universal mediator for long-range coupling between remote qubits. The proposal involves qubits interacting with a localized SAW phonon mode, defined by a high-quality resonator, which in turn can be coupled weakly to a SAW waveguide serving as a quantum bus (a concept well known from AMO physics with the role of photons replaced by SAW phonons). It is shown that the piezo-electric coupling between qubit and SAW phonon mode should enable a controlled mapping of the qubit state onto a coherent phonon superposition, which can then be converted to an itinerant SAW phonon in a waveguide, opening up the possibility to implement on-chip many quantum communication protocols well known in the context of optical quantum networks. The proposed combination of techniques and concepts known from quantum optics and quantum information, in conjunction with the technological expertise for SAW devices, should lead to further, rapid theoretical and experimental progress, opening up the avenue toward the widely anticipated field of quantum acoustics.

I do consider this thesis scientifically excellent and believe that the results motivate both (i) further research into what extent the proposed analogies between quantum optical and solid-state systems can be extended and generalized and (ii) the search for novel, emergent relations between these two subfields of modern quantum physics.

Garching, Germany   Prof. Ignacio Cirac
July 2016

Abstract

Electron spins confined in quantum dots (QDs) are among the leading contenders for implementing quantum information processing. In this thesis, we address two of the most significant technological challenges toward developing a scalable quantum information processor based on spins in quantum dots: (i) decoherence of the electronic spin qubit due to the surrounding nuclear spin bath and (ii) long-range spin–spin coupling between remote qubits. To this end, we develop novel strategies that turn the unavoidable coupling to the solid-state environment (in particular, nuclear spins and phonons) into a valuable asset rather than a liability.

In the first part of this thesis, we investigate electron transport through single and double QDs, with the aim of harnessing the (dissipative) coupling to the electronic degrees of freedom for the creation of coherence in both the transient- and steady-state behavior of the ambient nuclear spins. First, we theoretically show that intriguing features of coherent many-body physics can be observed in electron transport through a single QD. To this end, we first develop a master-equation-based formalism for electron transport in the Coulomb-blockade regime assisted by hyperfine (HF) interaction with the nuclear spin ensemble in the QD. This general tool is then used to study the leakage current through a single QD in a transport setting. When starting from an initially uncorrelated, highly polarized state, the nuclear system experiences a strong correlation buildup, due to the collective nature of the coupling to the central electron spin. We demonstrate that this results in a sudden intensity burst in the electronic tunneling current emitted from the QD system, which exceeds the maximal current of a corresponding classical system by several orders of magnitude. This gives rise to the new paradigm of electronic superradiance. Second, building upon the insight that the nuclear spin dynamics are governed by collective interactions giving rise to coherent effects such as superradiance, we propose a scheme for the deterministic generation of steady-state entanglement between the two nuclear spin ensembles in an electrically defined double quantum dot. Because of quantum interference in the collective coupling to the electronic degrees of freedom, the nuclear system is actively driven into a two-mode squeezed-like target state. The entanglement buildup is accompanied by a self-polarization of the nuclear spins toward large Overhauser field gradients. Moreover, the feedback between the electronic and nuclear dynamics is shown to lead to intriguing effects such as multistability and criticality in the steady-state solutions.

In the second part of this thesis, our focus turns toward the realization of long-range spin–spin coupling between remote qubits. We propose a universal, on-chip quantum transducer based on surface acoustic waves in piezo-active materials. Because of the intrinsic piezoelectric (and/or magnetostrictive) properties of the material, our approach provides a universal platform capable of coherently linking a broad array of qubits, including quantum dots, trapped ions, nitrogen-vacancy centers, or superconducting qubits. The quantized modes of surface acoustic waves lie in the gigahertz range, can be strongly confined close to the surface in phononic cavities and guided in acoustic waveguides. We show that this type of surface acoustic excitations can be utilized efficiently as a quantum bus, serving as an on-chip, mechanical cavity-QED equivalent of microwave photons and enabling long-range coupling of a wide range of qubits.

In summary, this thesis provides contributions toward developing a scalable quantum information processor based on spins in quantum dots in two different aspects. The first part is dedicated to a deeper understanding of the nuclear spin dynamics in quantum dots. In the second part, we put forward a novel sound-based strategy to realize long-range spin–spin coupling between remote qubits. This completes a broad picture of spin-based quantum information processing which integrates different perspectives, ranging from the single-qubit level to a broader quantum network level.

Publications Related to this Thesis

1. *Superradiance-like electron Transport through a Quantum Dot.* M. J. A. Schuetz, E. M. Kessler, J. I. Cirac, and G. Giedke, Phys. Rev. B **86**, 085322 (2012).

2. *Steady-State Entanglement in the Nuclear Spin Dynamics of a Double Quantum Dot.* M. J. A. Schuetz, E. M. Kessler, L. M. K. Vandersypen, J. I. Cirac, and G. Giedke, Phys. Rev. Lett. **111**, 246802 (2013).

3. *Nuclear Spin Dynamics in Double Quantum Dots: Multistability, Dynamical Polarization, Criticality, and Entanglement.* M. J. A. Schuetz, E. M. Kessler, L. M. K. Vandersypen, J. I. Cirac, and G. Giedke, Phys. Rev. B **89**, 195310 (2014).

4. *Universal Quantum Transducers based on Surface Acoustic Waves.* M. J. A. Schuetz, E. M. Kessler, G. Giedke, L. M. K. Vandersypen, M. D. Lukin, and J. I. Cirac, Phys. Rev. X **5**, 031031 (2015).

Acknowledgements

For the last four years, I did not have Pep Guardiola as my soccer coach. Nor did I have Boris Becker as my tennis trainer, or Butch Harmon as my golf teacher. But what I had was Ignacio Cirac as my Ph.D. supervisor. It has been an extremely exciting experience to work with him and I feel deeply indebted for all the guidance, support, patience, trust, and encouragement I have received. In particular, the latter two I cannot value high enough as a young researcher. Ignacio combines creativity, knowledge, work ethic, kindness, and humor in a very unique fashion, making him the best advisor and teacher I could have ever dreamed of. Moreover, as the head of a group with roughly 30 people, he manages to create an intellectually sparkling, friendly, open-door environment. It has been a pleasure and honor to be a part of it.

This work would not have been possible without the invaluable mentoring and support of my co-advisor Géza Giedke. Even though he is way too modest to say so himself, he has an incredibly wide knowledge spanning the most diverse fields of physics. On a personal level, I feel deeply indebted for all his patience, time, and continuous guidance. His contributions to our group (and the Max Planck Institute for quantum optics as a whole) go way beyond science, let alone our joint projects: He has been the center of social activities, organizational issues, etc. for years in our group. I wish him all the best for his new stage of life in the beautiful Basque country. I will always be grateful to him.

I would also like to thank Jan von Delft, who kindly agreed to be a referee of my Ph.D. thesis.

Further, I am indebted to Lieven Vandersypen and Misha Lukin for all their generous guidance, support, and hospitality in the course of several fruitful collaborations. I was able to learn a lot from their outstanding expertise, and my research visits to Delft and Boston with many stimulating discussions were not only a great experience for me, but also a pivotal element for this thesis.

Special thanks go to all people at MPQ for creating such a friendly and inspirational atmosphere and making MPQ such a great place to work. In particular, I would like to extend my thanks to the following (most likely incomplete) list of people: Eric Kessler has not only been a personal physics mentor with great pedagogical skills, but has become also a close friend. Some of our glorious victories together extend from the IPP pitch over Theresienwiese to our home turf, Klenze17, and will always be remembered; in case memories still exist. Gemma and Oriol have become some of my closest friends throughout the last years. Oriol's passion for science, sports (and actually many other things…) is unrivaled and will keep influencing me not only in physics questions. Officemates are an important part of a Ph.D. and I consider myself very fortunate to have shared my space with Michael Lubasch, Andras Molnar, and Alex Gonzalez-Tudela. Moreover, I also acknowledge Fernando Pastawski, Heike Schwager, Leonardo Mazza, Matteo Rizzi, Michelle Burello, Anika Pflanzer, Johannes Kofler, Lucas Clemente, Manuel Endres, Maarten van den Nest, Hong-Hao Tu, Roman Orus, and Mari-Carmen Banuls, and all my other coworkers and friends at the MPQ theory division for the always generous support in any physical questions, and the many fun hours together. Special thanks also go to Veronika Lechner and Andrea Kluth for invaluable support with any administrative matters.

Most importantly, I would like to thank my family, especially my mum and my sister for their unconditional support. I know I can always rely on them, and I am deeply thankful for that.

Finally, I would like to thank Steffi for all her support, ingenious humor, and love.

I gratefully acknowledge funding and support by the Max-Planck-Society and Sonderforschungsbereich 631 of the DFG.

Contents

Chapter 1
Introduction

Quantum mechanics describes our world at the microscopic level. Once seen as inaccessible to direct observation and irrelevant to any direct practical application, quantum physics has matured from a merely fundamental discipline to a vibrant research area where nowadays genuine quantum effects can be observed routinely in the lab. Over the past decades the quantum mechanical framework has been used extensively to explain and predict quantum phenomena of increasingly subtlety and complexity [1]. At the same time, it has also become the backbone for novel, revolutionary technological applications. Some early examples of quantum mechanical technology include: (i) the transistor [2], paving the way to modern electronics, (ii) the laser [3, 4], enabling diverse applications such as laser printers or laser surgeries, and (iii) superconducting magnets [5, 6], routinely used for example in MRI machines in hospitals.

This truly impressive development of ground-breaking technological applications operating on the laws of quantum physics is unlikely to have come to its end. On the contrary, the emerging field of quantum information science bears the potential to revolutionize the fields of communication and computation [7]. Borne out of a successful union of quantum mechanics and information science, this relatively young, interdisciplinary research field basically addresses one fundamental question: Can we gain some advantage by storing, transmitting and processing information encoded in systems that exhibit unique quantum properties such as superposition and entanglement [8]? Today it is understood that we can answer this question in the affirmative. This is the result of a broad range of fundamental discoveries that have been made throughout the past three decades: In 1981, Richard Feynman for the first time hypothesized that a device properly harnessing the laws of quantum physics could potentially outperform classical devices. In his seminal talk at the First Conference on the Physics of Computation he pointed out the manifest impossibility of efficiently simulating a generic quantum system on a classical device. The fundamental reason is the superposition principle together with the exponential growth of the dimension of the Hilbert space with the system size, which is believed to rule

© Springer International Publishing AG 2017
M.J.A. Schütz, *Quantum Dots for Quantum Information Processing: Controlling and Exploiting the Quantum Dot Environment*, Springer Theses,
DOI 10.1007/978-3-319-48559-1_1

out any efficient classical description: Already the description of a quantum system composed of only roughly 50 spins is practically impossible, and the description of 300 would require the simulation of more classical dimensions than there are particles in the universe. Yet, in the same talk Feynman showed how to circumvent this problem [9]. One needs to simulate the object of interest with a system of the same nature; in this case it implies the use of a (well-controlled) quantum device to simulate another quantum system. The concept of a quantum simulator was born. Shortly after, this idea was generalized in a seminal work by Deutsch [10], where he introduced the idea of a universal quantum computer, a device making direct use of quantum-mechanical phenomena, such as superposition and entanglement, to perform operations on data. These considerations, however, were largely considered a conceptual curiosity until the mid-1990s, when Peter Shor devised a quantum algorithm by which a quantum computer could factorize large numbers in a time exponentially faster than any known classical algorithms [11]. If implemented, Shor's algorithm would have profound implications in cryptography as it allows to crack today's standard cryptographic codes such as RSA [12] for which the difficulty of factoring constitutes the essential working principle. This breakthrough result was the starting signal for a worldwide sustained endeavor searching for feasible physical realizations of a quantum computing device. While similar technologies allowing (for example) for unconditional secure cryptography or ultra-sensitive measurements in the field of quantum metrology have become mature fields by now [13, 14], close to reaching the commercial level, the development of actual quantum computers is still in a very early stage of its development cycle.

The complexity of building a fully-fledged quantum computer is a fantastic challenge [15, 16]. The quantum computer has to be very well protected from its environment in order to suppress unwanted information leakage, a process called *decoherence*, which tends to corrupt and wash out the characteristic quantum properties that give rise to the power of quantum computation. At the same time, the building blocks of the device (quantum bits, in short qubits) typically need to be actively manipulated and read out at the end of the computation. Therefore, one arrives at the fundamental ambiguity that on the one hand isolation from the surrounding and on the other hand strong coupling to some classical control interface is required. In practice, no physical system is free of decoherence, but small amounts of information leakage may be tolerated thanks to various techniques subsumed under the name of quantum error correction [17–19]. Quantum error correction detects and corrects weak interactions with the environment or small quantum gate errors through redundant encoding of qubits. Therefore, experiments need not be perfect, and quantum computation is still feasible in the presence of noise, provided that the decoherence rate per qubit and qubit-operation is below a certain threshold. While more and more sophisticated quantum error correction schemes have been developed over the years [20–23], pushing this threshold to more practical values, decoherence is arguably still the strongest adversary to quantum information processing. This challenge together with some other general prerequisites for quantum computers has been identified by David DiVincenzo early on as follows [1] [24]: The target system should be controllable, i.e., it can be initialized, manipulated, and read out to achieve a computation. Moreover, it must be correctable (inevitable errors can be detected and compensated). Lastly, it should be scalable, meaning that a linear increase in effective system size should not require a corresponding exponential increase of required resources. Achieving these conditions for quantum computation has turned out to be extremely demanding. Nevertheless, as of today a whole plethora of physical systems has been put forward as possible candidate platforms for the realization of quantum information processing (QIP), spanning a truly fascinating range of systems [25] from elementary (quasi-)two level systems such as hyperfine levels in ions [26] and electron spins [27, 28], to more complex macroscopic structures like superconducting devices [29]. While each approach has its own benefits and a full comparison of the relative merits goes beyond the scope of this Thesis, broadly speaking, one can identify two different categories: Some systems comprise *natural* candidates, such as single atoms or ions, for which the manipulation of quantum states has a relatively long and rich history [8, 15]. Others are based on *artificial*, engineered systems such as phosphorus donors in silicon [30], superconducting circuits [31], or—in the focus of this Thesis—the promising systems of semiconductor quantum dots [27, 32]. While the former has served as prototypical quantum information testbeds, with unprecedented control of individual quantum systems, the latter candidates hold the promise of an automatized, large-scale manufacturing, using the well-established fabrication techniques of existing semiconductor industry [15, 16].

This Thesis investigates QIP related problems in concrete, experimentally accessible physical settings, with a primary focus on single localized electron spins in semiconductor quantum dots coupled to their naturally occurring nuclear spin environment. Against the background outlined above, we address two of the most significant challenges towards developing a scalable, solid-state spin based quantum information processor [28, 33]: (i) nuclear spin induced decoherence of the electronic spin qubit, and (ii) long-range coupling between remote spin qubits. Methodologically, our approach heavily relies on ideas, systems and techniques well-known in the context of quantum optics (such as atomic ensembles, cavity QED and quantum optical master equations) which turn out to be very useful in the theoretical description of this particular solid-state system. Therefore, to set the stage, the remainder of this chapter is structured as follows: In Sect. 1.1 we first touch upon quantum optical models and ideas (such as collective effects and quantum data buses) that serve as common background and inspirational resource for the research reported in this Thesis. Knowledge and experience already acquired in the realm of quantum optics will be actively transferred to the domain of semiconductor quantum dots throughout this Thesis. Furthermore, we also shed more light on the ambiguous relation between quantum coherence and dissipation and report on novel strategies that turn the unavoidable coupling to the environment into a resource rather than a liability. This relatively new paradigm in the quantum information toolbox will help to reverse

---

[1]Relatively new, complementary strategies based on engineered dissipation will be introduced in detail in Sect. 1.1.

the standard perception of nuclei in quantum dots, stimulating original insights into the role of nuclear spin dynamics in quantum dots. In Sect. 1.2 we then present a detailed theoretical and experimental account of this particular physical system in the focus of this Thesis. Its importance and prospects in the field of quantum physics, as well as the experimental state of the art are discussed. With this background knowledge established, Sect. 1.3 shortly outlines the particular research projects reported in this Thesis as well as the interdependencies between them.

1.1 Quantum Optical Review

Quantum optical setups provide unique quantum systems where complete control on the single quantum level can be realized routinely in the lab, with an unprecedented isolation from any uncontrolled environment [1]. Guided by theoretical proposals as reviewed in Refs. [13, 14], extraordinary progress has been achieved with (for example) trapped ions [34], cold atoms in optical lattices [35–38], cavity QED [39] and atomic ensembles [40]. Below we provide a summary of some general theoretical aspects of implementing QIP with quantum optical systems, that is tailored to the research objectives of this Thesis. For detailed reviews on the experimental state-of-the-art in these systems, we refer to Refs. [8, 15, 41–43]. In particular, we distinguish between approaches that rely on a high degree of isolation of the quantum system from its environment and strategies that actively harness the inevitable interaction between system and environment.

Coherent Engineering

The research area of atomic physics and quantum optics is home to some of the most advanced candidates for QIP [1]. Prominent realizations include trapped ions [34] and atoms in cavities [39]. These kind of systems share the feature that long-lived internal atomic states, such as atomic hyperfine ground states or metastable states, serve as quantum memory to store quantum information. Single qubit rotations can be performed by coupling the internal atomic levels to laser light for an appropriate period of time; this requires that single atoms can be addressed individually by laser light [13, 14]. Various schemes have also been developed for the realization of two-qubit gates. Here, one particularly promising approach relies on the concept of a **quantum data bus**: By coupling the qubits to a collective auxiliary quantum mode, entanglement of the qubits can be achieved by swapping qubits to excitations of the collective mode [13, 14]. For trapped ions, a single normal mode of the collective crystal motion serves as a quantum bus, allowing quantum information to be transferred between remote qubits in the crystal via the exchange of phonons, which is controlled by directing laser beams to each of the ions [26]. Similarly, photons constitute another natural candidate as carrier of quantum information [44–47], since they are highly coherent and can mediate interactions over very large distances [48, 49]. In order to create a photon-based quantum bus, the most common strategy is based on cavity QED techniques [50], where an atom interacts with a single, confined cavity mode only [39]. In the so-called strong coupling regime [50], this allows for the transfer of quantum information between an atom and a single photon. Formally, there is a strong connection between trapped ions and cavity QED, where the role of collective phonon modes of trapped ions is essentially taken by photons in high-quality cavity modes [13, 14]. Therefore, on the one hand, schemes developed for the entanglement generation in ion traps can readily be translated to their cavity QED counterparts. On the other hand, cavity QED provides a natural interconnect between stationary atomic qubits and photons as carriers of qubits for quantum communication, since photons (once they leave the cavity) can be guided in optical fibers to transport quantum information between remote locations [39, 49, 51]. In this way, small quantum computers can be wired together to make larger ones, allowing (for example) to spatially separate sensitive quantum memories from specific measurement hardware [15, 46]. Today, first experimental steps towards such a quantum network include the successful mapping of quantum information contained in an atomic superposition state onto the photon state (and vice versa) [52], quantum state transfer and entanglement generation between two quantum bits using both optical and microwave photons as intermediary [53–58].

The QIP schemes discussed above involve laser manipulation of *single* trapped particles. However, many QIP protocols can also be implemented by laser manipulation of atomic ensembles that contain *many* identical neutral atoms [13, 14]. Experimental candidate systems comprise both laser-cooled atoms [59–62] or room-temperature gas [63–66]. Very long coherence times of the relevant internal atomic levels have been observed in both kinds of systems [59–66]. The basic motivation of using a large number of atoms rather than single particles is three-folds [13, 14]: First, laser manipulation of atomic ensembles without the necessity of individual addressing of single particles is much easier to achieve experimentally. Second, quantum information encoded in atomic ensembles is more robust to some typical noise sources such as particle losses. Finally, and arguably most importantly, **collective effects** stemming from many-body coherence between the individual constituents allow for a strongly enhanced signal-to-noise ratio, that is the ratio between the rates of coherent information processing and of decoherence processes. For example, to achieve a high signal-to-noise ratio in cavity QED, commonly referred to as strong-coupling regime, one needs to place the atoms in high-finesse cavities [50]. By encoding quantum information into some collective excitations of the ensembles, however, QIP can be made possible with much more simplified experimental systems [8], such as atomic ensembles in weak-coupling cavities or even free space. This constitutes a critical improvement for the implementation of some quantum information applications [13, 14], such as quantum cryptography [67], quantum teleportation [68], and quantum memories [62, 64, 69–71].

Dissipative Engineering

The strategies outlined above rely on a high degree of isolation of the quantum system from its environment, in order to suppress decoherence. In this context, dissipative coupling to the environment appears exclusively as a malicious mechanism which tends to corrupt the genuine quantum properties of a quantum state, such as entanglement. As a consequence, for a long time the prime objective in the QIP community has been to develop more and more sophisticated methods to shield quantum systems from any environmental influence. However, with the advent of novel approaches such as engineering of dissipation [72], a new trend of actually harnessing dissipation has emerged in recent years. Here, the key idea is to properly engineer the continuous interaction of the system with its environment in order to create and utilize quantum coherences. In this way, dissipation—previously often viewed as a liability from a QIP perspective—can turn into a highly valuable asset and become the driving force behind the emergence of coherent quantum phenomena. For instance, theoretical and experimental works put forward their use for quantum state preparation [73–83], quantum computation [84], quantum memories and error correction [85, 86], as well as for open-system quantum simulators [87, 88]. The idea of actively using dissipation rather than relying on coherent evolution only extends the traditional DiVincenzo criteria [24] (which list pure, almost perfectly isolated qubits among the requirements for quantum computing; see the discussion above) to settings in which no unitary gates are available; also, it comes with potentially significant practical advantages, as dissipative methods are unaffected by timing and preparation errors and inherently robust against weak random perturbations [89, 90], allowing, in principle, to stabilize entanglement for arbitrary times. For example, in the case of quantum state engineering, the system is actively pumped into a (for instance highly entangled) steady state [91]. Recently, these concepts have been put into practice experimentally in different QIP architectures, namely atomic ensembles [92], trapped ions [87, 93] and superconducting qubits [94].

1.2   Semiconductor Quantum Dots

As illustrated in the previous Section, nature's own atoms and ions have pioneered the experimental investigation of quantum information processing, with many ground-breaking experiments demonstrating unprecedented control of individual quantum systems [95, 96]. This success story has fueled an impressive race to build "artificial atoms", tunable solid-state devices which behave much like their natural counterparts. Today, many approaches exist, employing different materials, temperatures, device structures etc., with all of them holding the promise of miniaturization and large-scale integration based on the well-established fabrication processes of our modern-day semiconductor industry [16, 28]. Furthermore, these on-chip devices typically operate at ultra-high clock speeds, much faster than their natural cousins, without the need for sophisticated cooling and trapping schemes [15, 97]. The spins of individual electrons and nuclei, in particular, have turned out to offer a promising combination of controllability and environmental isolation [16, 28]. One specific system that falls into this category are semiconductor quantum dots (QDs). In what follows, we first give an introduction to quantum dots and then briefly review recent progress towards full control of single and coupled spins in QDs [33], before turning the focus on the (ambiguous) role of nuclear spins in this context [98]. Finally, we outline the major challenges and goals ahead.

Quantum Dots

Quantum dots are nanoscale semiconductor heterostructures, which are able to confine charge carriers (conduction-band electrons or valence-band holes) in all three spatial dimensions, ultimately forming zero-dimensional boxes [16]. Due to strong confinement to length scales small compared to the particle's wavelength the energy spectrum is quantized. Many well-known atomic properties, such as shell structure and optical selection rules, have analogues in QDs; hence the informal nickname "artificial atom" [33]. Since a QD is such a general kind of system, there exist QDs of many different sizes and materials; compare for example Refs. [33, 97] and references therein. One of the most prominent examples are electrostatically defined quantum dots (EDQDs), also referred to as lateral quantum dots.[2] Lateral GaAs[3] quantum dots are fabricated from heterostructures of GaAs and AlGaAs grown by molecular beam epitaxy [33, 98]. With this method, semiconductor structures can be grown layer by layer, allowing to stack materials with different band gaps. Free charge carriers can be introduced into the system via doping, optical excitation or current injection. They accumulate at the GaAs/AlGaAs interface, typically ~50–100 nm below the surface [33], leading to the formation of a two-dimensional electron or hole gas (2DEG, 2DHG), respectively, a thin ~10 nm sheet of charge carriers that can move only along the interface. Surface gates on top of the heterostructure—so-called Schottky contacts—can then be used to locally deplete small regions of the 2DEG (typical dimensions of the resulting QD are ~40 nm in the plane and ~10 nm in the growth direction [98]) with an electric field, which facilitates control over both the electron number in the QDs and the tunnel-coupling between the QD and the adjacent reservoirs via by purely electrostatic means.

The electronic properties of QDs are governed by two basic effects [33]: First, as mentioned above, strong confinement in all three directions leads to a discrete energy spectrum. Second, the Coulomb repulsion between electrons on the dot leads to an energy penalty for adding an extra electron to the dot. Due to this charging energy tunneling of electrons to or from the reservoirs can be strongly suppressed at sufficiently low temperatures; this effect is referred to as Coulomb blockade (CB). To access the quantum regime, GaAs gated QDs have to be cooled down to temperatures well below 1 K; using dilution refrigerators, nowadays base temperatures of ~20 mK can be realized routinely in the lab [33]. In the CB-regime the charge on the QD is conserved, since no electrons can tunnel onto or off the dot. By applying gate voltages, the number of trapped electrons can then be controlled very well starting with 0, 1, 2 etc. Similarly, to study transport through the QD structure, one tunes the gate

---

[2] Yet another promising and actively pursued candidate for QIP are self-assembled quantum dots (SAQDs) [32, 97, 98]. They form spontaneously at random locations during the epitaxial growth process due to a lattice mismatch between the dot and substrate materials. Since they typically feature very large single-particle level spacing, they can be operated at elevated temperatures of ~4 K. In contrast to EDQDs, SAQDs allow for coherent optical manipulation, since they can trap both electrons and holes.

[3] Taking advantage of the well-developed growth of ultrapure GaAs/AlGaAs heterostructures, Gallium arsenide (GaAs) has been the working horse material for many years for these devices [33]. Alternative potential material choices such as silicon are discussed briefly below.

voltages to the boundary between two charge configurations such that electrons can be exchanged with the surrounding 2DEG via tunneling [99, 100]. Electron transport through the QD (or a series of several dots) can be used to probe the QD; alternatively, the charge state of the QD can be detected directly using a nearby electrometer, such as a quantum point contact (QPC).

Just like two (or more) atoms can form molecules, multiple near-by QDs with large tunnel-coupling (i.e., strong overlap of their electronic wavefunctions) can form "artificial molecules" [28]. Due to this covalent bonding, single-dot orbitals hybridize into molecular-like orbitals spanning multiple QDs. The best-studied QD molecule is the double quantum dot (DQD), which has served as the workhorse for many ground-breaking experiments with QDs [33]. It consists of just two EDQDs which are coupled by coherent tunneling. When operated in the (1, 1) charging region (that is, with a single electron on each dot) tunnel-coupling leads to an effective exchange interaction lifting the degeneracy in the four-dimensional subspace by splitting off the singlet from the triplet states [33]. As discussed in more detail below, this mechanism can be used to implement ultra-fast quantum gates in this system [101].

In what follows, we focus on single and coupled QDs in the few-electron regime, that is QDs containing only one or two electrons. These systems are of particular importance as they have been proposed as the basic building blocks of electron spin-based quantum information architectures to be discussed below.

QIP with Quantum Dots
One of the earliest theoretical proposals to turn quantum computing into reality dates back to the seminal work of Loss and DiVincenzo in 1998 [27]. They suggested to use single electron spins confined to EDQDs in order to encode quantum information. While single-qubit gates could be performed by locally controllable magnetic fields, two-qubit interactions were proposed to be realized by all-electrical control using the effective Heisenberg exchange interaction between neighbouring spins (the spins can be moved closer or further apart from each other simply by varying the tunnel barrier between the dots). Importantly, the proposal demonstrates that an array of EDQDs allows (in principle) for universal and scalable quantum computing, at ultrahigh clockspeed [97].

Within a few years, in a series of ground-breaking experiments, all major building blocks of the Loss-DiVincenzo proposal have been demonstrated in proof-of-principle experiments: initialization of the qubit, all-electrical single-shot read-out of spin states [102], coherent control of single spins [103], and two-qubit gates between two electron spins in a double dot system [104]. Within the past few years, the field of EDQD-based quantum computing has seen further tremendous progress. In the following we outline some of the most important theoretical and experimental achievements following Ref. [27] in more detail.

Qubit—In the original Loss-DiVincenzo proposal, the qubits are stored in the spin degree of freedom of single confined electrons: $|\uparrow\rangle \equiv |0\rangle$ and $|\downarrow\rangle \equiv |1\rangle$. This choice was made deliberately, since the spin (in contrast to the charge degree of freedom) is to first order insensitive to voltage fluctuations. In the presence of a magnetic field $\mathbf{B} = B\hat{z}$ the two spin states get split by the Zeeman-splitting ω_Z, according to

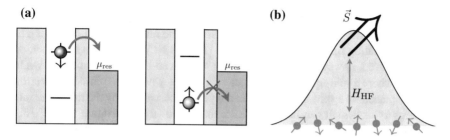

Fig. 1.1 a Spin-to-charge conversion: The energy difference between the spin states can be used to convert spin-state information into charge. In the configuration shown, an electron can tunnel from the dot to the reservoir only if it is in the spin-down state. Measurement of the charge on the dot yields the spin state. **b** An electron spin $\vec{S}$ confined in a GaAs EDQD interacts with $\sim 10^6$ nuclear spins (shown in *blue*) via the hyperfine interaction (H_{HF})

$\omega_Z = g\mu_B B$, with $g \approx -0.44$ the electron g-factor in GaAs [33] and μ_B the Bohr-magneton. Yet another prominent encoding scheme, typically referred to as singlet-triplet qubit, employs two-electron encoded qubits in DQD structures, in which logical qubits are encoded in a two-dimensional subspace of a four-dimensional two-electron spin system [105]. Within this approach, the qubit energy splitting is provided by the effective exchange coupling.

Single-shot read-out—Direct measurement of the electron spin is very challenging because of its tiny magnetic moment [33]. The standard approach to circumvent this problem very efficiently is based on spin-to-charge conversion [27]: Broadly speaking, the spin state is detected by correlating the spin states with different charge states and subsequently measuring the charge. The latter can be done with a nearby QPC, whose conductance changes due to the electric field provided by the electron charge on the EDQD [106, 107]. Several schemes have been proposed for spin-to-charge conversion [27, 30, 108–111]. Here, we shortly outline one particular method which was demonstrated first to work in a single-shot mode [102]: If the Zeeman splitting ω_Z exceeds well the thermal energy $k_B T$, with the Boltzmann constant k_B, the difference in energy between the two spin states can be used to make the tunneling rates of the electron from the dot strongly spin-dependent. For a schematic illustration, compare Fig. 1.1a. In this so-called energy-selective read-out scheme [33], for read-out the spin levels are pulsed to a position around the chemical potential of the electron reservoir μ_{res}, such that an electron with spin $\downarrow$ can tunnel off the dot, whereas an electron with spin $\uparrow$ is trapped. Therefore, if the charge measurement shows a change in occupation number, the state was spin $\downarrow$, and spin $\uparrow$ otherwise. This procedure allows for single-shot read-out, with demonstrated fidelities already exceeding 90 % [33]. We note that, with this setup, spin relaxation times T_1 of up to a second have been measured at magnetic fields of a few tesla [112]. For the singlet-triplet qubit in a DQD, adiabatic tuning of the gate voltages can be used to read-out information stored in the singlet-triplet basis [104, 113].

Initialization—At low temperatures and sufficiently high magnetic fields, the spin can be initialized into the pure state $|\uparrow\rangle$ in a brute-force approach simply by waiting long enough until energy relaxation will cause the spin on the dot to relax to the ground state $|\uparrow\rangle$. This procedure is known to be relatively slow ($\sim$10 ms), as the spin relaxation time T_1 can be very long [102, 112]. Very efficient initialization can be achieved, however, by placing the dot in the read-out configuration shown in Fig. 1.1a, where a spin-up electron will stay on the dot, whereas a spin-down electron will be replaced by a spin-up electron on a typical timescale set by the tunneling rate to the reservoir, which can be much faster than the energy relaxation time T_1 [33, 102, 114]. For initialization in the singlet-triplet qubit scheme, adiabatic tuning of the gate voltages can be used, similarly to read-out [104, 113].

Coherent control of single qubits—A broad range of techniques allows for coherent transitions between the Zeeman-split levels of a single electron; compare Ref. [33] and references therein. The most prominent one is electron spin resonance (ESR), whereby a rotating magnetic field $B_\perp$ is applied perpendicular to the static field B along $\hat{z}$, at a frequency $\nu = g\mu_B B/h$ matched to the spin-flip transition. ESR has been demonstrated first in a DQD device, tuned to the spin-blockade regime, where current through the device is blocked as soon as the DQD is occupied by two electrons with parallel spins (one in each dot) [103]. As shown in Ref. [103], the blockade can be lifted via ESR, resulting in periodic Rabi oscillations in the measured current through the dot as a function of the rf burst length. Here, already in this very first experimental demonstration of ESR, a fidelity of $\sim$75 % has been achieved for intended π-rotations. Alternatively, spin-rotations can also be induced electrically via spin-orbit coupling [115, 116] or inhomogeneous magnetic fields [117–119]. This technique is referred to as electric-dipole-induced spin resonance, in short EDSR. For typical GaAs QDs, EDSR allows for spin manipulation on a $\sim$10 ns timescale in state-of-the-art experiments [116], with the great advantage that spins can be addressed individually, since electric fields are much easier to confine to small regions than magnetic fields [28]. In the case of the singlet-triplet qubit, single qubit rotations can be performed using inhomogeneous magnetic fields [105, 120].

Coherent control over two-qubit states—Two-qubit operations can be performed all electrically by pulsing the electrostatic tunnel barrier between two adjacent spins in a DQD [27]. When the inter-dot barrier is high, the two spins are decoupled. When it is low, the two-electron wave functions overlap, resulting in a Heisenberg exchange coupling, described by the Hamiltonian

$$H(t) = J(t)\mathbf{S}_L \cdot \mathbf{S}_R,$$

with $\mathbf{S}_i$ referring to the spin in dot $i = L, R$ [121]. If the exchange Hamiltonian is turned on for a time τ such that $\int dt\, J(t)/\hbar = J_0\tau/\hbar = \pi$, the states of the two spins get exchanged. This is the so-called SWAP operation. Pulsing the exchange on for a time $\tau/2$ generates the $\sqrt{\text{SWAP}}$-operation, which can be used together with single-qubit gates in order to implement the controlled-NOT gate [27]. Typical gate operations times are extremely short, less than a nanosecond [28]. In the singlet-triplet encoding schemes, capacitive coupling due to the relative charge distributions of triplet and singlet states can be used in order to implement two-qubit gates [122, 123].

Several major challenges towards QD-based quantum computing have been identified and addressed already in the original work of Loss and DiVincenzo [27]. One of the major adversaries is decoherence (in particular, in the solid-state), that is the process by which quantum information is lost due to coupling of the system to other hard-to-control degrees of freedom in the environment, such as phonons, fluctuating charges, nuclear spins etc. In the following, we review our current understanding of the relevant decoherence processes in QDs and outline potential ways to surmount these challenges.

Quantum Dot Environment

Every experimental implementation aiming for QD-based quantum information processing will be (to some extent) subject to noise. Broadly speaking, one can distinguish between two types of noise sources [121]: (i) *extrinsic* noise sources, such as fluctuating magnetic fields and voltages, which may be seen as artifacts of a given experimental setup and may be mitigated with improved electronics, and (ii) on a more fundamental level, *intrinsic* noise sources, which warrant an in-depth analysis, since they cannot be unwound simply by improved electronics. For electron spins in semiconductor quantum dots, the most important intrinsic noise sources have been identified as the interactions with the environment via spin-orbit coupling and hyperfine coupling with the nuclei in the host environment [33, 97, 121].

Before we discuss these mechanisms in more detail, some short remarks on nomenclature are in order, since, throughout the following discussion, we will repeatedly refer to three characteristic timescales [124]: (i) Energy relaxation processes (i.e., random spin flips of the form $|\downarrow\rangle \rightsquigarrow |\uparrow\rangle$) are characterized by the spin relaxation time T_1. (ii) Phase randomization, describing the decay of a coherent superposition $\alpha |\downarrow\rangle + \beta |\uparrow\rangle$, is characterized by the spin coherence timescale T_2; by definition, T_1 sets a bound on T_2 such that $T_2 \leq 2T_1$. (iii) If a measurement samples many different environmental configurations, leading to random phase variations between successive measurements, the averaged result is characterized by the ensemble-averaged transverse spin decay time $T_2^\star$, which captures additional decoherence beyond that described by T_2.

Spin-orbit interaction and phonons—Spin-orbit interaction (SOI) couples the orbital motion of a charge to its spin [97]. This SOI mixes spin eigenstates by coupling states that contain both different orbital and different spin parts [125]; therefore, strictly speaking, what we usually call electronic spin-up and spin-down in a QD, are admixtures of spin and orbital states [33, 125]. Still, SOI on its own does not cause decoherence [124]. The character of the ground-state doublet, however, does change, since orbital and spin components get admixed. As a consequence, any fluctuations that couple to the orbital degree of freedom can lead to decoherence in combination with spin-orbital coupling [124, 126]. Here, the dominant contribution is due to phonons, while other processes are typically negligible [97, 125–129]. A large number of theoretical studies have shown that the resulting relaxation and dephasing times are of the same order (in contrast to the naively expected relation $T_2 \ll T_1$), due to the strong confinement extremely long and strongly suppressed for decreasing magnetic fields; see for example Refs. [33, 125, 126, 130]. These theoretical findings have been verified experimentally, with relaxation times ranging from $\sim$120 μs at 14 T to $\sim$170 ms at 1.75 T and following nicely the expected dependence on magnetic field strength ($T_1^{-1} \sim B^5$) over the applicable magnetic-field range [33, 131, 132]. Subsequently, the focus of the community turned towards the hyperfine interaction between electronic and nuclear spins as the remaining intrinsic source of noise [124].

Hyperfine interaction with nuclear spins—In analogy to the spin of an electron in an atom that interacts with its atomic nucleus through hyperfine coupling, an electron spin confined in a QD may interact with many (about 10^5–10^6 for typical GaAs EDQDs [33, 98]) nuclear spins in the host material. The dominant spin-spin coupling for this type of dot arises from the Fermi contact hyperfine interaction [133], given by

$$H_{\mathrm{HF}} = \frac{g}{2}\left(A^+ S^- + A^- S^+\right) + g A^z S^z. \qquad (1.1)$$

Here, S^μ and $A^\mu = \sum_{i=1}^{N} g_i \sigma_i^\mu$ with $\mu = +, -, z$ denote electron and collective nuclear spin operators, respectively, and σ_i^μ refers to the individual nuclear spin operators. The coupling coefficients can be normalized such that $\sum_i g_i^2 = 1$; g is related to the total HF coupling strength A_{HF} via $g = A_{\mathrm{HF}}/\sum_i g_i$. For a schematic illustration, compare Fig. 1.1b. The effect of the nuclear spins on the electron spin can be understood as an effective magnetic field $\mathbf{B}_N$, commonly referred to as Overhauser field (OF) [33, 98]. If the nuclear field $\mathbf{B}_N$ were fixed and precisely known, it would influence the electron spin dynamics in a systematic and known way [33]. In this case, there would be no nuclear contribution to the decoherence of the electronic spin [33]. However, both the orientation and magnitude of $\mathbf{B}_N$ fluctuate over time as a result of the redistribution of nuclear spin polarization due to dipolar coupling among the nuclei and/or virtual excitations of the electronic spin [33, 98]. The randomness in the nuclear field $\mathbf{B}_N$ directly translates to randomness in the time evolution of the electron spin via the hyperfine interaction [33]; therefore, during free evolution, the electron spin will pick up a random phase according to the random value of the OF, leading to a decay of the electronic spin coherence. In analogy to N coin tosses, in the limit of a large number of nuclei N, this effect may be accounted for by sampling the OF from a Gaussian distribution [98, 134, 135] with standard deviation $\sigma_{\mathrm{nuc}} = A/\sqrt{N}$, where the material-dependent parameter A gives the maximum hyperfine interaction strength for a fully polarized ensemble of nuclei[4]; for GaAs, $A \approx 100\,\mu$eV. For an electron confined in a GaAs QD that interacts with a typical number of $\sim$10^6 nuclei of spin-3/2, this results in a root-mean-square width of $\sigma_{\mathrm{nuc}} \approx 0.1\,\mu$eV. Note that both the maximum A and the typical interaction strength $A/\sqrt{N}$ are often expressed in terms of the equivalent magnetic field strength which amount to $A/(g\mu_B) \approx 4$ T and $A/(g\mu_B\sqrt{N}) \approx 4$ mT [33, 98, 136–138], respectively. The latter causes the phase of the electron spin to change by π in about 10 ns [28]. However, a measurement takes comparatively long, typically tens of seconds [28], during which time the nuclear spin

---

[4]Since the magnetic moment of a single nuclear spin is extremely small, nuclei are typically in an almost maximally mixed state with negligible thermal polarization even at the lowest temperatures and largest magnetic fields available in state-of-the-art experiments [28].

reconfigure themselves many times.[5] One time-averaged measurement of the electron spin precession therefore contains a spread of precession rates, yielding a decay of the envelope on a characteristic timescale $T_2^* \sim 1/\sigma_{nuc} \approx \hbar\sqrt{N}/A \approx 15\,\text{ns}$ [27, 98, 104, 134, 135, 141]. This rapid dephasing, resulting from the uncertainty in the nuclear field, has been verified experimentally; see for example Refs. [113, 142].

Broadly speaking, three main strategies to mitigate decoherence by the nuclear spin bath and therefore boost electronic spin coherence are actively pursued:

(1) *Dynamical decoupling*—Dephasing due to inhomogeneous broadening can be largely unwound via spin-echo pulses [104, 120, 141, 143–145]. In its simplest form, the Hahn echo, the random evolution of a certain time interval τ is reversed during a second time interval of the same duration by applying a π-pulse in between the two intervals [33]. This echo technique removes random dephasing due to the (quasi-static) OF to the extent that the random field is constant for the duration of the entire echo sequence [33]. The nuclear spin dynamics are sufficiently slow [98, 146] to allow for strongly prolonged electron coherence times in the presence of echo pulses, with experimentally demonstrated electronic dephasing times exceeding $200\,\mu\text{s}$ [120].

(2) *State preparation via polarization or measurement*—A common strategy to freeze the nuclear field fluctuations is to sharpen the OF distribution. One approach—referred to as dynamical nuclear polarization (DNP)—is to polarize the nuclear spins using the hyperfine interaction with the electron spin together with electrical or magnetic control over the electronic degrees of freedom [98, 101]. In this approach the electron spin is strongly polarized and the hyperfine coupling is used to transfer spin angular momentum from the electron to the nuclear spin. The basic motivation for this approach is that internuclear flip-flop processes can no longer take place [134, 147, 148], if all spins were aligned, which should result in a reduction of nuclear hyperfine field fluctuations. However, for this method to be effective, the nuclear spin polarization must be extremely close to unity [148]. Under the assumption that the nuclear ensemble is in a product state of individual spins with polarization p, polarizations above $99\,\%$ are required in order to achieve an improvement of coherence time by a factor of ten [97]. Since state-of-the-art experiments achieve nuclear polarizations $50\,\% < p < 80\,\%$ [98, 149–151], nuclear polarization on its own presently cannot contribute to a significant reduction of electron spin decoherence. However, besides polarizing, another attractive approach for lifetime prolongation is to narrow the nuclear spin distribution using either indirect measurement [152–154], or built-in feedback mechanisms in DNP schemes [155, 156], that lead to a narrow steady-state OF even at low nuclear polarizations. Here, nuclear spin induced dephasing is suppressed by preparing the nuclear spin bath in a narrowed, less noisy state [97]. Ultimately, in a finite magnetic field (where flip-flops between electron and nuclear spins are detuned), the coherence times can be improved by several orders of magnitude by preparing the system initially in an eigenstate to the field B_N^z, whereby

---

[5]The nuclei evolve on a typical timescale set by the dipolar correlation time $\sim$0.1 ms [139], as determined by NMR line width measurements [121, 140].

the coherence times scales $\sim N$ rather than $\sim \sqrt{N}$ [134, 147, 157–160]. In this limit, one recovers the (hyperfine-induced) electron decoherence time $T_2 \sim \hbar N/A$ [133, 134, 147] which arises from fluctuations of the OF due to the inhomogeneity of the hyperfine coupling constants α_j in $\mathbf{B}_N = \sum_j \alpha_j \mathbf{I}_j$, where $j = 1, \ldots, N$ labels the nuclear spins $\mathbf{I}_j$.

(3) *New materials*—In order to avoid decoherence due to hyperfine interaction with the nuclear spins, (isotopically purified) materials with smaller hyperfine interaction strength and/or low nuclear spin concentration have been proposed. A prominent example are silicon-based systems such as SiGe QDs [161–163], for which extremely long coherence times, with T_2^* exceeding $100 \, \mu s$ and $T_2 \approx 30 \, ms$, have recently been demonstrated experimentally [164, 165].

Beyond the active research into a quiescent and controllable magnetic environment for electron spins, nuclear spins themselves have been suggested as a useful resource that takes on an *active* role in quantum information processing [166]. The main motivation for this line of research are the strong coupling of the nuclei to the electron (but otherwise very good environmental isolation) and their ultra-long life- and coherence times [98]. Generally speaking, two ways of actively harnessing nuclear spins in QDs for QIP have been explored so far: First, nuclear spins can be used to perform quantum gates on the electron spin qubit, since they can provide a stable and *localized* effective magnetic field. The controlled, time-dependent application of strong magnetic fields acting on individual spin qubits in QDs is a pre-requisite to many single-qubit gates. This can be done by moving the electrons (via electrical control) in the inhomogeneous magnetic field generated by a close-by micro or nanomagnet [118, 119, 149, 167]. Alternatively, one may use the Overhauser field generated by polarized nuclear spins, since already a 10 % polarization difference between two dots in a GaAs DQD amounts to a few hundreds of milli-Tesla in effective field difference, much larger than typical OF fluctuations [97, 168]. Then, in order to implement a single-qubit gate, a spin qubit may be moved into a partially polarized QD for a suitable time period, while for singlet-triplet qubits OF gradients are routinely used to induce rotations between the computational states [168]. Moreover, OF gradients may also be used to realize individual addressing of single electrons by resonant microwave radiation [117]. This approach requires only a sufficiently narrow OF distribution and a sufficiently long nuclear relaxation time T_1; depending on the specific system, nuclear lifetimes ranging from second to hours have been observed [33, 98], all much longer than typical OF-based gate operation times (micro- to nanoseconds). Second, taking advantage of potentially long-lived nuclear states and in strong analogy to existing QIP proposals with atomic ensembles, nuclear spin ensembles in QDs have been considered as a quantum memory for the quantum information stored in the electron spin [169–171]. In contrast to our discussion so far, which can largely be understood in terms of an apparent *semiclassical* magnetic field $\mathbf{B}_N$, this approach involves the actual *quantum mechanical* degrees of freedom of the nuclear spin ensemble in order to coherently exchange a qubit state between the QD electron spin and a collective degree of freedom associated with the nuclear spin bath [170]. The nuclear spin state may live as long as the dipole-dipole correlation time $\sim 0.1 \, ms$ (in GaAs) or possibly even longer, if NMR pulses are applied in order to suppress the dipole-dipole interaction [98, 121, 166, 172].

Challenges and Opportunities
At the time of the first proposal for spin-based quantum computing in 1998 [27], the experimental situation was not very encouraging: single electrons had not been trapped in QDs, let alone the successful demonstration of single spin detection or manipulation [33]. Thanks to major theoretical and experimental breakthroughs in the field (such as advances in qubit design, nuclear spin preparation and the use of dynamical decoupling techniques), however, this has changed dramatically; for an extensive up-to-date review on the prospects of QD-based quantum computing compare Ref. [97]. Today, all standard building blocks for a quantum computer (single- and two-qubit gates, state preparation and measurement) have been demonstrated in a *single* device for both (all-electrical) spin-qubit [173] and singlet-triplet qubit designs [123]. The fidelities achieved to date are still moderate (for example, $F = 72\%$ for Bell state generation [123] and read-out fidelities of $\sim 86\%$ in Ref. [114]), yet calling for further significant improvements in order to approach the regime required for fault-tolerance. At present, the state of the art is still at the level of single and double QDs [33]. Large-scale quantum computers, however, need to reach a system size of several thousands of qubits or more [97]. Since the interactions between EDQDs are very short-range, enabling quantum computing architectures with nearest-neighbour interactions only, this poses serious architectural challenges, because a large amount of wiring and control electronics need to be accommodated on a very small scale. Therefore, a truly scalable design is likely to require coupling over distances of several micrometers [16, 174]. Accordingly, on a intermediate to long time scale, the major challenges towards QD-based QIP are believed to be (i) coherence and (ii) scalability [33]. With existing strategies to establish coherence as outlined above, the focus of the QD community is expected to shift from single-spin control towards the creation and manipulation of entangled states of two and more states over longer and longer distances [28]. The technological progress of the field has by far not come to an end yet and with ever improving control of electronic and nuclear degrees of freedom further thrilling breakthroughs can still be expected.

The hyperfine interaction in a QD realizes the so called central spin model, one of the hallmark models of modern quantum many-body physics [133, 175, 176], with theoretical and experimental evidence showing numerous compelling (dynamical) effects such as bistabilities, phase transitions and dragging [136, 155, 156, 177, 178]. Therefore, apart from the applied interests outlined above, the intriguing many-body interplay between slowly evolving nuclear spins and fast electronic degrees of freedom still offers many conceptually interesting questions and merits further investigations regardless of specific QIP applications.

1.3   Outline of This Thesis

Against the background described in the previous sections, the motivation for this
Thesis is basically two-fold: First, we propose schemes that actively harness the
environmental degrees of freedom that couple to semiconductor QDs; in this way,
we turn the presence of the solid-state environment from a liability to a potential
asset. Second, in doing so, we address two major challenges towards the realization
of spin-based quantum computing: (i) decoherence of the electronic spins due to
hyperfine coupling to the surrounding nuclear spin ensemble, and (ii) the realization
of long-range spin-spin coupling between remote qubits. Therefore, as schematically
depicted in Fig. 1.2, this Thesis naturally encompasses two major research blocks:
(1) First, on a single-node level, we investigate how coherent behaviour can emerge
in the *nuclear* spin dynamics from engineered (dissipative) coupling to the central
electronic spin. We develop a theoretical master-equation-based framework which
features coupled dynamics of electron and nuclear spins as a result of the hyperfine
interaction. Our analysis is based on the typical separation of time scales between
(fast) electron spin evolution and (slow) nuclear spin dynamics, yielding a coarse-
grained quantum master equation for the nuclear spins. This approach reverses the
standard perspective in which the nuclei are considered as an environment for the
electronic spins, but rather views the nuclear spins as the quantum system coupled
to an electronic environment with an exceptional degree of tunability. With this

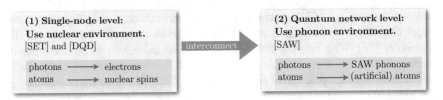

Fig. 1.2   The research goals of this Thesis can be grouped into two blocks. The first block investigates electron transport through single and double QDs, with the aim of harnessing the (dissipative)
coupling to the electronic degrees of freedom for the creation of coherence in both the transient
(see Chap. 2 [SET]) and steady-state (see Chap. 3 [DQD]) behaviour of the ambient nuclear spins.
In Chap. 2 [SET] we show that superradiance, i.e. spontaneous emergence of coherence, can be
observed in EDQDs in a transport setting. The signatures of nuclear coherence are imprinted on the
electronic current through the dot, giving rise to the new paradigm of electronic superradiance. In
Chap. 3 [DQD] we propose a scheme for the deterministic generation of (steady-state) entanglement
between the two nuclear spin ensembles of a DQD in the Pauli-blockade regime. On a qualitative
level, one can understand the main results of the first block within the following analogy between
mesoscopic solid-state physics and quantum optics: The nuclear spins surrounding a QD are identified with an atomic ensemble, individual nuclear spins corresponding to the internal levels of a
single atom and the electrons are associated with photons. In the second block we turn our focus
towards the question of how to interconnect individual qubits over large distances. To this end, we
propose surface acoustic waves (SAWs) in piezo-active materials as a universal quantum bus. In
close analogy to cavity QED, the main results of Chap. 4 [SAW] can be understood by identifying
SAW phonons with photons and artificial atoms with natural atoms

perspective, our framework provides novel insights into the complex, nonequilibrium many-body dynamics of localized electronic spins interacting with a mesoscopic number of nuclear spins. On the one hand, this may lead to a better quantum control over the nuclear spin bath and therefore improved schemes to coherently control electron spin qubits. On the other hand, our work may pave the way towards novel nuclear-spin-based information storage and manipulation protocols. (2) Second, on a broader quantum network level, we propose and analyze *phonon* modes associated with surface acoustic waves (SAWs) in piezo-active materials as a universal mediator for long-range coupling between remote qubits. Our approach involves qubits (as provided by quantum dots) interacting with a localized SAW phonon mode, defined by a high-quality resonator, which in turn can be coupled weakly to a SAW waveguide serving as a quantum bus. We show that the piezo-electric coupling between qubit and SAW phonon mode enables a controlled mapping of the qubit state onto a coherent phonon superposition, which can then be converted to an itinerant SAW phonon in a waveguide, opening up the possibility to implement on-chip many quantum communication protocols well known in the context of optical quantum networks.

The individual chapters of this Thesis are written in a self-contained style that allows the reader to study different chapters individually. In this introduction, the general context of the following chapters has been established. Each chapter provides an individual introduction into the specific topic, an executive summary of the main results and is supplemented by further concluding remarks and outlooks. The individual chapters of this Thesis are related by the focus on quantum information processing (QIP) in concrete, experimentally accessible physical systems and by the active transfer of methods and ideas from quantum optics to the solid-state system which is in the focus of this Thesis, namely single localized electron spins in semiconductor QDs coupled to their naturally occurring nuclear spin environment. In the following, we give a detailed overview over the individual chapters:

In Chap. 2 [**SET**] we present an experimentally feasible scheme where superradiance, i.e. spontaneous emergence of coherence, can be observed in the tunnel current through a single quantum dot, due to the build-up and reinforcement of strong internuclear correlations. The key experimental signature is a strong, sudden peak in the electronic current emitted form the quantum dot, giving rise to the new paradigm of electronic superradiance. Our scheme is based on a general, theoretical framework for the description of electron transport in the presence of hyperfine coupling to the ambient nuclear spins. The underlying connection between the fields quantum optics with atoms, where superradiance has been described first, and solid-state physics acts as a pivotal element for this research block.

Building upon the insight that the nuclear spin dynamics are governed by collective interactions giving rise to coherent effects such as superradiance, in Chap. 3 [**DQD**] we investigate the steady-state behaviour of a double quantum dot operated in the Pauli-blockade regime. Here, we propose the deterministic generation of an entangled state of two spatially separated nuclear spin ensembles in a double quantum dot. The mechanism is based on electron transport through the double quantum dot and relies on suitably engineering the effective electronic environment seen by the nuclear spins via external gate voltages. It ensures (i) a collective coupling of electrons and nuclei such that spin flips occurring in the course of electron transport can happen either in the left or right quantum dot and no which-way information is leaked, and (ii) that just two such processes with a common entangled stationary state are dominant. Rather than relying only on coherent evolution, the entanglement is actually stabilized by the dissipative dynamics unavoidably present in an open quantum system, making the proposed scheme inherently robust against weak random perturbations. We show that the nuclear spins are actively driven into an entangled state of the EPR (Einstein-Podolsky-Rosen) type, which is known to play a key role in continuous variable quantum information processing, quantum sensing and metrology. With this quantum control at hand, the nuclear spin bath could be turned from the dominant source of decoherence into a useful resource for manipulating an electron spin qubit. Our results may not only pave the way for improved schemes to coherently control electron spin qubits, but also, due to very long nuclear decoherence time scales, open up a new route towards nuclear spin based information storage and manipulation.

In Chap. 4 [**SAW**] we put forward a novel strategy based on phonon modes associated with surface acoustic waves (SAWs) in order to realize long-range spin-spin coupling between remote qubits. SAWs occupy a middle ground between previously investigated electromagnetic (transmission lines) and mechanical (fixed resonators) coupling mechanisms and naturally combine the advantageous properties of these systems. In piezo-active materials SAWs provide electromagnetic coupling but being mechanical, they can propagate in the substrate and are naturally coupled to charges/spins attached to or embedded in the material. Our setup opens up the route towards a novel sound-based quantum information architecture and comes with several compelling features: (1) Due to the plethora of physical properties associated with SAWs, our approach is accessible to a broad class of systems such as quantum dots, trapped ions, NV centers or superconducting qubits and is thus inherently universal. (2) We show that the proposed system bears striking similarities with the established fields of cavity- and circuit-QED, opening up the possibility to implement on-chip many quantum communication protocols well known from the context of optical quantum networks. (3) Since SAWs propagate elastically on the surface of a solid within a depth of approximately one wavelength, the mode volume is intrinsically confined in the direction normal to the surface. We show that further surface confinement yields large zero-point quantum fluctuations, enabling strong coupling on the single-phonon level. (4) Typical SAW frequencies lie in the gigahertz range, closely matching transition frequencies of artificial atoms and enabling ground state cooling by conventional cryogenic techniques. (5) Since SAW devices have been fabricated for decades and are ubiquitous in today's classical electronic devices, our scheme is built upon a well-established technology, where lithographic fabrication techniques provide almost arbitrary geometries with high precision, as evidenced by (for example) the experimental realization of high-quality SAW cavities and SAW waveguides. (6) For a given frequency in the gigahertz range, due to the slow speed of sound of approximately 10^3 m/s for typical materials, device dimensions are in micrometer range, which is convenient for fabrication and integration with semiconductor components, and about 10^5 times smaller than corresponding electromagnetic resonators. (7) Typically, for solid-state qubits coupling to phonons is considered as an undesired source of decoherence. Here, our scheme creates a new paradigm, where coupling to phonons becomes a valuable asset for coherent quantum control. (8) Our theoretical predictions indicate that the proposed SAW-based quantum state-transfer protocol (for coupling qubits over distances exceeding several millimeters) can be realized with existing experimental technology. (9) The proposed combination of techniques and concepts known from quantum optics and quantum information, in conjunction with the technological expertise for SAW devices, is likely to lead to further, rapid theoretical and experimental progress, opening up the avenue towards the widely anticipated field of quantum acoustics.

References

1. C. Gardiner, P. Zoller, *The Quantum World of Ultra-Cold Atoms and Light: Foundations of Quantum Optics* (Imperial College Press, London, 2014)
2. J. Bardeen, W.H. Brattain, The transistor, a semi-conductor triode. Phys. Rev. **74**, 230 (1948)
3. T.H. Maiman, Stimulated optical radiation in ruby. Nature **187**(4736), 493 (1960)
4. A. Javan, W.R. Bennett, D.R. Herriott, Population inversion and continuous optical maser oscillation in a gas discharge containing a He-Ne Mixture. Phys. Rev. Lett. **6**(3), 106 (1961)
5. D. van Delft, P. Kes, The discovery of superconductivity. Phys. Today **63**(9), 38 (2010)
6. J.E. Kunzler, E. Buehler, F.S.L. Hsu, J.H. Wernick, Superconductivity in Nb_3Sn at high current density in a magnetic field of 88 kgauss. Phys. Rev. Lett. **7**, 215 (1961)
7. M.A. Nielsen, I.L. Chuang, *Quantum Computation and Quantum Information* (Cambridge University Press, Cambridge, 2000)
8. C. Monroe, Quantum information processing with atoms and photons. Nature **416**(6877), 238 (2002)
9. R. Feynman, P.W. Shor, Simulating physics with computers. SIAM J. Comput. **26**, 1484 (1982)
10. D. Deutsch, Quantum theory, the church-turing principle and the universal quantum computer. Proc. R. Soc. A: Math. Phys. Eng. Sci. **400**(1818), 97 (1985)
11. P.W. Shor, Polynomial-time algorithms for prime factorization and discrete logarithms on a quantum computer. SIAM Rev. **41**(2), 303 (1999)
12. R.L. Rivest, A. Shamir, L. Adleman, A method for obtaining digital signatures and public-key cryptosystems. Commun. ACM **21**(2), 120 (1978)
13. J.I. Cirac, L.M. Duan, P. Zoller. Quantum optical implementation of quantum information processing, in F. Di Martini and C. Monroe (eds.), *"Experimental Quantum Computation and Information" Proceedings of the International School of Physics "Enrico Fermi", Course CXLVIII.* (IOS Press, Amsterdam, 2002), p. 263
14. P. Zoller, J.I. Cirac, L.M. Duan, J.J. Garcia-Ripoll, Quantum optical implementation of quantum information processing, in D. Esteve, J.M. Raymond, J. Dalibard (eds.), *"Quantum entanglement and Information Processing" Proceedings of the Les Houches Summer School, Session LXXIX.* (Elsevier, Amsterdam, 2004), p. 187
15. T.D. Ladd, F. Jelezko, R. Laflamme, Y. Nakamura, C. Monroe, J.L. O'Brien, Quantum computers. Nature **464**(7285), 45 (2010)
16. D.D. Awschalom, L.C. Bassett, A.S. Dzurak, E.L. Hu, J.R. Petta, Quantum spintronics: engineering and manipulating atom-like spins in semiconductors. Science **339**(6124), 1174 (2013)
17. A.M. Steane. *Quantum Computing and Error Correction.* arXiv:0304016 (2003)
18. P.W. Shor, Scheme for reducing decoherence in quantum computer memory. Phys. Rev. A **52**(4), R2493 (1995)
19. P.W. Shor, Fault-tolerant quantum computation, in *37th Annual Symposium on Foundations of Computer Science.* (IEEE Comput. Soc. Press, 1996), pp. 56–65

20. R. Raussendorf, J. Harrington, K. Goyal, Topological fault-tolerance in cluster state quantum computation. New J. Phys. **9**(6), 199 (2007)
21. R. Raussendorf, J. Harrington, Fault-tolerant quantum computation with high threshold in two dimensions. Phys. Rev. Lett. **98**(19), 190504 (2007)
22. D.S. Wang, A.G. Fowler, L.C.L. Hollenberg, Surface code quantum computing with error rates over 1 %. Phys. Rev. A **83**(2), 020302 (2011)
23. A.G. Fowler, A.M. Stephens, P. Groszkowski, High-threshold universal quantum computation on the surface code. Phys. Rev. A **80**(5), 052312 (2009)
24. D.P. DiVincenzo, The physical implementation of quantum computation. Fortschritte der Physik **48**(9–11), 771 (2000)
25. D. Bouwmeester, A. Ekert, A. Zeilinger, *The Physics of Quantum Information: Quantum Cryptography, Quantum Teleportation, Quantum Computation* (Springer, Berlin, 2000)
26. J.I. Cirac, P. Zoller, Quantum computations with cold trapped ions. Phys. Rev. Lett. **74**, 4091 (1995)
27. D. Loss, D.P. DiVincenzo, Quantum computation with quantum dots. Phys. Rev. A **57**, 120 (1998)
28. R. Hanson, D.D. Awschalom, Coherent manipulation of single spins in semiconductors. Nature **453**(7198), 1043 (2008)
29. Y. Makhlin, G. Schön, A. Shnirman, Quantum-state engineering with Josephson-junction devices. Rev. Mod. Phys. **73**, 357 (2001)
30. B.E. Kane, A silicon-based nuclear spin quantum computer. Nature **393**(6681), 133 (1998)
31. J. Clarke, F.K. Wilhelm, Superconducting quantum bits. Nat. Phys. **453**(7198), 1031 (2008)
32. B. Urbaszek, X. Marie, T. Amand, O. Krebs, P. Voisin, P. Maletinsky, A. Högele, A. Imamoglu, Nuclear spin physics in quantum dots: an optical investigation. Rev. Mod. Phys. **85**, 79 (2013)
33. R. Hanson, L.P. Kouwenhoven, J.R. Petta, S. Tarucha, L.M.K. Vandersypen, Spins in few-electron quantum dots. Rev. Mod. Phys. **79**, 1217 (2007)
34. R. Blatt, D. Wineland, Entangled states of trapped atomic ions. Nature **453**(7198), 1008 (2008)
35. D. Jaksch, C. Bruder, J.I. Cirac, C.W. Gardiner, P. Zoller, Cold bosonic atoms in optical lattices. Phys. Rev. Lett. **81**, 3108 (1998)
36. J.I. Cirac, P. Zoller, How to manipulate cold atoms. Science **301**, 176 (2003)
37. I. Bloch, Ultracold quantum gases in optical lattices. Nat. Phys. **1**, 23 (2005)
38. I. Bloch, J. Dalibard, W. Zwerger, Many-body physics with ultracold gases. Rev. Mod. Phys. **80**, 885 (2008)
39. J.M. Raimond, M. Brune, S. Haroche, Manipulating quantum entanglement with atoms and photons in a cavity. Rev. Mod. Phys. **73**, 565 (2001)
40. M.D. Lukin, Colloquium: trapping and manipulating photon states in atomic ensembles. Rev. Mod. Phys. **75**, 457 (2003)
41. L.-M. Duan, C. Monroe, Colloquium: quantum networks with trapped ions. Rev. Mod. Phys. **82**, 1209 (2010)
42. K. Hammerer, A. Sørensen, E.S. Polzik, Quantum interface between light and atomic ensembles. Rev. Mod. Phys. **82**, 1041 (2010)
43. N. Sangouard, C. Simon, H. de Riedmatten, N. Gisin, Quantum repeaters based on atomic ensembles and linear optics. Rev. Mod. Phys. **83**, 33 (2011)
44. L.-M. Duan, M.D. Lukin, J.I. Cirac, P. Zoller, Long-distance quantum communication with atomic ensembles and linear optics. Nature **414**, 413 (2001)
45. C.W. Chou, J. Laurat, H. Deng, K.S. Choi, H. de Riedmatten, D. Felinto, H.J. Kimble, Functional quantum nodes for entanglement distribution over scalable quantum networks. Science **316**(5829), 1316 (2007)
46. H.J. Kimble, The quantum internet. Nature **453**, 1023 (2008)
47. H.-J. Briegel, W. Dür, J.I. Cirac, P. Zoller, Quantum repeaters: the role of imperfect local operations in quantum communication. Phys. Rev. Lett. **81**, 5932 (1998)
48. T. Pellizzari, S.A. Gardiner, J.I. Cirac, P. Zoller, Decoherence, continuous observation, and quantum computing: a cavity QED model. Phys. Rev. Lett. **75**, 3788 (1995)

49. J.I. Cirac, P. Zoller, H.J. Kimble, H. Mabuchi, Quantum state transfer and entanglement distribution among distant nodes in a quantum network. Phys. Rev. Lett. **78**, 3221 (1997)
50. H. Mabuchi, A.C. Doherty, Cavity quantum electrodynamics: coherence in context. Science **298**, 1372 (2002)
51. T. Pellizzari, Quantum networking with optical fibres. Phys. Rev. Lett. **79**, 5242 (1997)
52. N. Kalb, A. Reiserer, S. Ritter, G. Rempe, Heralded storage of a photonic quantum bit in a single atom. Phys. Rev. Lett. **114**, 220501 (2015)
53. S. Ritter, C. Nölleke, C. Hahn, A. Reiserer, A. Neuzner, M. Uphoff, M. Mücke, E. Figueroa, J. Bochmann, G. Rempe, An elementary quantum network of single atoms in optical cavities. Nature **484**(7393), 195 (2012)
54. C. Nölleke, A. Neuzner, A. Reiserer, C. Hahn, G. Rempe, S. Ritter, Efficient teleportation between remote single-atom quantum memories. Phys. Rev. Lett. **110**, 140403 (2013)
55. J. Hofmann, M. Krug, N. Ortegel, L. Gérard, M. Weber, W. Rosenfeld, H. Weinfurter, An elementary quantum network of single atoms in optical cavities. Science **337**(6090), 72 (2012)
56. M.A. Sillanpää, J.I. Park, R.W. Simmonds, Coherent quantum state storage and transfer between two phase qubits via a resonant cavity. Nature **449**(7161), 438 (2007)
57. J. Majer et al., Coupling superconducting qubits via a cavity bus. Nature **449**(7161), 443 (2007)
58. R.J. Schoelkopf, S.M. Girvin, Wiring up quantum systems. Nature **451**(7179), 664 (2008)
59. J. Hald, J.L. Sørensen, C. Schori, E.S. Polzik, Spin squeezed atoms: a macroscopic entangled ensemble created by light. Phys. Rev. Lett. **83**, 1319 (1999)
60. J.-F. Roch, K. Vigneron, P. Grelu, A. Sinatra, J.-P. Poizat, P. Grangier, Quantum nondemolition measurements using cold trapped atoms. Phys. Rev. Lett. **78**, 634 (1997)
61. L.V. Hau, S.E. Harris, Z. Dutton, C.H. Behroozi, Light speed reduction to 17 metres per second in an ultracold atomic gas. Nature **397**(6720), 594 (1999)
62. C. Liu, Z. Dutton, C.H. Behroozi, L.V. Hau, Observation of coherent optical information storage in an atomic medium using halted light pulses. Nature **409**(6819), 490 (2001)
63. M.M. Kash, V.A. Sautenkov, A.S. Zibrov, L. Hollberg, G.R. Welch, M.D. Lukin, Y. Rostovtsev, E.S. Fry, M.O. Scully, Ultraslow Group velocity and enhanced nonlinear optical effects in a coherently driven hot atomic gas. Phys. Rev. Lett. **82**, 5229 (1999)
64. D.F. Phillips, A. Fleischhauer, A. Mair, R.L. Walsworth, M.D. Lukin, Storage of light in atomic vapor. Phys. Rev. Lett. **86**, 783 (2001)
65. B. Julsgaard, A. Kozhekin, E.S. Polzik, Experimental long-lived entanglement of two macroscopic objects. Nature **413**, 400 (2000)
66. B. Julsgaard, J. Sherson, J.I. Cirac, J. Fiurasek, E.S. Polzik, Experimental demonstration of quantum memory for light. Nature **432**, 482 (2004)
67. A.K. Ekert, Quantum cryptography based on Bell's theorem. Phys. Rev. Lett. **67**, 661 (1991)
68. C.H. Bennett, G. Brassard, C. Crépeau, R. Jozsa, A. Peres, W.K. Wootters, Teleporting an unknown quantum state via dual classical and Einstein-Podolsky-Rosen channels. Phys. Rev. Lett. **70**, 1895 (1993)
69. M.D. Lukin, S.F. Yelin, M. Fleischhauer, Entanglement of atomic ensembles by trapping correlated photon states. Phys. Rev. Lett. **84**, 4232 (2000)
70. M. Fleischhauer, M.D. Lukin, Dark-state polaritons in electromagnetically induced transparency. Phys. Rev. Lett. **84**, 5094 (2000)
71. A.E. Kozhekin, K. Mølmer, E. Polzik, Quantum memory for light. Phys. Rev. A **62**, 033809 (2000)
72. J.F. Poyatos, J.I. Cirac, P. Zoller, Quantum reservoir engineering with laser cooled trapped ions. Phys. Rev. Lett. **77**(23), 4728 (1996)
73. S. Diehl, A. Micheli, A. Kantian, B. Kraus, H.P. Büchler, P. Zoller, Quantum states and phases in driven open quantum systems with cold atoms. Nat. Phys. **4**, 878 (2008)
74. B. Kraus, H. Büchler, S. Diehl, A. Kantian, A. Micheli, P. Zoller, Preparation of entangled states by quantum Markov processes. Phys. Rev. A **78**(4), 042307 (2008)
75. J. Cho, S. Bose, M.S. Kim, Optical pumping into many-body entanglement. Phys. Rev. Lett. **106**(2), 020504 (2011)

76. A. Tomadin, S. Diehl, P. Zoller, Nonequilibrium phase diagram of a driven and dissipative many-body system. Phys. Rev. A **83**(1) (2011)
77. J.T. Barreiro, P. Schindler, O. Gühne, T. Monz, M. Chwalla, C.F. Roos, M. Hennrich, R. Blatt, Experimental multiparticle entanglement dynamics induced by decoherence. Nat. Phys. **6**(12), 943 (2010)
78. F. Benatti, R. Floreanini, U. Marzolino, Entangling two unequal atoms through a common bath. Phys. Rev. A **81**(1), 012105 (2010)
79. M.B. Plenio, S.F. Huelga, Entangled light from white noise. Phys. Rev. Lett. **88**(19), 197901 (2002)
80. S. Diehl, A. Tomadin, A. Micheli, R. Fazio, P. Zoller, Dynamical phase transitions and instabilities in open atomic many-body systems. Phys. Rev. Lett. **105**, 015702 (2010)
81. M.J. Kastoryano, F. Reiter, A.S. Sørensen, Dissipative preparation of entanglement in optical cavities. Phys. Rev. Lett. **106**(9), 090502 (2011)
82. B. Kraus, J.I. Cirac, Discrete entanglement distribution with squeezed light. Phys. Rev. Lett. **92**, 013602 (2004)
83. R. Sánchez, G. Platero, Dark Bell states in tunnel-coupled spin qubits. Phys. Rev. B **87**(8), 081305 (2013)
84. F. Verstraete, M.M. Wolf, J.I. Cirac, Quantum computation and quantum-state engineering driven by dissipation. Nat. Phys. **5**, 633 (2009)
85. F. Pastawski, L. Clemente, J.I. Cirac, Quantum memories based on engineered dissipation. Phys. Rev. A **83**(1), 012304 (2011)
86. J. Kerckhoff, H.I. Nurdin, D.S. Pavlichin, H. Mabuchi, Designing quantum memories with embedded control: photonic circuits for autonomous quantum error correction. Phys. Rev. Lett. **105**(4), 040502 (2010)
87. J.T. Barreiro, M. Müller, P. Schindler, D. Nigg, T. Monz, M. Chwalla, M. Hennrich, C.F. Roos, P. Zoller, R. Blatt, An open-system quantum simulator with trapped ions. Nat. Phys. **470**(7335), 486 (2011)
88. H. Weimer, M. Müller, I. Lesanovsky, P. Zoller, H.P. Büchler, A Rydberg quantum simulator. Nat. Phys. **6**(5), 382 (2010)
89. C.A. Muschik, E.S. Polzik, J.I. Cirac, Dissipatively driven entanglement of two macroscopic atomic ensembles. Phys. Rev. A **83**, 052312 (2011)
90. K.G.H. Vollbrecht, C.A. Muschik, J.I. Cirac, Entanglement distillation by dissipation and continuous quantum repeaters. Phys. Rev. Lett. **107**, 120502 (2011)
91. T. Ramos, H. Pichler, A.J. Daley, P. Zoller, Quantum spin dimers from chiral dissipation in cold-atom chains. Phys. Rev. Lett. **113**, 237203 (2014)
92. H. Krauter, C.A. Muschik, K. Jensen, W. Wasilewski, J.M. Petersen, J.I. Cirac, E.S. Polzik, Entanglement generated by dissipation and steady state entanglement of two macroscopic objects. Phys. Rev. Lett. **107**, 080503 (2011)
93. Y. Lin, J.P. Gaebler, F. Reiter, T.R. Tan, R. Bowler, A.S. Sørensen, D. Leibfried, D.J. Wineland, Dissipative production of a maximally entangled steady state of two quantum bits. Nature **504**, 415 (2013)
94. S. Shankar, M. Hatridge, Z. Leghtas, K.M. Sliwa, A. Narla, U. Vool, S.M. Girvin, L. Frunzio, M. Mirrahimi, M.H. Devoret, Autonomously stabilized entanglement between two superconducting quantum bits. Nature **504**(7480), 419 (2013)
95. S. Haroche, Nobel lecture: controlling photons in a box and exploring the quantum to classical boundary. Rev. Mod. Phys. **85**, 1083 (2013)
96. D.J. Wineland, Nobel lecture: superposition, entanglement, and raising Schrödinger's cat. Rev. Mod. Phys. **85**, 1103 (2013)
97. C. Kloeffel, D. Loss, Prospects for spin-based quantum computing in quantum dots. Ann. Rev. Condens. Matter Phys. **4**(1), 51 (2013)
98. E.A. Chekhovich, M.N. Makhonin, A.I. Tartakovskii, A. Yacoby, H. Bluhm, K.C. Nowack, L.M.K. Vandersypen, Nuclear spin effects in semiconductor quantum dots. Nat. Mater. **12**, 494 (2013)

99. L.P. Kouwenhoven, D.G. Austing, S. Tarucha, Few-electron quantum dots. Rep. Prog. Phys. **64**(6), 701 (2001)
100. S.M. Reimann, M. Manninen, Electronic structure of quantum dots. Rev. Mod. Phys. **74**, 1283 (2002)
101. G. Burkard, D. Loss, D.P. DiVincenzo, Coupled quantum dots as quantum gates. Phys. Rev. B **59**, 2070 (1999)
102. J.M. Elzerman, R. Hanson, L.H. Willems van Beveren, B. Witkamp, L.M.K. Vandersypen, L.P. Kouwenhoven, Single-shot read-out of an individual electron spin in a quantum dot. Nature **430**, 431 (2004)
103. F.H.L. Koppens, C. Buizert, K.J. Tielrooij, I.T. Vink, K.C. Nowack, T. Meunier, L.P. Kouwenhoven, L.M.K. Vandersypen, Driven coherent oscillations of a single electron spin in a quantum dot. Nature **442**, 766 (2006)
104. J.R. Petta, A.C. Johnson, J.M. Taylor, E.A. Laird, A. Yacoby, M.D. Lukin, C.M. Marcus, M.P. Hanson, A.C. Gossard, Coherent manipulation of coupled electron spins in semiconductor quantum dots. Science **309**(5744), 2180 (2005)
105. J. Levy, Universal quantum computation with Spin-1/2 Pairs and Heisenberg exchange. Phys. Rev. Lett. **89**, 147902 (2002)
106. M. Field, C.G. Smith, M. Pepper, D.A. Ritchie, J.E.F. Frost, G.A.C. Jones, D.G. Hasko, Measurements of Coulomb blockade with a noninvasive voltage probe. Phys. Rev. Lett. **70**, 1311 (1993)
107. J.M. Elzerman, R. Hanson, J.S. Greidanus, L.H. Willems van Beveren, S. De Franceschi, L.M.K. Vandersypen, S. Tarucha, L.P. Kouwenhoven, Few-electron quantum dot circuit with integrated charge read out. Phys. Rev. B **67**, 161308 (2003)
108. H.-A. Engel, V.N. Golovach, D. Loss, L.M.K. Vandersypen, J.M. Elzerman, R. Hanson, L.P. Kouwenhoven, Measurement efficiency and n-Shot readout of spin qubits. Phys. Rev. Lett. **93**, 106804 (2004)
109. M. Friesen, C. Tahan, R. Joynt, M.A. Eriksson, Spin readout and initialization in a semiconductor quantum dot. Phys. Rev. Lett. **92**, 037901 (2004)
110. A.D. Greentree, A.R. Hamilton, L.C.L. Hollenberg, R.G. Clark, Electrical readout of a spin qubit without double occupancy. Phys. Rev. B **71**, 113310 (2005)
111. R. Ionicioiu, A.E. Popescu, Single-spin measurement using spinorbital entanglement. New J. Phys. **7**(1), 120 (2005)
112. S. Amasha, K. MacLean, I.P. Radu, D.M. Zumbühl, M.A. Kastner, M.P. Hanson, A.C. Gossard, Electrical control of spin relaxation in a quantum dot. Phys. Rev. Lett. **100**, 046803 (2008)
113. A.C. Johnson, J.R. Petta, J.M. Taylor, A. Yacoby, M.D. Lukin, C.M. Marcus, M.P. Hanson, A.C. Gossard, Triplet-singlet spin relaxation via nuclei in a double quantum dot. Nature **435**, 925 (2005)
114. K.C. Nowack, M. Shafiei, M. Laforest, G.E.D.K. Prawiroatmodjo, L.R. Schreiber, C. Reichl, W. Wegscheider, L.M.K. Vandersypen, Single-shot correlations and two-qubit gate of solid-state spins. Science **333**(6047), 1269 (2011)
115. V.N. Golovach, M. Borhani, D. Loss, Electric-dipole-induced spin resonance in quantum dots. Phys. Rev. B **74**, 165319 (2006)
116. K.C. Nowack, F.H.L. Koppens, Y.V. Nazarov, L.M.K. Vandersypen, Coherent control of a single electron spin with electric fields. Science **318**(5855), 1430 (2007)
117. E.A. Laird, C. Barthel, E.I. Rashba, C.M. Marcus, M.P. Hanson, A.C. Gossard, Hyperfine-mediated gate-driven electron spin resonance. Phys. Rev. Lett. **99**, 246601 (2007)
118. M. Pioro-Ladrière, T. Obata, Y. Tokura, Y.S. Shin, T. Kubo, K. Yoshida, T. Taniyama, S. Tarucha, Electrically driven single-electron spin resonance in a slanting Zeeman field. Nat. Phys. **4**(10), 776 (2008)
119. Y. Tokura, W.G. van der Wiel, T. Obata, S. Tarucha, Coherent single electron spin control in a slanting zeeman field. Phys. Rev. Lett. **96**, 047202 (2006)
120. H. Bluhm, S. Foletti, I. Neder, M. Rudner, D. Mahalu, V. Umansky, A. Yacoby, Dephasing time of GaAs electron-spin qubits coupled to a nuclear bath exceeding 200μs. Nat. Phys. **7**(2), 109 (2010)

121. V. Cerletti, W.A. Coish, O. Gywat, D. Loss, Recipes for spin-based quantum computing. Nanotechnology **16**(4), R27 (2005)
122. J.M. Taylor, H.-A. Engel, W. Dür, A. Yacoby, C.M. Marcus, P. Zoller, M.D. Lukin, Fault-tolerant architecture for quantum computation using electrically controlled semiconductor spins. Nat. Phys. **1**, 177 (2005)
123. M.D. Shulman, O.E. Dial, S.P. Harvey, H. Bluhm, V. Umansky, A. Yacoby, Demonstration of entanglement of electrostatically coupled singlet-triplet qubits. Science **336**(6078), 202 (2012)
124. W.A. Coish, D. Loss, Quantum computing with spins in solids, in *Handbook of Magnetism and Advanced Magnetic Materials, 5*, vol. Set, ed. by H. Kronmüller, S. Parkin (Wiley, Hoboken, 2007)
125. A.V. Khaetskii, Y.V. Nazarov, Spin-flip transitions between Zeeman sublevels in semiconductor quantum dots. Phys. Rev. B **64**, 125316 (2001)
126. V.N. Golovach, A. Khaetskii, D. Loss, Phonon-induced decay of the electron spin in quantum dots. Phys. Rev. Lett. **93**, 016601 (2004)
127. D.V. Bulaev, D. Loss, Spin relaxation and anticrossing in quantum dots: Rashba versus Dresselhaus spin-orbit coupling. Phys. Rev. B **71**, 205324 (2005)
128. S.I. Erlingsson, Y.V. Nazarov, Hyperfine-mediated transitions between a Zeeman split doublet in GaAs quantum dots: The role of the internal field. Phys. Rev. B **66**, 155327 (2002)
129. V.N. Golovach, A. Khaetskii, D. Loss, Spin relaxation at the singlet-triplet crossing in a quantum dot. Phys. Rev. B **77**, 045328 (2008)
130. A.V. Khaetskii, Y.V. Nazarov, Spin relaxation in semiconductor quantum dots. Phys. Rev. B **61**, 12639 (2000)
131. S. Amasha, K. MacLean, I. Radu, D.M. Zumbuhl, M.A. Kastner, M.P. Hanson, A.C. Gossard. *Measurements of the spin relaxation rate at low magnetic fields in a quantum dot.* arXiv: cond-mat/0607110 (2006)
132. R. Hanson, B. Witkamp, L.M.K. Vandersypen, L.H.W. van Beveren, J.M. Elzerman, L.P. Kouwenhoven, Zeeman energy and spin relaxation in a one-electron quantum dot. Phys. Rev. Lett. **91**(19), 196802 (2003)
133. J. Schliemann, A. Khaetskii, D. Loss, Electron spin dynamics in quantum dots and related nanostructures due to hyperfine interaction with nuclei. J. Phys. Condens. Matter **15**(50), R1809 (2003)
134. A.V. Khaetskii, D. Loss, L. Glazman, Electron spin decoherence in quantum dots due to interaction with nuclei. Phys. Rev. Lett. **88**, 186802 (2002)
135. I.A. Merkulov, A.L. Efros, M. Rosen, Electron spin relaxation by nuclei in semiconductor quantum dots. Phys. Rev. B **65**, 205309 (2002)
136. A.I. Tartakovskii et al., Nuclear spin switch in semiconductor quantum dots. Phys. Rev. Lett. **98**, 026806 (2007)
137. P.-F. Braun, B. Urbaszek, T. Amand, X. Marie, O. Krebs, B. Eble, A. Lemaitre, P. Voisin, Bistability of the nuclear polarization created through optical pumping in $In_{1-x}Ga_xAs$ quantum dots. Phys. Rev. B **74**, 245306 (2006)
138. B. Urbaszek, P.-F. Braun, T. Amand, O. Krebs, T. Belhadj, A. Lemaître, P. Voisin, X. Marie, Efficient dynamical nuclear polarization in quantum dots: Temperature dependence. Phys. Rev. B **76**, 201301 (2007)
139. F.H.L. Koppens, D. Klauser, W.A. Coish, K.C. Nowack, L.P. Kouwenhoven, D. Loss, L.M.K. Vandersypen, Universal phase shift and nonexponential decay of driven single-spin oscillations. Phys. Rev. Lett. **99**, 106803 (2007)
140. D. Paget, G. Lampel, B. Sapoval, V.I. Safarov, Low field electron-nuclear spin coupling in gallium arsenide under optical pumping conditions. Phys. Rev. B **15**, 5780 (1977)
141. F.H.L. Koppens, K.C. Nowack, L.M.K. Vandersypen, Spin echo of a single electron spin in a quantum dot. Phys. Rev. Lett. **100**, 236802 (2008)
142. F.H.L. Koppens, J.A. Folk, J.M. Elzerman, R. Hanson, L.H. Willems van Beveren, I.T. Vink, H.-P. Tranitz, W. Wegscheider, L.P. Kouwenhoven, L.M.K. Vandersypen, Control and detection of singlet-triplet mixing in a random nuclear field. Science **309**, 1346 (2005)

143. C. Barthel, J. Medford, C.M. Marcus, M.P. Hanson, A.C. Gossard, Interlaced dynamical decoupling and coherent operation of a singlet-triplet qubit. Phys. Rev. Lett. **105**, 266808 (2010)

144. J. Medford, L. Cywiński, C. Barthel, C.M. Marcus, M.P. Hanson, A.C. Gossard, Scaling of dynamical decoupling for spin qubits. Phys. Rev. Lett. **108**, 086802 (2012)

145. G. de Lange, T. van der Sar, M.S. Blok, Z.H. Wang, V.V. Dobrovitski, R. Hanson, Controlling the quantum dynamics of a mesoscopic spin bath in diamond. Science **330**, 60 (2010)

146. A.E. Nikolaenko et al., Suppression of nuclear spin diffusion at a GaAs/$Al_x Ga_{1-x}$As interface measured with a single quantum-dot nanoprobe. Phys. Rev. B **79**, 081303 (2009)

147. A. Khaetskii, D. Loss, L. Glazman, Electron spin evolution induced by interaction with nuclei in a quantum dot. Phys. Rev. B **67**, 195329 (2003)

148. J. Schliemann, A.V. Khaetskii, D. Loss, Spin decay and quantum parallelism. Phys. Rev. B **66**, 245303 (2002)

149. G. Petersen, E.A. Hoffmann, D. Schuh, W. Wegscheider, G. Giedke, S. Ludwig, Large nuclear spin polarization in gate-defined quantum dots using a single-domain nanomagnet. Phys. Rev. Lett. **110**, 177602 (2013)

150. E.A. Chekhovich, M.N. Makhonin, K.V. Kavokin, A.B. Krysa, M.S. Skolnick, A.I. Tartakovskii, Pumping of nuclear spins by optical excitation of spin-forbidden transitions in a quantum dot. Phys. Rev. Lett. **104**, 066804 (2010)

151. F. Klotz, V. Jovanov, J. Kierig, E.C. Clark, M. Bichler, G. Abstreiter, M.S. Brandt, J.J. Finley, H. Schwager, G. Giedke, Asymmetric optical nuclear spin pumping in a single uncharged quantum dot. Phys. Rev. B **82**, 121307 (2010)

152. D. Klauser, W.A. Coish, D. Loss, Nuclear spin state narrowing via gate-controlled Rabi oscillations in a double quantum dot. Phys. Rev. B **73**, 205302 (2006)

153. G. Giedke, J.M. Taylor, D. D'Alessandro, M.D. Lukin, A. Imamoğlu, Quantum measurement of a mesoscopic spin ensemble. Phys. Rev. A **74**, 032316 (2006)

154. D. Stepanenko, G. Burkard, G. Giedke, A. Imamoglu, Enhancement of electron spin coherence by optical preparation of nuclear spins. Phys. Rev. Lett. **96**, 136401 (2006)

155. I.T. Vink, K.C. Nowack, F.H.L. Koppens, J. Danon, Y.V. Nazarov, L.M.K. Vandersypen, Locking electron spins into magnetic resonance by electron-nuclear feedback. Nat. Phys. **5**, 764 (2009)

156. J. Danon, I.T. Vink, F.H.L. Koppens, K.C. Nowack, L.M.K. Vandersypen, Y.V. Nazarov, Multiple nuclear polarization states in a double quantum dot. Phys. Rev. Lett. **103**, 046601 (2009)

157. W.A. Coish, D. Loss, Hyperfine interaction in a quantum dot: non-markovian electron spin dynamics. Phys. Rev. B **70**, 195340 (2004)

158. J. Fischer, M. Trif, W.A. Coish, D. Loss, Spin interactions, relaxation and decoherence in quantum dots. Solid State Commun. **149**, 1443 (2009)

159. W.A. Coish, J. Fischer, D. Loss, Exponential decay in a spin bath. Phys. Rev. B **77**, 125329 (2008)

160. W.A. Coish, J. Fischer, D. Loss, Free-induction decay and envelope modulations in a narrowed nuclear spin bath. Phys. Rev. B **81**, 165315 (2010)

161. A. Wild, J. Sailer, J. Ntzel, G. Abstreiter, S. Ludwig, D. Bougeard, Electrostatically defined quantum dots in a Si/SiGe heterostructure. New J. Phys. **12**(11), 113019 (2010)

162. C.B. Simmons et al., Tunable spin loading and T_1 of a silicon spin qubit measured by single-shot readout. Phys. Rev. Lett. **106**, 156804 (2011)

163. B.M. Maune et al., Coherent singlet-triplet oscillations in a silicon-based double quantum dot. Nature **481**, 344 (2012)

164. E. Kawakami, P. Scarlino, D.R. Ward, F.R. Braakman, D.E. Savage, M.G. Lagally, M. Friesen, S.N. Coppersmith, M.A. Eriksson, L.M.K. Vandersypen, Electrical control of a long-lived spin qubit in a Si/SiGe quantum dot. Nat. Nanotechnol. **9**, 666 (2014)

165. M. Veldhorst et al., An addressable quantum dot qubit with fault-tolerant control-fidelity. Nat. Nanotechnol. **9**, 981 (2014)

166. J.M. Taylor, G. Giedke, H. Christ, B. Paredes, J.I. Cirac, P. Zoller, M.D. Lukin, A. Imamoglu, *Quantum information processing using localized ensembles of nuclear spins.* arXiv:cond-mat/0407640 (2004)

167. T. Obata, M. Pioro-Ladrière, Y. Tokura, Y.-S. Shin, T. Kubo, K. Yoshida, T. Taniyama, S. Tarucha, Coherent manipulation of individual electron spin in a double quantum dot integrated with a micromagnet. Phys. Rev. B **81**, 085317 (2010)

168. S. Foletti, H. Bluhm, D. Mahalu, V. Umansky, A. Yacoby, Universal quantum control of two-electron spin quantum bits using dynamic nuclear polarization. Nat. Phys. **5**, 903 (2009)

169. J.M. Taylor, A. Imamoglu, M.D. Lukin, Controlling a mesoscopic spin environment by quantum bit manipulation. Phys. Rev. Lett. **91**, 246802 (2003)

170. J.M. Taylor, C.M. Marcus, M.D. Lukin, Long-lived memory for mesoscopic quantum bits. Phys. Rev. Lett. **90**, 206803 (2003)

171. W.M. Witzel, S. Das Sarma, Nuclear spins as quantum memory in semiconductor nanostructures. Phys. Rev. B **76**, 045218 (2007)

172. R. Takahashi, K. Kono, S. Tarucha, K. Ono, Voltage-selective bi-directional polarization and coherent rotation of nuclear spins in quantum dots. Phys. Rev. Lett. **107**, 026602 (2011)

173. R. Brunner, Y.S. Shin, T. Obata, M. Pioro-Ladrière, T. Kubo, K. Yoshida, T. Taniyama, Y. Tokura, S. Tarucha, Two-qubit gate of combined single-spin rotation and interdot spin exchange in a double quantum dot. Phys. Rev. Lett. **107**(14), 146801 (2011)

174. L.R. Schreiber, H. Bluhm, Quantum computation: silicon comes back. Nat. Nanotechnol. **9**, 966 (2014)

175. M. Bortz, J. Stolze, Spin and entanglement dynamics in the central-spin model with homogeneous couplings. J. Stat. Mech. Theory Exp. **2007**(06), P06018 (2007)

176. M. Gaudin, Diagonalisation d'une classe d'hamiltoniens de spin. J. Phys. France **37**(10), 1087 (1976)

177. E.M. Kessler, G. Giedke, A. Imamo?lu, S.F. Yelin, M.D. Lukin, J.I. Cirac, Dissipative phase transition in a central spin system. Phys. Rev. A **86**(1), 012116 (2012)

178. P. Maletinsky, C.W. Lai, A. Badolato, A. Imamoglu, Nonlinear dynamics of quantum dot nuclear spins. Phys. Rev. B **75**, 035409 (2007)

Chapter 2
Superradiance-like Electron Transport Through a Quantum Dot

In this chapter we theoretically show that intriguing features of coherent many-body physics can be observed in electron transport through a quantum dot (QD). We first derive a master equation based framework for electron transport in the Coulomb-blockade regime which includes hyperfine (HF) interaction with the nuclear spin ensemble in the QD. This general tool is then used to study the leakage current through a single QD in a transport setting. We find that, for an initially polarized nuclear system, the proposed setup leads to a strong current peak, in close analogy with superradiant emission of photons from atomic ensembles. This effect could be observed with realistic experimental parameters and would provide clear evidence of coherent HF dynamics of nuclear spin ensembles in QDs.

2.1 Introduction

Quantum coherence is at the very heart of many intriguing phenomena in today's nanostructures [1, 2]. For example, it is the essential ingredient to the understanding of the famous Aharonov-Bohm like interference oscillations of the conductance of metallic rings [3] or the well-known conductance steps in quasi one-dimensional wires [4, 5]. In particular, nonequilibrium electronic transport has emerged as a versatile tool to gain deep insights into the coherent quantum properties of mesoscopic solid-state devices [6, 7]. Here, with the prospect of spintronics and applications in quantum computing, a great deal of research has been directed towards the interplay and feedback mechanisms between electron and nuclear spins in gate-based semiconductor quantum dots [8–14]. Current fluctuations have been assigned to the random dynamics of the ambient nuclear spins [15] and/or hysteresis effects due to dynamic nuclear polarization [15–18]. Spin-flip mediated transport, realized in few-electron quantum dots in the so-called spin-blockade regime [19], has been shown to exhibit long time scale oscillations and bistability as a result of a buildup and relaxation of nuclear polarization [15, 16]. The nuclear spins are known to act collectively on the

© Springer International Publishing AG 2017

M.J.A. Schütz, *Quantum Dots for Quantum Information Processing: Controlling and Exploiting the Quantum Dot Environment*, Springer Theses,
DOI 10.1007/978-3-319-48559-1_2

electron spin via hyperfine interaction. In principle, this opens up an exciting testbed
for the observation of collective effects which play a remarkable role in a wide range
of many-body physics [20–22].

In Quantum Optics, the concept of superradiance (SR), describing the cooperative
emission of photons, is a paradigm example for a cooperative quantum effect [1, 23,
24]. Here, initially excited atoms emit photons collectively as a result of the buildup
and reinforcement of strong interatomic correlations. Its most prominent feature is
an emission intensity burst in which the system radiates much faster than an otherwise identical system of independent emitters. This phenomenon is of fundamental
importance in quantum optics and has been studied extensively since its first prediction by Dicke in 1954 [23]. Yet, in its original form the observation of optical
superradiance has turned out to be difficult due to dephasing dipole-dipole van der
Waals interactions, which suppress a coherence buildup in atomic ensembles.

As briefly sketched in the Introduction, this chapter is built upon analogies between
mesoscopic solid-state physics and Quantum Optics: the nuclear spins surrounding
a QD are identified with an atomic ensemble, individual nuclear spins corresponding
to the internal levels of a single atom and the electrons are associated with photons. Despite some fundamental differences—for example, electrons are fermions,
whereas photons are bosonic particles—this analogy stimulates conjectures about
the potential occurence of related phenomena in these two fields of physics. Led by
this line of thought, we will address the question if superradiant behaviour might
also be observed in a solid-state environment where the role of photons is played by
electrons. To this end, we analyze a gate-based semiconductor QD in the Coulomb
blockade regime, obtaining two main results, of both experimental and theoretical
relevance: First, in analogy to superradiant emission of photons, we show how to
observe superradiant emission of electrons in a transport setting through a QD. We
demonstrate that the proposed setup, when tuned into the spin-blockade regime,
carries clear fingerprints of cooperative emission, with no van der Waals dephasing
mechanism on relevant timescales. The spin-blockade is lifted by the HF coupling
which becomes increasingly more efficient as correlations among the nuclear spins
build up. This markedly enhances the spin-flip rate and hence the leakage current
running through the QD. Second, we develop a general theoretical master equation
framework that describes the nuclear spin mediated transport through a single QD.
Apart from the collective effects due to the HF interaction, the electronic tunneling
current is shown to depend on the internal state of the ambient nuclear spins through
the effective magnetic field (Overhauser field) produced by the hyperfine interaction.

2.2 Executive Summary: Reader's Guide

This chapter is structured as follows: In Sect. 2.3, we highlight our key findings
and provide an intuitive picture of our basic ideas, allowing the reader to grasp our
main results on a qualitative level. By defining the underlying Hamiltonian, Sect. 2.4
then describes the system in a more rigorous fashion. This enables us to present a

detailed derivation of the first main result of this chapter in Sect. 2.5: a general master equation for electron transport through a single QD which is coherently enhanced by the HF interaction with the ambient nuclear spins in the QD. It features both collective effects and feedback mechanisms between the electronic and the nuclear subsystem of the QD. Based on this theoretical framework, Sect. 2.6 puts forward the second main result, namely the observation of superradiant behavior in the leakage current through a QD. The qualitative explanations provided in Sect. 2.3 should allow to read this part independently of the derivation given in Sect. 2.5. Section 2.7 backs up our analytical predictions with numerical simulations. When starting from an initially polarized nuclear spin ensemble, the leakage current through the QD is shown to exhibit a strong peak whose relative height scales linearly with the number of nuclear spins, which we identify as the characteristic feature of superradiant behaviour. In Sect. 2.8 we draw conclusions and give an outlook on future directions of research.

2.3 Main Results

In this section we provide an intuitive exposition of our key ideas and summarize our main findings.

HF assisted electron transport.—We study a single electrically-defined QD in the Coulomb-blockade regime which is attached to two leads, as schematically depicted in Fig. 2.1. Formally, the Hamiltonian for the total system is given by

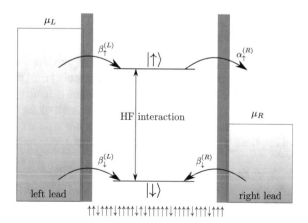

Fig. 2.1 Schematic illustration of the transport system: an electrically defined QD is tunnel-coupled to two electron reservoirs, the *left* and *right* lead respectively. A bias voltage $eV = \mu_L - \mu_R$ is applied between the two leads in order to induce a current through the QD. An external magnetic field is used to tune the system into the sequential-tunneling regime and the QD effectively acts as an spin-filter. The resulting spin-blockade can be lifted by the HF interaction between the QD electron and the nuclear spins in the surrounding host environment

$$H = H_Z + H_B + H_T + H_{\text{HF}}. \tag{2.1}$$

Here, H_Z describes the electronic level structure inside the QD in presence of an external magnetic field. Next, H_B refers to two independent reservoirs of non-interacting electrons, the left and right lead respectively. The coupling between these and the QD is described in terms of a tunneling Hamiltonian H_T and H_{HF} models the *collective* hyperfine interaction between an electron confined inside the QD and an ensemble of N proximal nuclear spins surrounding the QD. Note that the specific form of H will be given later on in Sect. 2.4.

Our analysis is built upon a quantum master equation approach, a technique originally rooted in the field of quantum optics. By tracing out the unobserved degrees of freedom of the leads we derive an effective equation of motion for the density matrix of the QD system ρ_S—describing the electron spin inside the QD as well as the nuclear spin ensemble—irreversibly coupled to source and drain electron reservoirs. In addition to the standard assumptions of a weak system-reservoir coupling (Born approximation), a flat reservoir spectral density, and a short reservoir correlation time (Markov approximation), we demand the hyperfine flip-flops to be strongly detuned with respect to the effective magnetic field seen by the electron throughout the dynamics. Under these conditions, the central master equation can be written as

$$
\begin{aligned}
\dot{\rho}_S(t) = & -i \left[H_Z + H_{\text{HF}}, \rho_S(t) \right] \\
& + \sum_{\sigma=\uparrow,\downarrow} \alpha_\sigma(t) \left[d_\sigma \rho_S(t) d_\sigma^\dagger - \frac{1}{2} \{ d_\sigma^\dagger d_\sigma, \rho_S(t) \} \right] \\
& + \sum_{\sigma=\uparrow,\downarrow} \beta_\sigma(t) \left[d_\sigma^\dagger \rho_S(t) d_\sigma - \frac{1}{2} \{ d_\sigma d_\sigma^\dagger, \rho_S(t) \} \right],
\end{aligned} \tag{2.2}
$$

where the tunneling rates $\alpha_\sigma(t)$ and $\beta_\sigma(t)$ describe dissipative processes by which an electron of spin σ tunnels from one of the leads into or out of the QD, respectively. Here, the fermionic operator $d_\sigma^\dagger$ creates an electron of spin σ inside the QD. While a detailed derivation of Eq. (2.2) along with the precise form of the tunneling rates is presented in Sect. 2.5, here we focus on a qualitative discussion of its theoretical and experimental implications. Essentially, our central master equation exhibits two core features:

Nuclear-state-dependent electronic dissipation.—First, dissipation only acts on the electronic subsystem with rates $\alpha_\sigma(t)$ and $\beta_\sigma(t)$ that depend dynamically on the state of the nuclear subsystem. This non-linear behavior potentially results in hysteretic behavior and feedback mechanisms between the two subsystems as already suggested theoretically [11, 14, 20, 21] and observed in experiments in the context of double QDs in the Pauli-blockade regime; see, e.g., Refs. [12, 13, 18]. On a qualitative level, this finding can be understood as follows: The nuclear spins provide an effective magnetic field for the electron spin, the Overhauser field, whose strength is proportional to the polarization of the nuclear spin ensemble. Thus, a changing nuclear polarization can either dynamically tune or detune the position of the electron levels inside the QD. This, in turn, can have a marked effect on the transport properties of the QD as they crucially depend on the position of these resonances with respect to the chemical potentials of the leads. In our model, this effect is directly captured by the tunneling rates dynamically depending on the state of the nuclei.

SR in electron transport.—Second, the collective nature of the HF interaction H_{HF} allows for the observation of coherent many-body effects. To show this, we refer to the following example: Consider a setting in which the bias voltage and an external magnetic field are tuned such that only one of the two electronic spin-components, say the level $|\uparrow\rangle$, lies inside the transport window. In this spin-blockade regime the electrons tunneling into the right lead are spin-polarized, i.e., the QD acts as an spin filter [25, 26]. If the HF coupling is sufficiently small compared to the external Zeeman splitting, the electron is predominantly in its $|\downarrow\rangle$ spin state allowing to adiabatically eliminate the electronic QD coordinates. In this way we obtain an effective equation of motion for the nuclear density operator μ only. It reads

$$\dot{\mu} = c_r \left[A^- \mu A^+ - \frac{1}{2} \{ A^+ A^-, \mu \} \right] + ic_i \left[A^+ A^-, \mu \right] + i\frac{g}{2} \left[A^z, \mu \right],$$ (2.3)

where $A^\mu = \sum_{i=1}^N g_i \sigma_i^\mu$ with $\mu = +, -, z$ are *collective* nuclear spin operators, composed of *all* N individual nuclear spin operators σ_i^μ, with g_i being proportional to the probability of the electron being at the location of the nucleus of site i. Again, we will highlight the core implications of Eq. (2.3) and for a full derivation thereof, including the definition of the effective rates c_r and c_i, we refer to Sect. 2.6. Most notably, Eq. (2.3) closely resembles the superradiance master equation which has been discussed extensively in the context of atomic physics [24] and therefore similar effects might be expected.

Superradiance is known as a macroscopic collective phenomenon which generalizes spontaneous emission from a single emitter to a many-body system of N atoms [1]. Starting from a fully polarized initial state the system evolves within a totally symmetric subspace under permutation and experiences a strong correlation build-up. As a consequence, the emission intensity is not of the usual exponentially decaying form, but conversely features a sudden peak occuring on a very rapid timescale $\sim 1/N$ with a maximum $\sim N^2$.

In this chapter, we show that the same type of cooperative emission can occur from an ensemble of nuclear spins surrounding an electrically-defined QD: The spin-blockade can be lifted by the HF interaction as the nuclei pump excitations into the electron. Starting from a highly polarized, weakly correlated nuclear state (which could be prepared by, e.g., dynamic polarization techniques [12, 13, 22]), this process becomes increasingly more efficient, as correlations among the nuclei build up due to the collective nature of the HF interaction. This results in an increased leakage current. Therefore, the current is collectively enhanced by the electron's HF interaction with the ambient nuclear spin ensemble giving rise to a superradiant-like

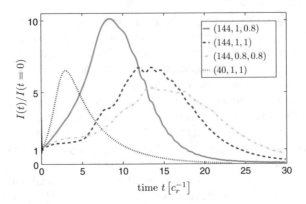

Fig. 2.2 Normalized leakage current through a QD in the spin-blockade regime for N nuclear spins, initial nuclear polarization p and external Zeeman splitting ω_0 in units of the total HF coupling constant $A_{HF} \approx 100\,\mu eV$, summarized as $(N, p, \omega_0/A_{HF})$. For homogeneous HF coupling the dynamics can be solved exactly (*black dotted line*). Compared to this idealized benchmark, the effects are reduced for realistic inhomogeneous HF coupling, but still present: the relative peak height becomes more pronounced for smaller detuning ω_0 or higher polarization p (*solid red line* compared to the *blue dashed* and *green dash-dotted line*, respectively). Even under realistic conditions, the relative peak height is found to scale linearly with N, corresponding to a strong enhancement for typically $N \approx 10^5$–10^6

effect in which the leakage current through the QD takes the role of the radiation field: To stress this relation, we also refer to this effect as *superradiant transport of electrons*.

Comparison to conventional SR.—Compared to its conventional atomic counterpart, our system incorporates two major differences: First, our setup describes superradiant behaviour from a *single* emitter, since in the strong Coulomb-blockade regime the electrons are emitted antibunched. As described above, the superradiant character is due to the nuclear spins acting collectively on the electron spin leading to an increased leakage current on timescales longer than single electron tunneling events. The second crucial difference is the *inhomogeneous* nature ($g_i \neq$ const.) of the collective operators A^μ. Accordingly, the collective spin is not conserved, leading to dephasing between the nuclei which in principle could prevent the observation of superradiant behavior. However, as exemplified in Fig. 2.2, we show that under realistic conditions—taking into account a finite initial polarization of nuclear spins p and dephasing processes due to the inhomogeneous nature of the HF coupling—the leakage current through the QD still exhibits the characteristic peak whose relative height scales linearly with the number of nuclear spins. Even though the effect is reduced compared to the ideal case, for an experimentally realistic number of nuclei $N \approx 10^5$–10^6 a strong increase is still predicted. The experimental key signature of this effect, the relative peak height of the leakage current, can be varied by either tuning the external Zeeman splitting or the initial polarization of the nuclear spins.

In the remainder of this chapter, Eqs. (2.2) and (2.3) are derived from first principles; in particular, the underlying assumptions and approximations are listed. Based
on this general theoretical framework, more results along with detailed discussions
will be presented. For both the idealized case of homogeneous HF coupling—in
which an exact solution is feasible even for relatively large N—and the more realistic inhomogeneous case, further numerical simulations prove the existence of a
strong superradiant peaking in the leakage current of single QD in the spin-blockade
regime.

2.4 The System

This section gives an in-depth description of the Hamiltonian under study, formally
introduced in Eq. (2.1). The system we consider consists a single electrically-
defined QD in a transport setting as schematically depicted in Fig. 2.1. Due to strong
confinement only a single orbital level is relevant. Moreover, the QD is assumed
to be in the strong Coulomb-blockade regime so that at maximum one electron
resides inside the QD. Therefore, the effective Hilbert-space of the QD electron is
span $\{|\uparrow\rangle, |\downarrow\rangle, |0\rangle\}$ where the lowest energy states for an additional electron in the
QD with spin $\sigma = \uparrow, \downarrow$ are split by an external magnetic field. The Hamiltonian for
the total system is given in Eq. (2.1).

Here, the first term

$$H_Z = \sum_\sigma \varepsilon_\sigma d_\sigma^\dagger d_\sigma \tag{2.4}$$

describes the electronic levels of the QD. The Zeeman splitting between the two
spin components is $\omega_0 = \varepsilon_\uparrow - \varepsilon_\downarrow$ (we set $\hbar = 1$) and the QD electron operators are
$d_\sigma^\dagger = |\sigma\rangle \langle 0|$, describing transitions from the state $|0\rangle$ with no electron inside the QD
to a state $|\sigma\rangle$ with one electron of spin σ inside the QD.

Electron transport through the QD is induced by attaching the QD to two electron
leads (labeled as L and R) which are in thermal equilibrium at chemical potentials μ_L
and μ_R, respectively. The leads themselves constitute reservoirs of non-interacting
electrons

$$H_B = \sum_{\alpha,k,\sigma} \varepsilon_{\alpha k} c_{\alpha k\sigma}^\dagger c_{\alpha k\sigma}, \tag{2.5}$$

where $c_{\alpha k\sigma}^\dagger$ ($c_{\alpha k\sigma}$) creates (annihilates) an electron in lead $\alpha = L, R$ with wavevector
k and spin σ. The operators $c_{\alpha k\sigma}^\dagger$ ($c_{\alpha k\sigma}$) fulfill the usual Fermi commutation relations:
$\{c_{\alpha k\sigma}^\dagger, c_{\alpha' k'\sigma'}^\dagger\} = \{c_{\alpha k\sigma}, c_{\alpha' k'\sigma'}\} = 0$ and $\{c_{\alpha k\sigma}^\dagger, c_{\alpha' k'\sigma'}\} = \delta_{\alpha,\alpha'}\delta_{k,k'}\delta_{\sigma,\sigma'}$. The effect of
the Coulomb interaction in the leads can be taken into account by renormalized
effective quasi-particle masses. A positive source-drain voltage $eV = \mu_L - \mu_R$ leads
to a dominant tunneling of electrons from left to right. Microscopically, the coupling
of the QD system to the electron reservoirs is described in terms of the tunneling
Hamiltonian

$$H_T = \sum_{\alpha,k,\sigma} T_{k,\sigma}^{(\alpha)} d_\sigma^\dagger c_{\alpha k \sigma} + \text{h.c.}, \tag{2.6}$$

with the tunnel matrix element $T_{k,\sigma}^{(\alpha)}$ specifying the transfer coupling between the lead $\alpha = L, R$ and the system. There is no direct coupling between the leads and electron transfer is only possible by charging and discharging the QD.

The cooperative effects are based on the collective hyperfine interaction of the electronic spin of the QD with N initially polarized nuclear spins in the host environment of the QD [27]. It is dominated by the isotropic contact term [28] given by

$$H_{\text{HF}} = \frac{g}{2}\left(A^+ S^- + A^- S^+\right) + g A^z S^z. \tag{2.7}$$

Here, S^μ and $A^\mu = \sum_{i=1}^N g_i \sigma_i^\mu$ with $\mu = +, -, z$ denote electron and collective nuclear spin operators, respectively. The coupling coefficients are normalized such that $\sum_i g_i^2 = 1$ and individual nuclear spin operators σ_i^μ are assumed to be spin $1/2$ for simplicity; g is related to the total HF coupling strength A_{HF} via $g = A_{\text{HF}} / \sum_i g_i$. We neglect the typically very small nuclear Zeeman and nuclear dipole-dipole terms [28]. For simplicity, we also restrict our analysis to one nuclear species only. These simplifications will be addressed in more detail in Sect. 2.7.

The effect of the HF interaction with the nuclear spin ensemble is two-fold: The first part of the above Hamiltonian $H_{\text{ff}} = \frac{g}{2}\left(A^+ S^- + A^- S^+\right)$ is a Jaynes-Cummings-type interaction which exchanges excitations between the QD electron and the nuclei. The second term $H_{\text{OH}} = g A^z S^z$ constitutes a quantum magnetic field, the Overhauser field, for the electron spin generated by the nuclei. If the Overhauser field is not negligible compared to the external Zeeman splitting, it can have a marked effect on the current by (de)tuning the hyperfine flip-flops.

2.5 Generalized Quantum Master Equation

Electron transport through a QD can be viewed as a tool to reveal the QD's non-equilibrium properties in terms of the current-voltage I/V characteristics. From a theoretical perspective, a great variety of methods such as the scattering matrix formalism [29] and non-equilibrium Green's functions [7, 30] have been used to explore the I/V characteristics of quantum systems that are attached to two metal leads. Our analysis is built upon the master equation formalism, a tool widely used in quantum optics for studying the irreversible dynamics of quantum systems coupled to a macroscopic environment.

In what follows, we employ a projection operator based technique to derive an effective master equation for the QD system—comprising the QD electron spin as well as the nuclear spins—which experiences dissipation via the electron's coupling to the leads. This dissipation is shown to dynamically depend on the state of the nuclear system potentially resulting in feedback mechanisms between the two subsystems. We derive conditions which allow for a Markovian treatment of the problem and list the assumptions our master equation based framework is based on.

2.5.1 Superoperator Formalism—Nakajima-Zwanzig Equation

The state of the global system that comprises the QD as well as the environment is represented by the full density matrix $\rho(t)$. However, the actual states of interest are the states of the QD which are described by the reduced density matrix $\rho_S = \text{Tr}_B[\rho]$, where $\text{Tr}_B \ldots$ averages over the unobserved degrees of freedom of the Fermi leads. We derive a master equation that governs the dynamics of the reduced density matrix ρ_S using the superoperator formalism. We start out from the von Neumann equation for the full density matrix

$$\dot{\rho} = -i\,[H\,(t)\,,\rho]\,, \tag{2.8}$$

where $H(t)$ can be decomposed into the following form which turns out to be convenient later on

$$H\,(t) = H_0\,(t) + H_1\,(t) + H_T. \tag{2.9}$$

Here, $H_0(t) = H_Z + H_B + g\,\langle A^z \rangle_t\,S^z$ comprises the Zeeman splitting caused by the external magnetic field via H_Z and the Hamiltonian of the non-interacting electrons in the leads H_B; moreover, the time-dependent expectation value of the Overhauser field has been absorbed into the definition of $H_0(t)$. The HF interaction between the QD electron and the ensemble of nuclear spins has been split up into the flip-flop term H_{ff} and the Overhauser field H_{OH}, that is $H_{HF} = H_{OH} + H_{ff}$. The term $H_1(t) = H_{\Delta OH}(t) + H_{ff}$ comprises the Jaynes-Cummings-type dynamics H_{ff} and fluctuations due to deviations of the Overhauser field from its expectation value, i.e., $H_{\Delta OH}(t) = g\delta A^z S^z$, where $\delta A^z = A^z - \langle A^z \rangle_t$.

The introduction of superoperators—operators acting on the space of linear operators on the Hilbert space—allows for a compact notation. The von Neumann equation is written as $\dot{\rho} = -i\mathcal{L}(t)\,\rho$, where $\mathcal{L}(t) = \mathcal{L}_0(t) + \mathcal{L}_1(t) + \mathcal{L}_T$ is the Liouville superoperator defined via $\mathcal{L}_\alpha \cdot = [H_\alpha, \cdot]$. Next, we define the superoperator $\mathcal{P}$ as a projector onto the relevant subspace

$$\mathcal{P}\rho\,(t) = \text{Tr}_B\,[\rho\,(t)] \otimes \rho_B^0 = \rho_S\,(t) \otimes \rho_B^0, \tag{2.10}$$

where ρ_B^0 describes separate thermal equilibria of the two leads whose chemical potentials are different due to the bias voltage $eV = \mu_L - \mu_R$. Essentially, $\mathcal{P}$ maps a density operator onto one of product form with the environment in equilibrium but still retains the relevant information on the system state. The complement of $\mathcal{P}$ is $\mathcal{Q} = 1 - \mathcal{P}$.

By inserting $\mathcal{P}$ and $\mathcal{Q}$ in front of both sides of the von Neumann equation one can derive a closed equation for the projection $\mathcal{P}\rho(t)$, which for factorized initial condition, where $\mathcal{Q}\rho(0) = 0$, can be rewritten in the form of the generalized Nakajima-Zwanzig master equation

$$\frac{d}{dt}\mathcal{P}\rho = -i\mathcal{P}\mathcal{L}\mathcal{P}\rho$$
$$- \int_0^t dt'\, \mathcal{P}\mathcal{L}\mathcal{Q}\,\hat{T}e^{-i\int_{t'}^t d\tau \mathcal{Q}\mathcal{L}(\tau)}\mathcal{Q}\mathcal{L}\mathcal{P}\rho(t'), \qquad (2.11)$$

which is non-local in time and contains all orders of the system-leads coupling [31]. Here, $\hat{T}$ denotes the chronological time-ordering operator. Since $\mathcal{P}$ and $\mathcal{Q}$ are projectors onto orthogonal subspaces that are only connected by $\mathcal{L}_T$, this simplifies to

$$\frac{d}{dt}\mathcal{P}\rho = -i\mathcal{P}\mathcal{L}\mathcal{P}\rho - \int_0^t dt'\mathcal{P}\mathcal{L}_T\hat{T}e^{-i\int_{t'}^t d\tau \mathcal{Q}\mathcal{L}(\tau)}\mathcal{L}_T\mathcal{P}\rho(t'). \qquad (2.12)$$

Starting out from this exact integro-differential equation, we introduce some approximations: In the weak coupling limit we neglect all powers of $\mathcal{L}_T$ higher than two (Born approximation). Consequently, we replace $\mathcal{L}(\tau)$ by $\mathcal{L}(\tau) - \mathcal{L}_T$ in the exponential of Eq. (2.12). Moreover, we make use of the fact that the nuclear spins evolve on a time-scale that is very slow compared to all electronic processes: In other words, the Overhauser field is quasi-static on the timescale of single electronic tunneling events [22, 32]. That is, we replace $\langle A^z \rangle_\tau$ by $\langle A^z \rangle_t$ in the exponential of Eq. (2.12) which removes the explicit time dependence in the kernel. By taking the trace over the reservoir and using $\text{Tr}_B\left[\mathcal{P}\dot{\rho}(t)\right] = \dot{\rho}_S(t)$, we get

$$\dot{\rho}_S(t) = -i\left(\mathcal{L}_Z + \mathcal{L}_{HF}\right)\rho_S(t)$$
$$- \int_0^t d\tau\, \text{Tr}_B\left(\mathcal{L}_T e^{-i[\mathcal{L}_0(t) + \mathcal{L}_1(t)]\tau}\mathcal{L}_T\mathcal{P}\rho(t-\tau)\right). \qquad (2.13)$$

Here, we also used the relations $\mathcal{P}\mathcal{L}_T\mathcal{P} = 0$ and $\mathcal{L}_B\mathcal{P} = 0$ and switched the integration variable to $\tau = t - t'$. Note that, for notational convenience, we suppress the explicit time-dependence of $\mathcal{L}_{0(1)}(t)$ in the following. In the next step, we iterate the Schwinger-Dyson identity

$$e^{-i(\mathcal{L}_0 + \mathcal{L}_1)\tau} = e^{-i\mathcal{L}_0\tau}$$
$$-i\int_0^\tau d\tau'\, e^{-i\mathcal{L}_0(\tau - \tau')}\mathcal{L}_1 e^{-i(\mathcal{L}_0 + \mathcal{L}_1)\tau'}. \qquad (2.14)$$

In what follows, we keep only the first term of this infinite series (note that the next two leading terms are explicitly calculated in Appendix 2.9.1). In quantum optics, this simplification is well known as approximation of independent rates of variation [33]. In our setting it is valid, if $\mathcal{L}_1(t)$ is small compared to $\mathcal{L}_0(t)$ and if the bath correlation time τ_c is short compared to the HF dynamics, $A_{HF} \ll 1/\tau_c$. Pictorially, this means that during the correlation time τ_c of a tunneling event, there is not sufficient time for the Rabi oscillation with frequency $g \lesssim A_{\mathrm{HF}}$ to occur. For typical materials [34], the relaxation time τ_c is in the range of $\sim 10^{-15}$ s corresponding to a relaxation rate $\Gamma_c = \tau_c^{-1} \approx 10^5$ μeV. Indeed, this is much faster than all other relevant processes. In this limit, the equation of motion for the reduced density matrix of the system simplifies to

$$\dot{\rho}_S(t) = -i\left(\mathcal{L}_Z + \mathcal{L}_{\mathrm{HF}}\right)\rho_S(t)$$
$$- \int_0^t d\tau\, \mathsf{Tr}_{\mathsf{B}}\left(\mathcal{L}_T e^{-i\mathcal{L}_0(t)\tau}\mathcal{L}_T\rho_S(t-\tau) \otimes \rho_B^0\right). \tag{2.15}$$

Note, however, that this master equation is not Markovian as the rate of change of $\rho_S(t)$ still depends on its past. Conditions which allow for a Markovian treatment of the problem will be addressed in the following.

2.5.2 Markov Approximation

Using the general relation $e^{-i\mathcal{L}_0\tau}\mathcal{O} = e^{-iH_0\tau}\mathcal{O}e^{iH_0\tau}$ for any operator $\mathcal{O}$, we rewrite Eq. (2.15) as

$$\dot{\rho}_S(t) = -i\left[H_Z + H_{\mathrm{HF}}, \rho_S(t)\right] - \int_0^t d\tau\, \mathsf{Tr}_{\mathsf{B}}\left(\left[H_T, \left[\tilde{H}_T(\tau), e^{-iH_0\tau}\rho_S(t-\tau)e^{iH_0\tau} \otimes \rho_B^0\right]\right]\right). \tag{2.16}$$

In accordance with the previous approximations, we replace $e^{-iH_0\tau}\rho_S(t-\tau)e^{iH_0\tau}$ by $\rho_S(t)$ which is approximately the same since any correction to H_0 would be of higher order in perturbation theory [35, 36]. In other words, the evolution of $\rho_S(t-\tau)$ is approximated by its unperturbed evolution which is legitimate provided that the relevant timescale for this evolution τ_c is very short (Markov approximation). This step is motivated by the typically rapid decay of the lead correlations functions [35]; the precise validity of this approximation is elaborated below. In particular, this simplification disregards dissipative effects induced by H_T which is valid self-consistently provided that the tunneling rates are small compared to the dynamics generated by H_0.

Moreover, in Eq. (2.16) we introduced the tunneling Hamiltonian in the interaction picture as $\tilde{H}_T(\tau) = e^{-iH_0\tau}H_T e^{iH_0\tau}$. For simplicity, we will only consider one lead for now and add the terms referring to the second lead later on. Therefore, we can disregard an additional index specifying the left or right reservoir and write explicitly

$$\tilde{H}_T(\tau) = \sum_{k,\sigma} T_{k,\sigma}e^{-i[\varepsilon_\sigma(t) - \varepsilon_k]\tau}d_\sigma^\dagger c_{k\sigma} + \text{h.c.} \tag{2.17}$$

Here, the resonances $\varepsilon_\sigma(t)$ are explicitly time-dependent as they dynamically depend on the polarization of the nuclear spins

$$\varepsilon_{\uparrow(\downarrow)}(t) = \varepsilon_{\uparrow(\downarrow)} \pm \frac{g}{2} \langle A^z \rangle_t . \tag{2.18}$$

The quantity

$$\omega = \varepsilon_{\uparrow}(t) - \varepsilon_{\downarrow}(t) = \omega_0 + g \langle A^z \rangle_t \tag{2.19}$$

can be interpreted as an effective Zeeman splitting which incorporates the external magnetic field as well as the mean magnetic field generated by the nuclei.

Since the leads are assumed to be at equilibrium, their correlation functions are given by

$$\text{Tr}_\text{B}\left[c_{k\sigma}^\dagger(\tau) c_{k'\sigma'} \rho_B^0 \right] = \delta_{\sigma,\sigma'} \delta_{k,k'} e^{-i\varepsilon_k \tau} f_k \tag{2.20}$$

$$\text{Tr}_\text{B}\left[c_{k\sigma}(\tau) c_{k'\sigma'}^\dagger \rho_B^0 \right] = \delta_{\sigma,\sigma'} \delta_{k,k'} e^{i\varepsilon_k \tau} (1 - f_k) , \tag{2.21}$$

where the Fermi function $f_k = (1 + \exp[\beta(\varepsilon_k - \mu)])^{-1}$ with inverse temperature $\beta = 1/(k_B T)$ gives the thermal occupation number of the respective lead in equilibrium. Note that all terms comprising two lead creation $c_{k\sigma}^\dagger$ or annihilation operators $c_{k\sigma}$ vanish since ρ_B^0 contains states with definite electron number only [35]. The correlation functions are diagonal in spin space and the tunneling Hamiltonian preserves the spin projection; therefore only co-rotating terms prevail. If we evaluate all dissipative terms appearing in Eq. (2.16), due to the conservation of momentum and spin in Eqs. (2.20) and (2.21), only a single sum over k, σ survives. Here, we single out one term explicitly, but all other terms follow analogously. We obtain

$$\dot{\rho}_S(t) = \cdots + \sum_\sigma \int_0^t d\tau \, \mathcal{C}_\sigma(\tau) d_\sigma^\dagger e^{-iH_0\tau} \rho_S(t - \tau) e^{iH_0\tau} d_\sigma, \tag{2.22}$$

where the correlation time of the bath τ_c is determined by the decay of the noise correlations

$$\mathcal{C}_\sigma(\tau) = \sum_k |T_{k,\sigma}|^2 f_k e^{i[\varepsilon_\sigma(t) - \varepsilon_k]\tau}$$

$$= \int_0^\infty d\varepsilon \, J_\sigma(\varepsilon) e^{i[\varepsilon_\sigma(t) - \varepsilon]\tau}. \tag{2.23}$$

Here, we made use of the fact that the leads are macroscopic and therefore exhibit a continuous density of states per spin $n(\varepsilon)$. On top of that, we have introduced the spectral density of the bath as

$$J_\sigma(\varepsilon) = D_\sigma(\varepsilon) f(\varepsilon), \tag{2.24}$$

where $D_\sigma(\varepsilon) = n(\varepsilon) |T_\sigma(\varepsilon)|^2$ is the effective density of states. The Markovian treatment manifests itself in a self-consistency argument: We assume that the spectral density of the bath $J_\sigma(\varepsilon)$ is flat around the (time-dependent) resonance $\varepsilon_\sigma(t)$ over a range set by the characteristic width Γ_d. Typically, both the tunneling matrix elements $T_\sigma(\varepsilon)$ as well as the density of states $n(\varepsilon)$ are slowly varying functions of energy. In the so-called wide-band limit the effective density of states $D_\sigma(\varepsilon)$ is assumed to be constant so that the self-consistency argument will exclusively concern the behaviour of the Fermi function $f(\varepsilon)$ which is intimately related to the temperature of the bath T. Under the condition, that $J_\sigma(\varepsilon)$ behaves flat on the scale Γ_d, it can be replaced by its value at $\varepsilon_\sigma(t)$, and the noise correlation simplifies to

$$\mathcal{C}_\sigma(\tau) = J_\sigma(\varepsilon_\sigma(t))\, e^{i\varepsilon_\sigma(t)\tau} \int_0^\infty d\varepsilon\, e^{-i\varepsilon\tau}. \tag{2.25}$$

Using the relation

$$\int_0^\infty d\varepsilon\, e^{-i\varepsilon\tau} = \pi\delta(\tau) - i\mathbb{P}\frac{1}{\tau}, \tag{2.26}$$

with $\mathbb{P}$ denoting Cauchy's principal value, we find that the Markov approximation $\mathrm{Re}\,[\mathcal{C}_\sigma(\tau)] \propto \delta(\tau)$ is fulfilled provided that the self-consistency argument holds. This corresponds to the white-noise limit where the correlation-time of the bath is $\tau_c = 0$. Pictorially, the reservoir has no memory and instantaneously relaxes to equilibrium. We can then indeed replace $e^{-iH_0\tau}\rho_S(t-\tau)e^{iH_0\tau}$ by $\rho_S(t)$ and extend the integration in Eq. (2.16) to infinity, with negligible contributions due to the rapid decay of the memory kernel. In the following, we derive an explicit condition for the self-consistency argument to be satisfied.

Let us first consider the limit $T = 0$: As schematically depicted in Fig. 2.3, in this case $f(\varepsilon)$ behaves perfectly flat except for $\varepsilon = \mu$ where the self-consisteny argument is violated. Therefore, the Markovian approximation is valid at $T = 0$ given that the

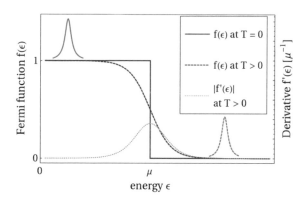

Fig. 2.3 Fermi function for finite temperature (*dashed blue line*) and in the limit $T = 0$ (*solid blue line*). The absolute value of the derivative of the Fermi function $f'(\varepsilon)$ (*dotted orange line* for finite temperature) is maximized at the chemical potential μ and tends to a delta function in the limit $T \to 0$. The Markovian description is valid provided that the Fermi function is approximately constant around the resonances $\varepsilon_\sigma(t)$ on a scale of the width of these resonances, schematically shown in red (*solid line* for $\varepsilon_\sigma(t) < \mu$ and *dashed line* for $\varepsilon_\sigma(t) > \mu$)

condition $|\varepsilon_\sigma(t) - \mu| \gg \Gamma_d$ is fulfilled. In this limit, all tunneling rates are constant over time and effectively decoupled from the nuclear dynamics. Note that for the observation of superradiant transport it is sufficient to restrict oneself to this case.

For a more general analysis, we now turn to the case of finite temperature $T > 0$. We require the absolute value of the relative change of the Fermi function around the resonance $\varepsilon_\sigma(t)$ over a range of the characteristic width Γ_d to be much less than unity, that is

$$\left| \frac{\partial f(\varepsilon)}{\partial \varepsilon} \right|_{\varepsilon_\sigma(t)} \Gamma_d \ll 1. \tag{2.27}$$

An upper bound for the first factor can easily be obtained as this quantity is maximized at the chemical potential μ, for all temperatures. Evaluating the derivative at $\varepsilon_\sigma(t) = \mu$ results in the compact condition

$$\Gamma_d \ll 4k_B T. \tag{2.28}$$

Thus, finite temperature $T > 0$ washes out the rapid character of $f(\varepsilon)$ at the chemical potential μ and, provided that Eq. (2.28) is fulfilled, allows for an Markovian treatment.

Two distinct mechanisms contribute to the width Γ_d: dissipation due to coupling to the leads and the effect of $H_1(t)$. Both of them have been neglected self-consistently in the memory kernel when going from Eqs. (2.12) to (2.15). Typically, the tunneling rates are of the order of ~ 5–$20\,\mu eV$, depending on the transparency of the tunnel-barrier. Regarding the contribution due to $H_1(t)$, we first consider two limits of particular importance: For a completely mixed state the fluctuation of the nuclear field around its zero expectation value is of the order of $\sim A_{HF}/\sqrt{N} \approx 0.1\,\mu eV$. In contrast, for a fully polarized state these fluctuations can be neglected whereas the effective strength of the flip-flop dynamics is $\sim A_{HF}/\sqrt{N}$ as well. Therefore, in both limits considered here, the dominant contribution to Γ_d is due to the coupling to the leads and the self-consistency condition could still be met with cryostatic temperatures $k_B T \gtrsim 10\,\mu eV$, well below the orbital level spacing. However, we note that in the course of a superradiant evolution, where strong correlations among the nuclei build up, the dominant contribution to Γ_d may come from the flip-flop dynamics, which are $A_{HF}/4 \approx 25\,\mu eV$ at maximum for homogeneous coupling. For realisitic conditions, though, this effect is significantly reduced, as demonstrated in our simulations in Sect. 2.7.

2.5.3 General Master Equation for Nuclear Spin Assisted Transport

Assuming that the self-consistency argument for a Markovian treatment is satisfied, we now apply the following modifications to Eq. (2.16): First, we neglect level shifts due to the coupling to the continuum states which can be incorporated by replacing the bare frequencies $\varepsilon_\sigma(t)$ with renormalized frequencies. Second, one adds the second electron reservoir that has been omitted in the derivation above. Lastly, one performs a suitable transformation into a frame rotating at the frequency $\bar{\varepsilon} = (\varepsilon_\uparrow + \varepsilon_\downarrow)/2$ leaving all terms invariant but changing H_Z from $H_Z = \varepsilon_\uparrow d_\uparrow^\dagger d_\uparrow + \varepsilon_\downarrow d_\downarrow^\dagger d_\downarrow$ to $H_Z = \omega_0 S^z$. After these manipulations one arrives at the central master equation as stated in Eq. (2.2) where the tunneling rates with $\alpha_\sigma(t) = \sum_{x=L,R} \alpha_\sigma^{(x)}(t)$, $\beta_\sigma(t) = \sum_{x=L,R} \beta_\sigma^{(x)}(t)$ and

$$\frac{\alpha_\sigma^{(x)}(t)}{2\pi} = n_x(\varepsilon_\sigma(t)) \left|T_\sigma^{(x)}(\varepsilon_\sigma(t))\right|^2 [1 - f_x(\varepsilon_\sigma(t))]$$

$$\frac{\beta_\sigma^{(x)}(t)}{2\pi} = n_x(\varepsilon_\sigma(t)) \left|T_\sigma^{(x)}(\varepsilon_\sigma(t))\right|^2 f_x(\varepsilon_\sigma(t)) \qquad (2.29)$$

govern the dissipative processes in which the QD system exchanges single electrons with the leads. The tunneling rates, as presented here, are widely used in nanostructure quantum transport problems [35, 37, 38]. However, in our setting they are evaluated at the resonances $\varepsilon_\sigma(t)$ which dynamically depend on the polarization of the nuclear spins; see Eq. (2.18). Note that Eq. (2.2) incorporates finite temperature effects via the Fermi functions of the leads. This potentially gives rise to feedback mechansims between the electronic and the nuclear dynamics, since the purely electronic diffusion markedly depends on the nuclear dynamics.

Since Eq. (2.2) marks our first main result, at this point we quickly reiterate the assumptions our master equation treatment is based on:

- The system-lead coupling is assumed to be weak and therefore treated perturbatively up to second order (Born-approximation).
- In particular, the tunneling rates are small compared to the effective Zeeman splitting ω.
- Level shifts arising from the coupling to the continuum states in the leads are merely incorporated into a redefinition of the QD energy levels $\varepsilon_\sigma(t)$.
- There is a separation of timescales between electron-spin dynamics and nuclear-spin dynamics. In particular, the Overhauser field $g\langle A^z\rangle_t$ evolves on a timescale that is slow compared to single electron tunneling events.
- The HF dynamics generated by $H_1(t) = H_{\mathrm{ff}} + H_{\mathrm{\Delta OH}}(t)$ is (i) sufficiently weak compared to H_0 and (ii) slow compared to the correlation time of the bath τ_c, that is $A_{\mathrm{HF}}\tau_c \ll 1$ (approximation of independent rates of variation). Note that the flip-flop dynamics can become very fast as correlations among the nuclei build up culminating in a maximum coupling strength of $A_{\mathrm{HF}}/4$ for homogeneous coupling. This potentially drives the system into the strong coupling regime where condition (i), that is $\omega \gg \|H_1(t)\|$, might be violated. However, under realistic conditions of inhomogeneous coupling this effect is significantly reduced.
- The effective density of states $D_\sigma(\varepsilon) = n(\varepsilon)|T_\sigma(\varepsilon)|^2$ is weakly energy-dependent (wide-band limit). In particular, it is flat on a scale of the characteristic widths of the resonances.

- The Markovian description is valid provided that either the resonances are far away from the chemical potentials of the leads on a scale set by the characteristic widths of the resonances or the temperature is sufficiently high to smooth out the rapid character of the Fermi functions of the leads. This condition is quantified in Eq. (2.28).

In summary, we have derived a Quantum master equation describing electronic transport through a single QD which is collectively enhanced due to the interaction with a large ancilla system, namely the nuclear spin ensemble in the host environment. Equation (2.2) incorporates two major intriguing features both of theoretical and experimental relevance: Due to a separation of timescales, only the electronic subsystem experiences dissipation with rates that depend dynamically on the state of the ancilla system. This non-linearity gives rise to feedback mechanisms between the two subsystems as well as hysteretic behavior. Moreover, the collective nature of the HF interaction offers the possibility to observe intriguing coherent many-body effects. Here, one particular outcome is the occurence of superradiant electron transport, as shown in the remainder of this chapter.

Note that in the absence of HF interaction between the QD electron and the proximal nuclear spins, i.e., in the limit $g \to 0$, our results agree with previous theoretical studies [36].

2.6 Superradiance-like Electron Transport

Proceeding from our general theory derived above, this section is devoted to the prediction and analysis of superradiant behavior of nuclear spins, evidenced by the strongly enhanced leakage current through a single QD in the Coulomb-blockade regime; see Fig. 2.1 for the scheme of the setup. A pronounced peak in the leakage current will serve as the main evidence for SR behaviour in this setting.

We note that, in principle, an enhancement seen in the leakage current could also simply arise from the Overhauser field dynamically tuning the hyperfine flip-flops. However, we can still ensure that the measured change in the leakage current through the QD is due to cooperative emission only by dynamically compensating the Overhauser field. This can be achieved by applying a time-dependent magnetic or spin-dependent AC Stark field such that $H_{comp}(t) = -g \langle A^z \rangle_t S^z$ which is done in most of our simulations below to clearly prove the existence of superradiant behaviour in this setting. Consequently, in our previous analysis $H_0(t)$ is replaced by $H_0 = H_0(t) - g \langle A^z \rangle_t S^z = H_Z + H_B$ so that the polarization dependence of the tunneling rates is removed and we can drop the explicit time-dependence of the resonances $\varepsilon_\sigma(t) \to \varepsilon_\sigma$. Under this condition, the master equation for the reduced system density operator can be written as

$$\dot{\rho}_S\left(t\right) = -i\left[\omega_0 S^z + H_{\mathrm{HF}} + H_{\mathrm{comp}}\left(t\right), \rho_S\left(t\right)\right]$$

$$+ \sum_{\sigma=\uparrow,\downarrow} \alpha_\sigma \left[d_\sigma \rho_S\left(t\right) d_\sigma^\dagger - \frac{1}{2}\left\{d_\sigma^\dagger d_\sigma, \rho_S\left(t\right)\right\}\right]$$

$$+ \sum_{\sigma=\uparrow,\downarrow} \beta_\sigma \left[d_\sigma^\dagger \rho_S\left(t\right) d_\sigma - \frac{1}{2}\left\{d_\sigma d_\sigma^\dagger, \rho_S\left(t\right)\right\}\right]. \qquad (2.30)$$

In accordance with our previous considerations, in this specific setting the Markovian treatment is valid provided that the spectral density of the reservoirs varies smoothly around the (time-independent) resonances ε_σ on a scale set by the natural widths of the level and the fluctuations of the dynamically compensated Overhauser field. More specifically, throughout the whole evolution the levels are assumed to be far away from the chemical potentials of the reservoirs [39, 40]; for an illustration see Fig. 2.3. In this wide-band limit, the tunneling rates α_σ, β_σ are independent of the state of the nuclear spins. The master equation is of Lindblad form which guarantees the complete positivity of the generated dynamics. Equation (2.30) agrees with previous theoretical results [36] except for the appearance of the collective HF interaction between the QD electron and the ancilla system in the Hamiltonian dynamics of Eq. (2.30).

To some extent, Eq. (2.30) bears some similarity with the quantum theory of the laser. While in the latter the atoms interact with bosonic reservoirs, in our transport setting the QD is pumped by the nuclear spin ensemble and emits fermionic particles [30, 38].

If the HF dynamics is the slowest timescale in the problem, Eq. (2.30) can be recast into a form which makes its superradiant character more apparent. In this case, the system is subject to the slaving principle [30]. The dynamics of the whole system follow that of the subsystem with the slowest time constant allowing to adiabatically eliminate the electronic QD coordinates and to obtain an effective equation of motion for the nuclear spins. In this limit, the Overhauser field is much smaller than the Zeeman splitting so that a dynamic compensation of the OH can be disregarded for the moment. For simplicity, we consider a transport setting in which only four tunneling rates are different from zero, see Fig. 2.1. The QD can be recharged from the left and the right lead, but only electrons with spin projection $\sigma = \uparrow$ can tunnel out of the QD into the right lead. We define the total recharging rate $\beta = \beta_\downarrow + \beta_\uparrow = \beta_\downarrow^{(L)} + \beta_\downarrow^{(R)} + \beta_\uparrow^{(L)}$ and for notational convenience unambiguously set $\alpha = \alpha_\uparrow^{(R)}$. First, we project Eq. (2.30) onto the populations of the electronic levels and the coherences in spin space according to $\rho_{mn} = \langle m|\rho_S|n\rangle$, where $m, n = 0, \uparrow, \downarrow$. This yields

$$\dot{\rho}_{00} = \alpha\rho_{\uparrow\uparrow} - \beta\rho_{00}, \tag{2.31}$$

$$\dot{\rho}_{\uparrow\uparrow} = -i\frac{g}{2}\left[A^z, \rho_{\uparrow\uparrow}\right] - i\frac{g}{2}\left(A^-\rho_{\downarrow\uparrow} - \rho_{\uparrow\downarrow}A^+\right) - \alpha\rho_{\uparrow\uparrow} + \beta_\uparrow\rho_{00}, \tag{2.32}$$

$$\dot{\rho}_{\downarrow\downarrow} = +i\frac{g}{2}\left[A^z, \rho_{\downarrow\downarrow}\right] - i\frac{g}{2}\left(A^+\rho_{\uparrow\downarrow} - \rho_{\downarrow\uparrow}A^-\right) + \beta_\downarrow\rho_{00}, \tag{2.33}$$

$$\dot{\rho}_{\uparrow\downarrow} = -i\omega_0\rho_{\uparrow\downarrow} - i\frac{g}{2}\left(A^z\rho_{\uparrow\downarrow} + \rho_{\uparrow\downarrow}A^z\right) - i\frac{g}{2}\left(A^-\rho_{\downarrow\downarrow} - \rho_{\uparrow\uparrow}A^-\right) - \frac{\alpha}{2}\rho_{\uparrow\downarrow}. \tag{2.34}$$

We can retrieve an effective master equation for the regime in which on relevant timescales the QD is always populated by an electron. This holds for a sufficiently strong recharging rate, that is in the limit $\beta \gg \alpha$, which can be implemented experimentally by making the left tunnel barrier more transparent than the right one. Then, the state $|0\rangle$ is populated negligibly throughout the dynamics and can be eliminated adiabatically according to $\rho_{00} \approx \frac{\alpha}{\beta}\rho_{\uparrow\uparrow}$. In analogy to the Anderson impurity model, in the following this limit will be referred to as *local moment regime*. The resulting effective master equation reads

$$\begin{aligned}\dot{\rho}_S = &-i\left[\omega_0 S^z + H_{\text{HF}}, \rho_S\right] \\ &+ \gamma\left[S^-\rho_S S^+ - \frac{1}{2}\left\{S^+ S^-, \rho_S\right\}\right] \\ &+ \Gamma\left[S^z\rho_S S^z - \frac{1}{4}\rho_S\right],\end{aligned} \tag{2.35}$$

where

$$\gamma = \frac{\beta_\downarrow}{\beta}\alpha \tag{2.36}$$

is an effective decay rate and

$$\Gamma = \frac{\beta_\uparrow}{\beta}\alpha \tag{2.37}$$

represents an effective electronic dephasing rate. This situation is schematized in Fig. 2.4. The effective decay (dephasing) describes processes in which the QD is recharged with a spin down (up) electron after a spin up electron has tunneled out of the QD. As demonstrated in Ref. [41], additional electronic dephasing mechanisms only lead to small corrections to the dephasing rate Γ and are therefore neglected in Eq. (2.35).

In the next step we aim for an effective description that contains only the nuclear spins: Starting from a fully polarized state, SR is due to the increase in the operative HF matrix element $\langle A^+ A^-\rangle$. The scale of the coupling is set by the total HF coupling constant $A_{\text{HF}} = g\sum_i g_i$. For a sufficiently small *relative coupling strength* [27]

$$\varepsilon = A_{\text{HF}}/(2\Delta), \tag{2.38}$$

Fig. 2.4 The electronic QD system in the local moment regime after the adiabatic elimination of the $|0\rangle$ level including the relevant dissipative processes. Within the effective system (*box*) we encounter an effective decay term and an effective pure dephasing term, with the rates γ and Γ, respectively. This simplification is possible for fast recharging of the QD, i.e., $\beta \gg \alpha$

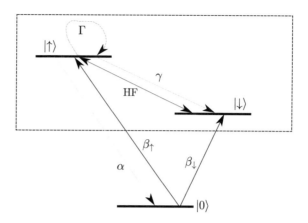

where

$$\Delta = |\alpha/2 + i\omega_0|, \tag{2.39}$$

the electron is predominantly in its $|\downarrow\rangle$ spin state and we can project Eq. (2.35) to the respective subspace. As shown in detail in Appendix 2.9.2, in this limit the master equation for the reduced nuclear density operator $\mu = \mathsf{Tr}_{\mathsf{el}}\left[\rho_S\right]$ is given by Eq. (2.3), where the effective coefficients read

$$c_r = \frac{g^2\alpha}{4\Delta^2}, \tag{2.40}$$

$$c_i = \frac{g^2\omega_0}{4\Delta^2}. \tag{2.41}$$

This master equation is our second main result. In an optical setting, it has previously been predicted theoretically to exhibit strong SR signatures [27]. Conceptually, its superradiant character can be understood immediately in the ideal case of homogeneous coupling in which the collective state of all nuclear spins can be described in terms of Dicke states $|J, m\rangle$: The enhancement of the HF interaction is directly associated with the transition through nuclear Dicke states $|J, m\rangle$, $m \ll J$. In this idealized setting, the angular momentum operator $\mathbf{I} = \sqrt{N}\mathbf{A}$ of the nuclear spin ensemble obeys the SU(2) Lie algebra, from which one can deduce the ladder operator relation $I^-|J, m\rangle = \sqrt{J(J + 1) - m(m - 1)}\,|J, m - 1\rangle$. This means that, starting from an initially fully polarized state $|J = N/2, m = N/2\rangle$, the nuclear system cascades down the Dicke-ladder with an effective rate

$$\tilde{\Gamma}_{m \to m-1} = \frac{c_r}{N}(N/2 + m)(N/2 - m + 1), \tag{2.42}$$

since, according to the first term in Eq. (2.3), the populations of the Dicke states evolve as

$$\dot{\mu}_{m,m} = -\frac{c_r}{N} \left(N/2 + m \right) \left(N/2 - m + 1 \right) \mu_{m,m}$$
$$+ \frac{c_r}{N} \left(N/2 + m + 1 \right) \left(N/2 - m \right) \mu_{m+1,m+1}. \qquad (2.43)$$

While the effective rate is $\tilde{\Gamma}_{N/2 \rightarrow N/2-1} = c_r$ at the very top of the the ladder it increases up to $\tilde{\Gamma}_{|m| \ll N/2} \approx \frac{c_r}{4} N$ at the center of the Dicke ladder. This implies the characteristic intensity peaking as compared to the limit of independent classical emitters the emission rate of which would be $\tilde{\Gamma}_{\text{cl}} = \frac{c_r}{N} N_\uparrow = \frac{c_r}{N} \left(N/2 + m \right)$.

However, there is also a major difference compared to the superradiant emission of photons from atomic ensembles: In contrast to its atomic cousin, the prefactor $c_r/N \propto 1/N^2$ is N-dependent, resulting in an overall time of the SR evolution $\langle t_D \rangle$ which increases with N. By linearizing Eq. (2.42) for the beginning of the superradiant evolution [24] as $\tilde{\Gamma}_{m \rightarrow m-1} \approx c_r(s+1)$, where $s = N/2 - m$ gives the number of nuclear flips, one finds that the first flip takes place in an average time c_r^{-1}, the second one in a time $(2c_r)^{-1}$ and so on. The summation of all these elementary time intervals gives an upper bound estimate for the process duration till the SR peaking as

$$\langle t_D \rangle \lesssim \frac{2}{c_r} \left[1 + \frac{1}{2} + \cdots + \frac{1}{N/2} \right] \approx \frac{2 \ln(N/2)}{c_r}, \qquad (2.44)$$

which, indeed, increases with the number of emitters as $\sim N \ln(N)$, whereas one obtains $\langle t_D \rangle \sim \ln(N)/N$ for ordinary superradiance [24]. Accordingly, in our solid-state system the characteristic SR peak appears at later times for higher N. The underlying reason for this difference is that in the atomic setting each new emitter adds to the overall coupling strength, whereas in the central spin setting a fixed overall coupling strength A_{HF} is distributed over an increasing number of particles. Note that in an actual experimental setting N is not a tunable parameter, of course. For our theoretical discussion, though, it is convenient to fix the total HF coupling strength A_{HF} and to extrapolate from our findings to an experimentally relevant number of nuclear spins N.

For large relative coupling strength $\varepsilon \gg 1$ the QD electron saturates and super-radiant emission is capped by the decay rate $\alpha/2$, prohibiting the observation of a strong intensity peak. In order to circumvent this bottleneck regime, one has to choose a detuning ω_0 such that $0 < \varepsilon \leq 1$. However, to realize the spin-blockade regime, where the upper spin manifold is energetically well separated from the lower spin manifold, the Zeeman splitting has to be of the order of $\omega_0 \sim A_{\text{HF}}$ which guarantees $\varepsilon < 1$. In this parameter range, the early stage of the evolution—in which the correlation buildup necessary for SR takes place [24]—is well described by Eq. (2.3).

The inhomogeneous nature ($g_i \neq$ const) of the collective operators A^μ leads to dephasing between the nuclei, possibly preventing the phased emission necessary for the observation of SR [24, 27, 42, 43]. The inhomogeneous part of the last term in Eq. (2.3)—the electron's Knight field—causes dephasing [44] $\propto g \sqrt{\text{Var}(g_i)}/2$, possibly leading to symmetry reducing transitions $J \rightarrow J - 1$. Still, it has been shown that SR is also present in realistic inhomogeneous systems [27], since the system evolves in a many-body protected manifold (MPM): The second term in Eq. (2.3) energetically separates different total nuclear spin-J manifolds, protecting the correlation build-up for large enough ε.

The superradiant character of Eq. (2.3) suggests the observation of its prominent intensity peak in the leakage current through the QD in the spin-blockade regime. We have employed the method of Full-Counting-Statistics (FCS) [45, 46] in order to obtain an expression for the current and find (setting the electron's charge $e = 1$)

$$I(t) = \alpha \rho_{\uparrow\uparrow} - \beta_{\downarrow}^{(R)} \rho_{00}. \tag{2.45}$$

This result is in agreement with previous theoretical findings: The current through the device is completely determined by the occupation of the levels adjacent to one of the leads [29, 37, 39]. The first term describes the accumulation of electrons with spin $\sigma = \uparrow$ in the right lead, whereas the second term describes electrons with $\sigma = \downarrow$ tunneling from the right lead into the QD. As done before [27], we take the ratio of the maximum current to the initial current (the maximum for independent emitters) $I_{\text{coop}}/I_{\text{ind}}$ as our figure of merit: a relative intensity peak height $I_{\text{coop}}/I_{\text{ind}} > 1$ indicates cooperative effects. One of the characteristic features of SR is that this quantity scales linearly with the number of spins N.

In the local-moment regime, described by Eq. (2.35), the expression for the current simplifies to $I(t) = (1 - \beta_{\downarrow}^{(R)}/\beta)\alpha \langle S^+ S^- \rangle_t \propto \langle S^+ S^- \rangle_t$ showing that it is directly proportional to the electron inversion. This, in turn, increases as the nuclear system pumps excitations into the electronic system. A compact expression for the relation between the current and the dynamics of the nuclear system can be obtained immediately in the case of homogeneous coupling

$$\frac{d}{dt} \langle S^+ S^- \rangle_t = -\frac{d}{dt} \langle I^z \rangle_t - \gamma \langle S^+ S^- \rangle_t. \tag{2.46}$$

Since the nuclear dynamics are in general much slower than the electron's dynamics, the approximate solution of this equation is $\langle S^+ S^- \rangle_t \approx -\frac{d}{dt} \langle I^z \rangle_t / \gamma$. As a consequence, the current $I(t)$ is proportional to the time-derivative of the nuclear polarization

$$I(t) \propto -\frac{d}{dt} \langle I^z \rangle_t. \tag{2.47}$$

Still, no matter how strong the cooperative effects are, on a timescale of single electron tunneling events, the electrons will always be emitted antibunched, since in the strong Coulomb-blockade regime the QD acts as a single-electron emitter [47]. Typically, the rate for single-electron emission events is even below the tunneling rate α due to the spin-blockade. On electronic timescales $\sim 1/\alpha$, the SR mechanism manifests in lifting this blockade; as argued above, the efficiency of this process is significantly enhanced by collective effects.

Before we proceed with an in-depth analysis of the current $I(t)$, we note that an intriguing extension of the present work would be the study of fluctuations thereof (see for example Ref. [48] for studies of the shot noise spectrum in a related system). Insights into the statistics of the current could be obtained by analyzing two-time correlation functions such as $\langle n_\uparrow(t+\tau)n_\uparrow(t)\rangle$, where $n_\uparrow = d_\uparrow^\dagger d_\uparrow$. This can conveniently be done via the Quantum Regression Theorem [49] which yields the formal result $\langle n_\uparrow(t+\tau)n_\uparrow(t)\rangle = \text{Tr}_S\left[n_\uparrow e^{\mathcal{W}\tau}\left(n_\uparrow \rho_S(t)\right)\right]$. Here, $\mathcal{W}$ denotes the Liouvillian governing the system's dynamics according to $\dot\rho_S = \mathcal{W}\rho_S$ (see Eq. 2.35) and $\text{Tr}_S[\ldots]$ refers to the trace over the system's degree of freedoms. This procedure can be generalized to higher order correlation functions and full evaluation of the current statistics might reveal potential connections between current fluctuations and cooperative nuclear dynamics.

2.7 Analysis and Numerical Results

2.7.1 Experimental Realization

The proposed setup described here may be realized with state-of-the-art experimental techniques. First, the Markovian regime, valid for sufficiently large bias eV, is realized if the Fermi functions of the leads are smooth on a scale set by the natural widths of the levels and residual fluctuations due to the dynamically compensated Overhauser field. Since for typical materials [8] the hyperfine coupling constant is $A_{HF} = 1-100\,\mu\text{eV}$ and tunneling rates are typically [9] of the order of $\sim 10\,\mu\text{eV}$, this does not put a severe restriction on the bias voltage which is routinely [18, 19] in the range of hundreds of μV or mV. Second, in order to tune the system into the spin-blockade regime, a sufficiently large external magnetic field has to be applied. More precisely, the corresponding Zeeman splitting ω_0 energetically separates the upper and lower manifolds in such a way that the Fermi function of the right lead drops from one at the lower manifold to zero at the upper manifold. Finite temperature T smears out the Fermi function around the chemical potential by approximately $\sim k_B T$. Accordingly, with cryostatic temperatures of $k_B T \sim 10\,\mu\text{eV}$ being routinely realized in the lab [10], this condition can be met by applying an external magnetic field of $\sim 5-10$ T which is equivalent to $\omega_0 \approx 100-200\,\mu\text{eV}$ in GaAs [8, 50]. Lastly, the charging energy U, typically $\sim 1-4$ meV [9, 19], sets the largest energy scale in the problem justifying the Coulomb-blockade regime with negligible double occupancy of the QD provided that the chemical potential of the left lead is well below the doubly occupied level. Lastly, we note that similar setups to the one proposed here have previously been realized experimentally by, e.g., Hanson et al. [26, 50].

Proceeding from these considerations, we now show by numerical simulation that an SR peaking of several orders of magnitude can be observed for experimentally relevant parameters in the leakage current through a quantum dot in the spin-blockade regime. We first consider the idealized case of homogeneous coupling for which an exact numerical treatment is feasible even for a larger number of coupled nuclei. Then, we continue with the more realistic case of inhomogeneous coupling for which an approximative scheme is applied. Here, we also study scenarios in which the nuclear spins are not fully polarized initially. Moreover, we discuss intrinsic nuclear dephasing effects and undesired cotunneling processes which have been omitted in our simulations. In particular, we show that the inhomogeneous nature of the HF coupling accounts for the strongest dephasing mechanism in our system. We note that this effect is covered in the second set of our simulations. Finally, we self-consistently justify the perturbative treatment of the Overhauser-field fluctuations as well as the HF flip-flop dynamics.

2.7.2 Superradiant Electron Transport

Idealized Setting
The homogeneous case allows for an exact treatment even for a relatively large number of nuclei as the system evolves within the totally symmetric low-dimensional subspace $\{|J, m\rangle,\ m = -J, \ldots, J\}$. Starting from a fully-polarized state, a strong intensity enhancement is observed; typical results obtained from numerical simulations of Eq. (2.30) are depicted in Fig. 2.5 for $N = 60$ and $N = 100$ nuclear spins. The corresponding relative peak heights display a linear dependence with N, cf. Fig. 2.6, which we identify as the characteristic feature of superradiance. Here, we have used the numerical parameters $A_{HF} = 1$, $\omega_0 = 1$ and $\alpha = \beta_\uparrow^{(L)} = \beta_\downarrow^{(L)} = \beta_\downarrow^{(R)} = 0.1$ in units of $\sim 100\,\mu eV$, corresponding to a relative coupling strength $\varepsilon = 0.5$.

Before we proceed, some further remarks on the dynamic compensation of the Overhauser field seem appropriate: We have merely introduced it in our analysis

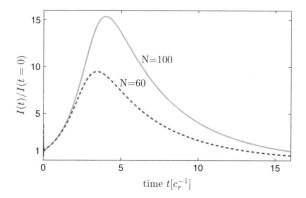

Fig. 2.5 Typical time-evolution of the normalized current for homogeneous coupling under dynamical compensation of the Overhauser field and a relative coupling strength of $\varepsilon = 0.5$, shown here for $N = 60$ and $N = 100$ nuclear spins. The characteristic feature of superradiance, a pronounced peak in the leakage current proportional to N, is clearly observed

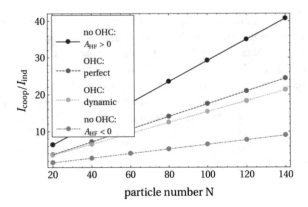

Fig. 2.6 Ratio of the maximum current to the initial current I_{coop}/I_{ind} as a function of the number of nuclear spins N for homogeneous coupling and a relative coupling strength of $\varepsilon = 0.5$: results for perfect compensation (*dashed line*) are compared to the case of dynamic compensation (*dotted line*) of the Overhauser field (OHC). Simulations without compensation of the Overhauser field set bounds for the enhancement of the leakage current, depending on the sign of the HF coupling constant A_{HF}; *solid* and *dash-dotted line* for $A_{HF} > 0$ and $A_{HF} < 0$, respectively

in order to provide a clear criterion for the presence of purely collective effects, given by $I_{coop}/I_{ind} > 1$. In other words, dynamic compensation of the Overhauser field is not a necessary requirement for the observation of collective effects, but it is rather an adequate tool to display them clearly. From an experimental point of view, the dynamic compensation of the Overhauser field might be challenging as it requires accurate knowledge about the evolution of the nuclear spins. Therefore, we also present results for the case in which the external magnetic field is constant and no compensation is applied. Here, we can distinguish two cases: Depending on the sign of the HF coupling constant A_{HF}, the time-dependence of the effective Zeeman-splitting ω can either give rise to an additional enhancement of the leakage current ($A_{HF} > 0$) or it can counteract the collective effects ($A_{HF} < 0$). As shown in Fig. 2.6, this sets lower and upper bounds for the observed enhancement of the leakage current.

In Fig. 2.6 we also compare the results obtained for dynamic compensation of the Overhauser field to the idealized case of perfect compensation in which the effect of the Overhauser term is set to zero, i.e., $H_{OH} = g A^z S^z = 0$. Both approaches display the same features justifying our approximation of neglecting residual (de)tuning effects of the dynamically compensated Overhauser field w.r.t. the external Zeeman splitting ω_0. This is also discussed in greater detail below.

Beyond the Idealized Setting
Inhomogeneous HF coupling.—In principle, the inhomogeneous HF coupling could prevent the phasing necessary for SR. However, as shown below, SR is still present in realistically inhomogeneous systems. In contrast to the idealized case of homogeneous coupling, the dynamics cannot be restricted to a low-dimensional subspace so that an exact numerical treatment is not feasible due to the large number of nuclei. We

therefore use an approximate approach which has previously been shown to capture
the effect of nuclear spin coherences while allowing for a numerical treatment of
hundreds of spins [22, 27]. For simplicity, we restrict ourselves to the local moment
regime in which the current can be obtained directly from the electron inversion
$I(t) \propto \langle S^+ S^- \rangle_t$. By Eq. (2.35), this expectation value is related to a hierachy of correlation terms involving both the electron and nuclear spins. Based on a Wick type
factorization scheme, higher order expressions are factorized in terms of the covari-
ance matrix $\gamma_{ij}^+ = \langle \sigma_i^+ \sigma_j^- \rangle$ and the "mediated covariance matrix" $\gamma_{ij}^- = \langle \sigma_i^+ S^z \sigma_j^- \rangle$.
For further details, see Refs. [22, 27].

The coupling constants g_j have been obtained from the assumption of a two-
dimensional Gaussian spatial electron wavefunction of width $\sqrt{N}/2$. Specifically,
we will present results for two sets of numerical parameters, corresponding to a
relative coupling strength of $\varepsilon = 0.5$, where $A_{HF} = 1$, $\omega_0 = 1$, $\gamma = 0.1$ and
$\Gamma = 0.08$, and $\varepsilon = 0.55$ with $A_{HF} = 1$, $\omega_0 = 0.9$, $\gamma = 0.1$ and $\Gamma = 0.067$.

As shown in Figs. 2.7 and 2.8, the results obtained with these methods demonstrate clear SR signatures. In comparison to the ideal case of homogeneous coupling,
the relative height is reduced, but for a *fully polarized* initial state we still find a
linear enhancement $I_{coop}/I_{ind} \approx 0.043\,N$ ($\varepsilon = 0.5$); therefore, as long as this linear
dependence is valid, for typically $N \approx 10^5$–10^6 a strong intensity enhancement of
several orders of magnitude is predicted $(\sim 10^3$–$10^4)$.

Imperfect initial polarization.—If the initial state is not fully polarized, SR effects
are reduced: However, when starting from a mixture of symmetric Dicke states $|J, J\rangle$
with polarization $p = 80(60)\,\%$, we find that the linear N dependence is still present:
$I_{coop}/I_{ind} \approx 0.0075(0.0025)\,N$ for $\varepsilon = 0.5$, i.e., the scaling is about a factor of

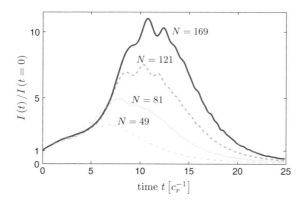

Fig. 2.7 Typical time-evolution of the normalized current for inhomogeneous coupling, shown
here for up to $N = 13^2$ nuclear spins and a relative coupling strength $\varepsilon = 0.55$. Compared to the
idealized case of homogeneous coupling, the SR effects are reduced, but still clearly present. A
Gaussian spatial electron wave function has been assumed and the Overhauser field is compensated
dynamically

Fig. 2.8 Ratio of the maximum current to the initial current $I_{\mathrm{coop}}/I_{\mathrm{ind}}$ as a function of the number of nuclear spins N for relative coupling strengths $\varepsilon = 0.5$ and $\varepsilon = 0.55$: results for inhomogeneous coupling. The linear dependence is still present when starting from a nuclear state with finite polarization $p = 0.8$

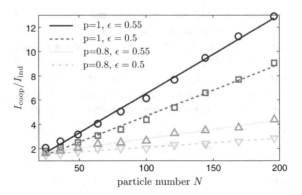

$\sim$5(15) weaker than for full polarization.[1] Still, provided the linear scaling holds up to an experimentally realistic number of nuclei $N \approx 10^5 - 10^6$, this amounts to a relative enhancement of the order of $I_{\mathrm{coop}}/I_{\mathrm{ind}} \sim 10^2 - 10^3$. To clearly resolve this peak experimentally, any spurious current should not be larger than the initial HF-mediated leakage current. As we argue below, this condition can be fulfilled in our setup, since the main spurious mechanism, cotunneling, is strongy suppressed.

Nuclear Zeeman term and species inhomogeneity.—In our simulations we have disregarded the nuclear Zeeman energies. For a single nuclear species, this term plays no role in the SR dynamics. However, in typical QDs several nuclear species with different g factors are present ("species inhomogeneity"). In principle, these are large enough to cause additional dephasing between the nuclear spins, similar to the inhomogeneous Knight field [22]. However, this dephasing mechanism only applies to nuclei which belong to different species [22]. This leads to few (in GaAs three) mutually decohered subsystems each of which is described by our theory.

Nuclear interactions.—Moreover, we have neglected the dipolar and quadrupolar interactions among the nuclear spins. First, the nuclear dipole-dipole interaction can cause diffusion and dephasing processes. Diffusion processes that can change A^z are strongly detuned by the Knight field and therefore of minor importance, as corroborated by experimentally measured spin diffusion rates [51, 52]. Resonant processes such as $\propto I_i^z I_j^z$ can lead to dephasing similar to the inhomogeneous Knight shift. This competes with the phasing necessary for the observation of SR as expressed by the first term in Eq. (2.3). The SR process is the weakest at the very beginning of the evolution where we estimate its strength as $c_r^{\mathrm{min}} \approx 10 \, \mu\mathrm{eV}/N$. An upper bound for the dipole-dipole interaction in GaAs has been given in Ref. [28] as $\sim 10^{-5} \, \mu\mathrm{eV}$, in agreement with values given in Refs. [32, 41]. Therefore, the nuclear dipole-dipole interaction can safely be neglected for $N \lesssim 10^5$. In particular, its dephasing effect should be further reduced for highly polarized ensembles.

Second, the nuclear quadrupolar interactions can have two origins—strain (largely absent in electrically defined QDs) and electric field gradients originating from the

---

[1] For finite polarization the initial covariance matrix has been determined heuristically from the dark state condition $\langle A^- A^+ \rangle = 0$ in the homogeneous limit.

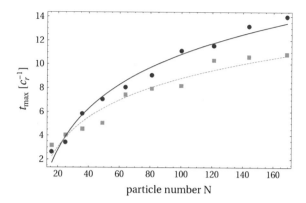

Fig. 2.9 Total time till the observation of the characteristic SR peaking $t_{\max}$ for $\varepsilon = 0.5$ (*blue dots*) and $\varepsilon = 0.55$ (*orange squares*). Based on Eq. (2.44), logarithmic fits are obtained from which we estimate $t_{\max}$ for experimentally realistic number of nuclear spins $N \approx 10^5$

electron. These have been estimated for typical electrically defined QDs in Ref. [41] to lead to an additional nuclear level splitting on the order of $\sim 10^{-5}$ µeV. Moreover, they are absent for nuclear spin $I = 1/2$ (e.g. CdSe QDs). To summarize, the additional dephasing mechanisms induced by nuclear interactions are much smaller than the terms arising from the inhomogeneous Knight field [32]. As argued above and confirmed by our simulations, the latter does not prevent the observation of SR behavior due to the presence of the MPM-term in Eq. (2.3).

Quantitative Aspects

Initially, the HF mediated superradiance dynamics is rather slow, with its characteristic time scale set by c_r^{-1}; for experimentally realistic parameters—in what follows we use the parameter set $\left(\varepsilon = 0.5, \ \alpha \approx 10\,\mu\text{eV}, \ N \approx 10^5 \right)$ for numerical estimates— this corresponds to $c_r^{-1} \approx 10\,\mu\text{s}$. Based on fits as shown in Fig. 2.9, we then estimate for the SR process duration $\langle t_D \rangle \approx 50\,c_r^{-1} \approx 500\,\mu\text{s}$ which is still smaller than recently reported [53] nuclear decoherence times of ~ 1 m s. Therefore, it should be possible to observe the characteristic enhancement of the leakage current before the nuclear spins decohere.

Leakage current.—Accordingly, in the initial phasing stage, the HF mediated lifting of the spin-blockade is rather weak resulting in a low leakage current, approximatively given by $I\left(t = 0 \right) / (e\hbar^{-1}) \approx \varepsilon^2 \alpha / N$. Therefore, the initial current due to HF processes is inversely proportional to the number of nuclear spins N. However, as correlations among the nuclei build up, the HF mediated lifting becomes more efficient culminating in a maximum current of $I_{\max} / (e\hbar^{-1}) \approx \varepsilon^2 \alpha$, independent of N. For realistic experimental values—also taking into account the effects of inhomogeneous HF coupling and finite initial polarization $p \approx 0.6$—we estimate the initial (maximum) leakage current to be of the order of $I\left(t = 0 \right) \approx 6\,\text{fA}$ ($I_{\max} \approx 10\,\text{pA}$). Leakage currents in this range of magnitudes have already been detected in single QD spin-filter experiments [26] as well as double QD Pauli-blockade experiments [15, 16, 18, 19]; here, leakage currents below 10 and 150 fA, respectively, have been attributed explicitly to other spurious processes [18, 26]. These are addressed in greater detail in the following.

Our transport setting is tuned into the sequential tunneling regime and therefore we have disregarded cotunneling processes which are fourth order in H_T. In principle, cotunneling processes could lift the spin-blockade and add an extra contribution to the leakage current that is independent of the HF dynamics. However, note that cotunneling current scales as $I_{ct} \propto \alpha^2$, whereas sequential tunneling current $I \propto \alpha$; accordingly, cotunneling current can always be suppressed by making the tunnel barriers less transparent [26]. Moreover, inelastic cotunneling processes exciting the QD spin can be ruled out for $eV, k_B T < \omega_0$ due to energy conservation [25]. The effectiveness of a single quantum dot to act as an electrically tunable spin filter has also been demonstrated experimentally [26]: The spin-filter efficiency was measured to be nearly 100 %, with I_{ct} being smaller than the noise floor ~ 10 fA. Its actual value has been calculated as $\sim 10^{-4}$ fA, from which we roughly estimate $I_{ct} \sim 10^{-2}$ fA in our setting. This is smaller than the initial HF mediated current $I (t = 0)$ and considerably smaller than I_{max}, even for an initially not fully polarized nuclear spin ensemble. Still, if one is to explore the regime where cotunneling cannot be neglected, phenomenological dissipative terms—effectively describing the corresponding spin-flip and pure dephasing mechanisms for inelastic and elastic processes respectively—should be added to Eq. (2.30).

Self-consistency

In our simulations we have self-consistently verified that the fluctuations of the Overhauser field, defined via

$$\Delta_{OH} (t) = g\sqrt{\langle A_z^2 \rangle_t - \langle A_z \rangle_t^2},\qquad(2.48)$$

are indeed small compared to the external Zeeman splitting ω_0 throughout the entire evolution. This ensures the validity of our perturbative approach and the realization of the spin-blockade regime. From atomic superradiance it is known that in the limit of homogeneous coupling large fluctuations can build up, since in the middle of the emission process the density matrix becomes a broad distribution over the Dicke states [24]. Accordingly, in the idealized, exactly solvable case of homogeneous coupling we numerically find rather large fluctuations of the Overhauser field; as demonstrated in Fig. 2.10, this holds independently of N. In particular, for a relative coupling strength $\varepsilon = 0.5$ the fluctuations culminate in $\max [\Delta_{OH}]/\omega_0 \approx 0.35$. However, in the case of inhomogeneneous HF coupling the Overhauser field fluctuations are found to be smaller as the build-up of these fluctuations is hindered by the Knight term causing dephasing among the nuclear spins. As another limiting case, we also estimate the fluctuations for completely independent homogeneously coupled nuclear spins via the binominal distribution as $\max [\Delta_{OH}] \sim 0.5 A_{HF}/\sqrt{N}$.[2]

Moreover, we have also ensured self-consistently the validity of the perturbative treatment of the flip-flop dynamics; that is, throughout the entire evolution, even for

---

[2]This limit is realized if strong nuclear dephasing processes prevent the coherence build-up of the SR evolution.

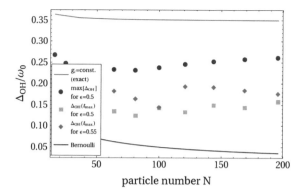

Fig. 2.10 Fluctuations of the Overhauser field relative to the external Zeeman splitting ω_0. In the limit of homogeneous HF coupling, strong fluctuations build up towards the middle of the emission process (*red line*, $\varepsilon = 0.5$). For inhomogeneous coupling this build-up of fluctuations is hindered by the dephasing between the nuclear spins, resulting in considerably smaller fluctuations: the value of the Overhauser fluctuations is shown at the time of the SR peak t_{max} for $\varepsilon = 0.5$ (*orange squares*) and $\varepsilon = 0.55$ (*green diamonds*). The Overhauser fluctuations reach a maximum value later than t_{max}, see *blue dots* for $\varepsilon = 0.5$. For independent homogeneously coupled nuclear spins, one can estimate the fluctuations via the binominal distribution (*black line*)

maximum operative matrix elements $\langle A^+ A^- \rangle_t$, the strength of the flip-flop dynamics $\|H_{ff}\|$ was still at least five times smaller than ω_0.

2.8 Conclusion and Outlook

In summary, we have developed a master equation based theoretical framework for nuclear-spin-assisted transport through a QD. Due to the collective nature of the HF interaction, it incorporates intriguing many-body effects as well as feedback mechanisms between the electron spin and nuclear spin dynamics. As a prominent application, we have shown that the current through a single electrically defined QD in the spin-blockade regime naturally exhibits superradiant behavior. This effect stems from the collective hyperfine interaction between the QD electron and the nuclear spin ensemble in the QD. Its most striking feature is a lifting of the spin-blockade and a pronounced peak in the leakage current. The experimental observation of this effect would provide clear evidence of coherent HF dynamics of nuclear spin ensembles in QDs.

Finally, we highlight possible directions of research going beyond our present work: Apart from superradiant electron transport, the setup proposed here is inherently well suited for other experimental applications like dynamic polarization of nuclear spins (DNP): In analogy to optical pumping, Eq. (2.3) describes *electronic* pumping of the nuclear spins. Its steady states are eigenstates of A^z, which lie in the kernel of the collective jump-operator A^-. In particular, for a completely inhomogeneous system the only steady state is the fully polarized one, the ideal initial state required for the observation of SR effects. When starting from a completely unpolarized nuclear state, the uni-directionality of Eq. (2.3)—electrons with one spin orientation exchange excitations with the nuclear spins, while electrons of opposite spin primarily do not—implies that the rather warm electronic reservoir can still extract entropy out of the nuclear system. More generally, the transport setting studied here possibly opens up the route towards the (feedback-based) electronic preparation of particular nuclear states in single QDs. This is in line with similar ideas previously developed in double QD settings, see e.g., Refs. [12, 15, 18, 20, 53].

In this work we have specialized on a single QD. However, our theory could be extended to a double QD (DQD) setting which is likely to offer even more possibilities. DQDs are routinely operated in the Pauli-blockade regime where despite the presence of an applied source-drain voltage the current through the device is blocked whenever the electron tunneling into the DQD has the same spin orientation as the one already present. The DQD parameters and the external magnetic field can be tuned such that the role of the states $|\sigma\rangle, \sigma = \downarrow, \uparrow$, in our model is played by a pair of singlet and triplet states, while all other states are off-resonant. Then, along the lines of our study, non-linearities appear due to dependencies between the electronic and nuclear subsystems and collective effects enter via the HF-mediated lifting of the spin-blockade.

While we have focused on the Markovian regime and the precise conditions for its validity, Eq. (2.15) offers a starting point for studies of non-Markovian effects in the proposed transport setting. All terms appearing in the memory kernel of Eq. (2.15) are quadratic in the fermionic creation and annihilation operators allowing for an efficient numerical simulation, without having to explicitly invoke the flatness of the spectral density of the leads. This should then shed light on possibly abrupt changes in the QD transport properties due to feedback mechanism between the nuclear spin ensemble and the electron spin.

Lastly, our work also opens the door towards studies of dissipative phase transitions in the transport setting: when combined with driving, the SR dynamics can lead to a variety of strong-correlation effects, non-equilibrium and dissipative phase transitions [1, 54–56], which could now be studied in a mesoscopic solid state system, complementing other approaches to dissipative phase transitions in QDs [57–60].

2.9 Appendix to Chap. 2

2.9.1 Microscopic Derivation of the Master Equation

In this Appendix we provide some details regarding the derivation of the master equations as stated in Eqs. (2.2) and (2.30). It comprises the effect of the HF dynamics in the memory kernel of Eq. (2.13) and the subsequent approximation of independent rates of variation.

In the following, we will show that it is self-consistent to neglect the effect of the HF dynamics $\mathcal{L}_1(t)$ in the memory-kernel of Eq. (2.13) provided that the bath correlation time τ_c is short compared to the Rabi flips produced by the HF dynamics. This needs to be addressed as cooperative effects potentially drive the system from a weakly coupled into a strongly coupled regime. First, we reiterate the Schwinger-Dyson identity in Eq. (2.14) as an infinite sum over time-ordered nested commutators

$$e^{-i(\mathcal{L}_0+\mathcal{L}_1)\tau} = e^{-i\mathcal{L}_0\tau} \sum_{n=0}^{\infty} (-i)^n \int_0^{\tau} d\tau_1 \int_0^{\tau_1} d\tau_2 \dots \int_0^{\tau_{n-1}} d\tau_n \, \tilde{\mathcal{L}}_1(\tau_1) \tilde{\mathcal{L}}_1(\tau_2) \dots \tilde{\mathcal{L}}_1(\tau_n),$$

(2.49)

where for any operator X

$$\tilde{\mathcal{L}}_1(\tau) X = e^{i\mathcal{L}_0\tau} \mathcal{L}_1 e^{-i\mathcal{L}_0\tau} X = \left[e^{iH_0\tau} H_1 e^{-iH_0\tau}, X \right] = \left[\tilde{H}_1(\tau), X \right]. \quad (2.50)$$

More explicitly, up to second order Eq. (2.49) is equivalent to

$$e^{-i(\mathcal{L}_0+\mathcal{L}_1)\tau} X = e^{-i\mathcal{L}_0\tau} X - i e^{-i\mathcal{L}_0\tau} \int_0^{\tau} d\tau_1 \left[\tilde{H}_1(\tau_1), X \right]$$

$$- e^{-i\mathcal{L}_0\tau} \int_0^{\tau} d\tau_1 \int_0^{\tau_1} d\tau_2 \left[\tilde{H}_1(\tau_1), \left[\tilde{H}_1(\tau_2), X \right] \right] + \cdots \quad (2.51)$$

Note that the time-dependence of $\tilde{H}_1(\tau)$ is simply given by

$$\tilde{H}_1(\tau) = e^{i\omega\tau} H_+ + e^{-i\omega\tau} H_- + H_{\Delta\text{OH}}, \qquad H_{\pm} = \frac{g}{2} S^{\pm} A^{\mp}, \quad (2.52)$$

where the effective Zeeman splitting $\omega = \omega_0 + g \langle A^z \rangle_t$ is time-dependent. Accordingly, we define $\tilde{\mathcal{L}}_1(\tau) = \tilde{\mathcal{L}}_+(\tau) + \tilde{\mathcal{L}}_-(\tau) + \tilde{\mathcal{L}}_{\Delta\text{OH}}(\tau) = e^{i\omega\tau} \mathcal{L}_+ + e^{-i\omega\tau} \mathcal{L}_- + \mathcal{L}_{\Delta\text{OH}}$, where $\mathcal{L}_{x\cdot} = [H_x, \cdot]$ for $x = \pm, \Delta\text{OH}$. In the next steps, we will explicitly evaluate the first two contributions to the memory kernel that go beyond $n = 0$ and then generalize our findings to any order n of the Schwinger-Dyson series.

First order correction

The first order contribution $n = 1$ in Eq. (2.13) is given by

$$\Xi^{(1)} = i \int_0^t d\tau \int_0^{\tau} d\tau_1 \text{Tr}_B \left(\mathcal{L}_T e^{-i\mathcal{L}_0\tau} \left[\tilde{H}_1(\tau_1), X \right] \right). \quad (2.53)$$

Performing the integration in τ_1 leads to

$$\Xi^{(1)} = \int_0^t d\tau \left\{ \frac{g}{2\omega} \left(1 - e^{-i\omega\tau} \right) \text{Tr}_B \left(\mathcal{L}_T \left[S^+ A^-, \tilde{X}_\tau \right] \right) \right.$$

$$+ \frac{g}{2\omega} \left(e^{i\omega\tau} - 1 \right) \text{Tr}_B \left(\mathcal{L}_T \left[S^- A^+, \tilde{X}_\tau \right] \right)$$

$$\left. + i g \tau \text{Tr}_B \left(\mathcal{L}_T \left[\left(A^z - \langle A^z \rangle_t \right) S^z, \tilde{X}_\tau \right] \right) \right\} \qquad (2.54)$$

where, for notational convenience, we introduced the operators $X = \mathcal{L}_T \rho_S (t - \tau) \rho_B^0$ and $\tilde{X}_\tau = e^{-iH_0\tau} \left[H_T, \rho_S (t - \tau) \rho_B^0 \right] e^{iH_0\tau} \approx \left[\tilde{H}_T (\tau), \rho_S (t) \rho_B^0 \right]$. In accordance with previous approximations, we have replaced $e^{-iH_0\tau} \rho_S (t - \tau) e^{iH_0\tau}$ by $\rho_S (t)$ since any additional term besides H_0 would be of higher order in perturbation theory [35, 36]. In particular, this disregards dissipative effects: In our case, this approximation is valid self-consistently provided that the tunneling rates are small compared to effective Zeeman splitting ω. The integrand decays on the leads-correlation timescale τ_c which is typically much faster than the timescale set by the effective Zeeman splitting, $\omega\tau_c \ll 1$. This separation of timescales allows for an expansion in the small parameter $\omega\tau$, e.g. $\frac{g}{\omega} \left(e^{i\omega\tau} - 1 \right) \approx i g\tau$. We see that the first order correction can be neglected if the the bath correlation time τ_c is sufficiently short compared to the timescale of the HF dynamics, that is $g\tau_c \ll 1$. The latter is bounded by the total hyperfine coupling constant A_{HF} (since $||gA^x|| \leq A_{HF}$) so that the requirement for disregarding the first order term reads $A_{HF}\tau_c \ll 1$.

Second order correction

The contribution of the second term $n = 2$ in the Schwinger-Dyson expansion can be decomposed into

$$\Xi^{(2)} = \Xi_{zz}^{(2)} + \Xi_{ff}^{(2)} + \Xi_{fz}^{(2)}. \qquad (2.55)$$

The first term $\Xi_{zz}^{(2)}$ contains contributions from $H_{\Delta OH}$ only

$$\Xi_{zz}^{(2)} = \int_0^t d\tau \int_0^\tau d\tau_1 \int_0^{\tau_1} d\tau_2 \text{Tr}_B \left(\mathcal{L}_T e^{-i\mathcal{L}_0\tau} \left[\tilde{H}_{\Delta OH} (\tau_1), \left[\tilde{H}_{\Delta OH} (\tau_2), X \right] \right] \right) \quad (2.56)$$

$$= - \int_0^t d\tau \, (g\tau)^2 \, \text{Tr}_B \left[\mathcal{L}_T \left(\delta A^z S^z \tilde{X}_\tau \delta A^z S^z - \frac{1}{2} \left\{ \delta A^z S^z \delta A^z S^z, \tilde{X}_\tau \right\} \right) \right] \quad (2.57)$$

Similarly, $\Xi_{ff}^{(2)}$ which comprises contributions from H_{ff} only is found to be

$$\Xi_{ff}^{(2)} = \frac{g^2}{4\omega^2} \int_0^t d\tau \left\{ \left(1 + i\omega\tau - e^{i\omega\tau} \right) \text{Tr}_B \left[\mathcal{L}_T \left(S^+ S^- A^- A^+ \tilde{X}_\tau + \tilde{X}_\tau S^- S^+ A^+ A^- \right) \right] \right.$$

$$\left. + \left(1 - i\omega\tau - e^{-i\omega\tau} \right) \text{Tr}_B \left[\mathcal{L}_T \left(S^- S^+ A^+ A^- \tilde{X}_\tau + \tilde{X}_\tau S^+ S^- A^- A^+ \right) \right] \right\}. \quad (2.58)$$

Here, we have used the following simplification: The time-ordered products which include flip-flop terms only can be simplified to two possible sequences in which $\mathcal{L}_+$ is followed by $\mathcal{L}_-$ and vice versa. This holds since

$$\mathcal{L}_\pm \mathcal{L}_\pm X = \left[H_\pm, \left[H_\pm, X\right]\right] = H_\pm H_\pm X + X H_\pm H_\pm - 2 H_\pm X H_\pm = 0. \quad (2.59)$$

Here, the first two terms drop out immediately since the electronic jump-operators $S^\pm$ fulfill the relation $S^\pm S^\pm = 0$. In the problem at hand, also the last term gives zero because of particle number superselection rules: In Eq. (2.13) the time-ordered product of superoperators acts on $X = \left[H_T, \rho_S\left(t - \tau\right) \rho_B^0\right]$. Thus, for the term $H_\pm X H_\pm$ to be nonzero, coherences in Fock space would be required which are consistently neglected; compare Ref. [36]. This is equivalent to ignoring coherences between the system and the leads. Note that the same argument holds for any combination $H_\mu X H_\nu$ with $\mu, \nu = \pm$.

Similar results can be obtained for $\Xi_{fz}^{(2)}$ which comprises $H_\pm$ as well as $H_{\Delta OH}$ in all possible orderings. Again, using that the integrand decays on a timescale τ_c and expanding in the small parameter $\omega\tau$ shows that the second order contribution scales as $\sim (g\tau_c)^2$. Our findings for the first and second order correction suggest that the n-th order correction scales as $\sim (g\tau_c)^n$. This will be proven in the following by induction.

n-th order correction

The scaling of the n-th term in the Dyson series is governed by the quantities of the form

$$\xi_{+-\dots}^{(n)}\left(\tau\right) = g^n \int_0^\tau d\tau_1 \int_0^{\tau_1} d\tau_2 \dots \int_0^{\tau_{n-1}} d\tau_n e^{i\omega\tau_1} e^{-i\omega\tau_2} \dots, \quad (2.60)$$

where the index suggests the order in which $H_\pm$ (giving an exponential factor) and $H_{\Delta OH}$ (resulting in a factor of 1) appear. Led by our findings for $n = 1, 2$, we claim that the expansion of $\xi_{+-\dots}^{(n)}\left(\tau\right)$ for small $\omega\tau$ scales as $\xi_{+-\dots}^{(n)}\left(\tau\right) \sim (g\tau)^n$. Then, the $(n+1)$-th terms scale as

$$\xi_{-(\Delta OH)+-\dots}^{(n+1)}\left(\tau\right) = g^{n+1} \int_0^\tau d\tau_1 \int_0^{\tau_1} d\tau_2 \dots \int_0^{\tau_{n-1}} d\tau_n \int_0^{\tau_n} d\tau_{n+1} \begin{pmatrix} e^{-i\omega\tau_1} \\ 1 \end{pmatrix} e^{+i\omega\tau_2} \dots$$
$$(2.61)$$

$$= g \int_0^\tau d\tau_1 \begin{pmatrix} e^{-i\omega\tau_1} \\ 1 \end{pmatrix} \xi_{+-\dots}^{(n)}\left(\tau_1\right) \quad (2.62)$$

$$\sim (g\tau)^{n+1}. \quad (2.63)$$

Since we have already verified this result for $n = 1, 2$, the general result follows by induction. This completes the proof.

2.9.2 Adiabatic Elimination of the QD Electron

For a sufficiently small relative coupling strength ε the nuclear dynamics are slow compared to the electronic QD dynamics. This allows for an adiabatic elimination of the electronic degrees of freedom yielding an effective master equation for the nuclear spins of the QD.

Our analysis starts out from Eq. (2.35) which we write as

$$\dot{\rho} = \mathcal{W}_0 \rho + \mathcal{W}_1 \rho, \tag{2.64}$$

where

$$\mathcal{W}_0 \rho = -i\left[\omega_0 S^z, \rho\right] + \gamma\left[S^- \rho S^+ - \frac{1}{2}\left\{S^+ S^-, \rho\right\}\right] + \Gamma\left[S^z \rho S^z - \frac{1}{4}\rho\right] \tag{2.65}$$

$$\mathcal{W}_1 \rho = -i\left[H_{\text{HF}}, \rho\right]. \tag{2.66}$$

Note that the superoperator $\mathcal{W}_0$ only acts on the electronic degrees of freedom. It describes an electron in an external magnetic field that experiences a decay as well as a pure dephasing mechanism. In zeroth order of the coupling parameter ε the electronic and nuclear dynamics of the QD are decoupled and SR effects cannot be expected. These are contained in the interaction term $\mathcal{W}_1$.

Formally, the adiabatic elimination of the electronic degrees of freedom can be achieved as follows [61]. To zeroth order in ε the eigenvectors of $\mathcal{W}_0$ with zero eigenvector $\lambda_0 = 0$ are

$$\mathcal{W}_0 \mu \otimes \rho_{ss} = 0, \tag{2.67}$$

where $\rho_{ss} = |\downarrow\rangle \langle\downarrow|$ is the stationary solution for the electronic dynamics and μ describes some arbitrary state of the nuclear system. The zero-order Liouville eigenstates corresponding to $\lambda_0 = 0$ are coupled to the subspaces of "excited" nonzero (complex) eigenvalues $\lambda_k \neq 0$ of $\mathcal{W}_0$ by the action of $\mathcal{W}_1$. Physically, this corresponds to a coupling between electronic and nuclear degrees of freedom. In the limit where the HF dynamics are slow compared to the electronic frequencies, i.e. the Zeeman splitting ω_0, the decay rate γ and the dephasing rate Γ, the coupling between these blocks of eigenvalues and Liouville subspaces of $\mathcal{W}_0$ is weak justifying a perturbative treatment. This motivates the definition of a projection operator P onto the subspace with zero eigenvalue $\lambda_0 = 0$ of $\mathcal{W}_0$ according to

$$P\rho = \text{Tr}_{\text{el}}\left[\rho\right] \otimes \rho_{ss} = \mu \otimes |\downarrow\rangle \langle\downarrow|, \tag{2.68}$$

where $\mu = \text{Tr}_{\text{el}}\left[\rho\right]$ is a density operator for the nuclear spins, $\text{Tr}_{\text{el}} \ldots$ denotes the trace over the electronic subspace and by definition $\mathcal{W}_0 \rho_{ss} = 0$. The complement of P is $Q = 1 - P$. By projecting the master equation on the P subspace and tracing over the electronic degrees of freedom we obtain an effective master equation for the nuclear spins in second order perturbation theory

$$\dot{\mu} = \mathrm{Tr}_{\mathrm{el}}\left[P\mathcal{W}_1 P\rho - P\mathcal{W}_1 Q\mathcal{W}_0^{-1} Q\mathcal{W}_1 P\rho\right]. \tag{2.69}$$

Using $\mathrm{Tr}_{\mathrm{el}}\left[S^z \rho_{SS}\right] = -1/2$, the first term is readily evaluated and yields the Knight shift seen by the nuclear spins

$$\mathrm{Tr}_{\mathrm{el}}\left[P\mathcal{W}_1 P\rho\right] = +i\frac{g}{2}\left[A^z, \mu\right]. \tag{2.70}$$

The derivation of the second term is more involved. It can be rewritten as

$$-\mathrm{Tr}_{\mathrm{el}}\left[P\mathcal{W}_1 Q\mathcal{W}_0^{-1} Q\mathcal{W}_1 P\rho\right]$$
$$= -\mathrm{Tr}_{\mathrm{el}}\left[P\mathcal{W}_1 (1-P)\mathcal{W}_0^{-1}(1-P)\mathcal{W}_1 P\rho\right] \tag{2.71}$$
$$= \int_0^\infty d\tau\,\mathrm{Tr}_{\mathrm{el}}\left[P\mathcal{W}_1 e^{\mathcal{W}_0\tau}\mathcal{W}_1 P\rho\right] - \int_0^\infty d\tau\,\mathrm{Tr}_{\mathrm{el}}\left[P\mathcal{W}_1 P\mathcal{W}_1 P\rho\right]. \tag{2.72}$$

Here, we used the Laplace transform $-\mathcal{W}_0^{-1} = \int_0^\infty d\tau\, e^{\mathcal{W}_0\tau}$ and the property $e^{\mathcal{W}_0\tau} P = P e^{\mathcal{W}_0\tau} = P$.

Let us first focus on the first term in Eq. (2.72). It contains terms of the form

$$\mathrm{Tr}_{\mathrm{el}}\left[P\left[A^+ S^-, e^{\mathcal{W}_0\tau}\left[A^- S^+, \mu \otimes \rho_{SS}\right]\right]\right] = \mathrm{Tr}_{\mathrm{el}}\left[S^- e^{\mathcal{W}_0\tau}\left(S^+ \rho_{SS}\right)\right] A^+ A^- \mu \tag{2.73}$$
$$- \mathrm{Tr}_{\mathrm{el}}\left[S^- e^{\mathcal{W}_0\tau}\left(S^+ \rho_{SS}\right)\right] A^- \mu A^+ \tag{2.74}$$
$$+ \mathrm{Tr}_{\mathrm{el}}\left[S^- e^{\mathcal{W}_0\tau}\left(\rho_{SS} S^+\right)\right] \mu A^- A^+ \tag{2.75}$$
$$- \mathrm{Tr}_{\mathrm{el}}\left[S^- e^{\mathcal{W}_0\tau}\left(\rho_{SS} S^+\right)\right] A^+ \mu A^- \tag{2.76}$$

This can be simplified using the following relations: Since $\rho_{SS} = |\downarrow\rangle\langle\downarrow|$, we have $S^- \rho_{SS} = 0$ and $\rho_{SS} S^+ = 0$. Moreover, $|\uparrow\rangle\langle\downarrow|$ and $|\downarrow\rangle\langle\uparrow|$ are eigenvectors of $\mathcal{W}_0$ with eigenvalues $-(i\omega_0 + \alpha/2)$ and $+(i\omega_0 - \alpha/2)$, where $\alpha = \gamma + \Gamma$, yielding

$$e^{\mathcal{W}_0\tau}\left(S^+ \rho_{SS}\right) = e^{-(i\omega_0 + \alpha/2)\tau}\,|\uparrow\rangle\langle\downarrow| \tag{2.77}$$
$$e^{\mathcal{W}_0\tau}\left(\rho_{SS} S^-\right) = e^{+(i\omega_0 - \alpha/2)\tau}\,|\downarrow\rangle\langle\uparrow|. \tag{2.78}$$

This leads to

$$\mathrm{Tr}_{\mathrm{el}}\left[P\left[A^+ S^-, e^{\mathcal{W}_0\tau}\left[A^- S^+, \mu \otimes \rho_{SS}\right]\right]\right] = e^{-(i\omega_0 + \alpha/2)\tau}\left(A^+ A^- \mu - A^- \mu A^+\right). \tag{2.79}$$

Similarly, one finds

$$\mathrm{Tr}_{\mathrm{el}}\left[P\left[A^- S^+, e^{\mathcal{W}_0\tau}\left[A^+ S^-, \mu \otimes \rho_{SS}\right]\right]\right] = e^{+(i\omega_0 - \alpha/2)\tau}\left(\mu A^+ A^- - A^- \mu A^+\right). \tag{2.80}$$

Analogously, one can show that terms containing two flip or two flop terms give zero. The same holds for mixed terms that comprise one flip-flop and one Overhauser term with $\sim A^z S^z$. The term consisting of two Overhauser contributions gives

$$\text{Tr}_{\text{el}}\left[P\left[A^z S^z, e^{\mathcal{W}_0 \tau}\left[A^z S^z, \mu \otimes \rho_{SS}\right]\right]\right] = -\frac{1}{4}\left[2 A^z \mu A^z - \left[A^z A^z, \mu\right]\right]. \quad (2.81)$$

However, this term exactly cancels with the second term from Eq. (2.72). Thus we are left with the contributions coming from Eqs. (2.79) and (2.80). Restoring the prefactors of $-ig/2$, we obtain

$$\text{Tr}_{\text{el}}\left[P\mathcal{W}_1 Q\left(-\mathcal{W}_0^{-1}\right) Q\mathcal{W}_1 P\rho\right] = \frac{g^2}{4}\int_0^\infty d\tau \left[e^{-(i\omega_0 + \alpha/2)\tau}\left(A^- \mu A^+ - A^+ A^- \mu\right)\right.$$
$$\left. + e^{+(i\omega_0 - \alpha/2)\tau}\left(A^- \mu A^+ - \mu A^+ A^-\right)\right]. \quad (2.82)$$

Performing the integration and separating real from imaginary terms yields

$$\text{Tr}_{\text{el}}\left[P\mathcal{W}_1 Q\left(-\mathcal{W}_0^{-1}\right) Q\mathcal{W}_1 P\rho\right] = c_r \left[A^- \mu A^+ - \frac{1}{2}\left\{A^+ A^-, \mu\right\}\right] + i c_i \left[A^+ A^-, \mu\right],$$
$$(2.83)$$

where $c_r = g^2/\left(4\omega_0^2 + \alpha^2\right)\alpha$ and $c_i = g^2/\left(4\omega_0^2 + \alpha^2\right)\omega_0$. Combining Eq. (2.70) with Eq. (2.83) directly gives the effective master equation for the nuclear spins given in Eq. (2.3) in the main text.

References

1. T. Brandes, Coherent and collective quantum optical effects in mesoscopic systems. Phys. Reports **408**(5–6), 315 (2005)
2. D.D. Awschalom, N. Samarath, D. Loss, *Semiconductor Spintronics and Quantum Computation* (Springer, Berlin, 2002)
3. D.Y. Sharvin, Y.V. Sharvin, Magnetic-flux quantization in a cylindrical film of a normal metal. JETP Lett. **34**(7), 272 (1981)
4. B.J. van Wees, H. van Houten, C.W.J. Beenakker, J.G. Williamson, L.P. Kouwenhoven, D. van der Marel, C.T. Foxon, Quantized conductance of point contacts in a two-dimensional electron gas. Phys. Rev. Lett. **60**(9), 848 (1988)
5. D.A. Wharam, T.J. Thornton, R. Newbury, M. Pepper, H. Ahmed, J.E.F. Frost, D.G. Hasko, D.C. Peacock, D.A. Ritchie, G.A.C. Jones, One-dimensional transport and the quantisation of the ballistic resistance. J. Phys. C Solid State Phys. **21**(8), L209 (2000)
6. Y.V. Nazarov, Y.M. Blanter, *Quantum Transport* (Cambridge University Press, Cambrigde, 2009)
7. S. Datta, *Electronic Transport in Mesoscopic Systems* (Cambridge University Press, Cambridge, 1997)
8. R. Hanson, L.P. Kouwenhoven, J.R. Petta, S. Tarucha, L.M.K. Vandersypen, Spins in few-electron quantum dots. Rev. Mod. Phys. **79**, 1217 (2007)
9. W.G. van der Wiel, S. De Franceschi, J.M. Elzerman, T. Fujisawa, S. Tarucha, L.P. Kouwenhoven, Electron transport through double quantum dots. Rev. Mod. Phys. **75**, 1 (2002)
10. A.C. Johnson, J.R. Petta, J.M. Taylor, A. Yacoby, M.D. Lukin, C.M. Marcus, M.P. Hanson, A.C. Gossard, Triplet-singlet spin relaxation via nuclei in a double quantum dot. Nature **435**, 925 (2005)
11. O.N. Jouravlev, Y.V. Nazarov, Electron transport in a double quantum dot governed by a nuclear magnetic field. Phys. Rev. Lett. **96**, 176804 (2006)

12. J. Baugh, Y. Kitamura, K. Ono, S. Tarucha, Large nuclear Overhauser fields detected in vertically-coupled double quantum dots. Phys. Rev. Lett. **99**, 096804 (2007)
13. J.R. Petta, J.M. Taylor, A.C. Johnson, A. Yacoby, M.D. Lukin, C.M. Marcus, M.P. Hanson, A.C. Gossard, Dynamic nuclear polarization with single electron spins. Phys. Rev. Lett. **100**, 067601 (2008)
14. J. Iñarrea, G. Platero, A.H. MacDonald, Electronic transport through a double quantum dot in the spin-blockade regime: theoretical models. Phys. Rev. B **76**(8), 085329 (2007)
15. F.H.L. Koppens, J.A. Folk, J.M. Elzerman, R. Hanson, L.H. Willems van Beveren, I.T. Vink, H.-P. Tranitz, W. Wegscheider, L.P. Kouwenhoven, L.M.K. Vandersypen, Control and detection of singlet-triplet mixing in a random nuclear field. Science **309**, 1346 (2005)
16. K. Ono, S. Tarucha, Nuclear-spin-induced oscillatory current in spin-blockaded quantum dots. Phys. Rev. Lett. **92**, 256803 (2004)
17. A. Pfund, I. Shorubalko, K. Ensslin, R. Leturcq, Suppression of spin relaxation in an in as nanowire double quantum dot. Phys. Rev. Lett. **99**(3), 036801 (2007)
18. T. Kobayashi, K. Hitachi, S. Sasaki, K. Muraki, Observation of hysteretic transport due to dynamic nuclear spin polarization in a GaAs lateral double quantum dot. Phys. Rev. Lett. **107**(21), 216802 (2011)
19. K. Ono, D.G. Austing, Y. Tokura, S. Tarucha, Current rectification by Pauli exclusion in a weakly coupled double quantum dot system. Science **297**(5585), 1313 (2002)
20. M.S. Rudner, L.S. Levitov, Self-polarization and cooling of spins in quantum dots. Phys. Rev. Lett. **99**, 036602 (2007)
21. M. Eto, T. Ashiwa, M. Murata, Current-induced entanglement of nuclear spins in quantum dots. J. Phys. Soc. Jpn. **73**(2), 307 (2004)
22. H. Christ, J.I. Cirac, G. Giedke, Quantum description of nuclear spin cooling in a quantum dot. Phys. Rev. B **75**, 155324 (2007)
23. R.H. Dicke, Coherence in spontaneous radiation processes. Phys. Rev. **93**, 99 (1954)
24. M. Gross, S. Haroche, Superradiance: an essay on the theory of collective spontaneous emission. Phys. Reports **93**, 301 (1982)
25. P. Recher, E.V. Sukhorukov, D. Loss, Quantum dot as spin filter and spin memory. Phys. Rev. Lett. **85**(9), 1962 (2000)
26. R. Hanson, L.M.K. Vandersypen, L.H.W. van Beveren, J.M. Elzerman, I.T. Vink, L.P. Kouwenhoven, Semiconductor few-electron quantum dot operated as a bipolar spin filter. Phys. Rev. B **70**(24), 241304 (2004)
27. E.M. Kessler, S. Yelin, M.D. Lukin, J.I. Cirac, G. Giedke, Optical superradiance from nuclear spin environment of single photon emitters. Phys. Rev. Lett. **104**, 143601 (2010)
28. J. Schliemann, A. Khaetskii, D. Loss, Electron spin dynamics in quantum dots and related nanostructures due to hyperfine interaction with nuclei. J. Phys. Condens. Matter **15**(50), R1809 (2003)
29. H. Bruus, K. Flensberg, *Many-body Quantum Theory in Condensed Matter Physics* (Oxford University Press, New York, 2006)
30. Y. Yamamoto, A. Imamoglu, *Mesoscopic Quantum Optics* (Wiley, New York, 1999)
31. S. Welack, M. Esposito, U. Harbola, S. Mukamel, Interference effects in the counting statistics of electron transfers through a double quantum dot. Phys. Rev. B **77**(19), 195315 (2008)
32. J.M. Taylor, J.R. Petta, A.C. Johnson, A. Yacoby, C.M. Marcus, M.D. Lukin, Relaxation, dephasing, and quantum control of electron spins in double quantum dots. Phys. Rev. B **76**, 035315 (2007)
33. C. Cohen-Tannoudji, J. Dupont-Roc, G. Grynberg, *Atom-Photon Interactions: Basic Processes and Applications* (Wiley, New York, 1992)
34. C. Timm, *Private Communication* (2011)
35. C. Timm, Tunneling through molecules and quantum dots: master-equation approaches. Phys. Rev. B **77**(19), 195416 (2008)
36. U. Harbola, M. Esposito, S. Mukamel, Quantum master equation for electron transport through quantum dots and single molecules. Phys. Rev. B **74**(23), 235309 (2006)

37. H.-A. Engel, D. Loss, Single-spin dynamics and decoherence in a quantum dot via charge transport. Phys. Rev. B **65**(19), 195321 (2002)
38. N. Zhao, J.-L. Zhu, R.-B. Liu, C.P. Sun, Quantum noise theory for quantum transport through nanostructures. New J. Phys. **13**(1), 013005 (2011)
39. S.A. Gurvitz, Y.S. Prager, Microscopic derivation of rate equations for quantum transport. Phys. Rev. B **53**(23), 15932 (1996)
40. S.A. Gurvitz, Rate equations for quantum transport in multidot systems. Phys. Rev. B **57**(11), 6602 (1998)
41. H. Bluhm, S. Foletti, I. Neder, M. Rudner, D. Mahalu, V. Umansky, A. Yacoby, Dephasing time of GaAs electron-spin qubits coupled to a nuclear bath exceeding 200 μs. Nat. Phys. **7**(2), 109 (2010)
42. G.S. Agarwal, Master-equation approach to spontaneous emission. III. Many-body aspects of emission from two-level atoms and the effect of inhomogeneous broadening. Phys. Rev. A **4**, 1791 (1971)
43. C. Leonardi, A. Vaglica, Superradiance and inhomogeneous broadening. II: spontaneous emission by many slightly detuned sources. Nuovo Cimento B Serie **67**, 256 (1982)
44. V.V. Temnov, U. Woggon, Superradiance and subradiance in an inhomogeneously broadened ensemble of two-level systems coupled to a low-Q cavity. Phys. Rev. Lett. **95**, 243602 (2005)
45. D.A. Bagrets, Y.V. Nazarov, Full counting statistics of charge transfer in Coulomb blockade systems. Phys. Rev. B **67**(8), 085316 (2003)
46. M. Esposito, U. Harbola, S. Mukamel, Nonequilibrium fluctuations, fluctuation theorems, and counting statistics in quantum systems. Rev. Mod. Phys. **81**(4), 1665 (2009)
47. C. Emary, C. Pöltl, A. Carmele, J. Kabuss, A. Knorr, T. Brandes, Bunching and antibunching in electronic transport. Phys. Rev. B **85**(16), 165417 (2012)
48. L.D. Contreras-Pulido, R. Aguado, Shot noise spectrum of artificial single-molecule magnets: measuring spin relaxation times via the Dicke effect. Phys. Rev. B **81**(16), 161309(R) (2010)
49. H.J. Carmichael, *Statistical Methods in Quantum Optics 1* (Springer, Berlin, 1999)
50. R. Hanson, B. Witkamp, L.M.K. Vandersypen, L.H.W. van Beveren, J.M. Elzerman, L.P. Kouwenhoven, Zeeman energy and spin relaxation in a one-electron quantum dot. Phys. Rev. Lett. **91**(19), 196802 (2003)
51. D. Paget, Optical detection of NMR in high-purity GaAs: direct study of the relaxation of nuclei close to shallow donors. Phys. Rev. B **25**, 4444 (1982)
52. T. Ota, G. Yusa, N. Kumada, S. Miyashita, T. Fujisawa, Y. Hirayama, Decoherence of nuclear spins due to direct dipole-dipole interactions probed by resistively detected nuclear magnetic resonance. Appl. Phys. Lett. **91**, 193101 (2007)
53. R. Takahashi, K. Kono, S. Tarucha, K. Ono, Voltage-selective bi-directional polarization and coherent rotation of nuclear spins in quantum dots. Phys. Rev. Lett. **107**, 026602 (2011)
54. H.J. Carmichael, Analytical and numerical results for the steady state in cooperative resonance fluorescence. J. Phys. B Atom. Mol. Phys. **13**(18), 3551 (1980)
55. S. Morrison, A.S. Parkins, Collective spin systems in dispersive optical cavity-QED: quantum phase transitions and entanglement. Phys. Rev. A **77**(4), 043810 (2008)
56. E.M. Kessler, G. Giedke, A. Imamoglu, S.F. Yelin, M.D. Lukin, J.I. Cirac, Dissipative phase transition in a central spin system. Phys. Rev. A **86**(1), 012116 (2012)
57. C.-H. Chung, K. Le Hur, M. Vojta, P. Wölfle, Nonequilibrium transport at a dissipative quantum phase transition. Phys. Rev. Lett. **102**, 216803 (2009)
58. A.J. Leggett, S. Chakravarty, A.T. Dorsey, M.P.A. Fisher, A. Garg, W. Zwerger, Dynamics of the dissipative two-state system. Rev. Mod. Phys. **59**, 1 (1987)
59. M.S. Rudner, L.S. Levitov, Phase transitions in dissipative quantum transport and mesoscopic nuclear spin pumping. Phys. Rev. B **82**, 155418 (2010)
60. L. Borda, G. Zarand, D. Goldhaber-Gordon, Dissipative quantum phase transition in a quantum dot. arXiv:cond-mat/0602019 (2006)
61. J.I. Cirac, R. Blatt, P. Zoller, W.D. Phillips, Laser cooling of trapped ions in a standing wave. Phys. Rev. A **46**(5), 2668 (1992)

Chapter 3
Nuclear Spin Dynamics in Double Quantum Dots: Multi-stability, Dynamical Polarization, Criticality and Entanglement

In the previous chapter we have investigated the transient creation of nuclear coherence as a result of electron transport through a single quantum dot. In this chapter we theoretically study the nuclear spin dynamics driven by electron transport and hyperfine interaction in an electrically defined double quantum dot in the Pauli-blockade regime. We derive a master-equation-based framework and show that the coupled electron-nuclear system displays an instability towards the buildup of large nuclear spin polarization gradients in the two quantum dots. In the presence of such inhomogeneous magnetic fields, a quantum interference effect in the collective hyperfine coupling results in sizable nuclear spin entanglement between the two quantum dots in the steady state of the evolution. We investigate this effect using analytical and numerical techniques, and demonstrate its robustness under various types of imperfections.

3.1 Introduction

The prospect of building devices capable of quantum information processing (QIP) has fueled an impressive race to implement well-controlled two-level quantum systems (qubits) in a variety of physical settings [1]. For any such system, generating and maintaining entanglement—one of the most important primitives of QIP—is a hallmark achievement. It serves as a benchmark of experimental capabilities and enables essential information processing tasks such as the implementation of quantum gates and the transmission of quantum information [2].

In the solid state, electron spins confined in electrically defined semiconductor quantum dots have emerged as a promising platform for QIP [3–6]: Essential ingredients such as initialization, single-shot readout, universal quantum gates and, quite recently, entanglement have demonstrated experimentally [7–12]. In this context, nuclear spins in the surrounding semiconductor host environment have attracted

© Springer International Publishing AG 2017

M.J.A. Schütz, *Quantum Dots for Quantum Information Processing: Controlling and Exploiting the Quantum Dot Environment*, Springer Theses,
DOI 10.1007/978-3-319-48559-1_3

considerable theoretical [13–19] and experimental [20–25] attention, as they have been identified as the main source of electron spin decoherence due to the relatively strong hyperfine (HF) interaction between the electronic spin and $N \sim 10^6$ nuclei [4]. However, it has also been noted that the nuclear spin bath itself, with nuclear spin coherence times ranging from hundreds of microseconds to a millisecond [4, 26], could be turned into an asset, for example, as a resource for quantum memories or quantum computation [27–31]. Since these applications require yet unachieved control of the nuclear spins, novel ways of understanding and manipulating the dynamics of the nuclei are called for. The ability to control and manipulate the nuclei will open up new possibilities for nuclear spin-based information storage and processing, but also directly improve electron spin decoherence timescales [32–34].

Dissipation has recently been identified as a novel approach to control a quantum system, create entangled states or perform quantum computing tasks [35–39]. This is done by properly engineering the continuous interaction of the system with its environment. In this way, dissipation—previously often viewed as a vice from a QIP perspective—can turn into a virtue and become the driving force behind the emergence of coherent quantum phenomena. The idea of actively using dissipation rather than relying on coherent evolution extends the traditional DiVincenzo criteria [40] to settings in which no unitary gates are available; also, it comes with potentially significant practical advantages, as dissipative methods are inherently robust against weak random perturbations, allowing, in principle, to stabilize entanglement for arbitrary times. Recently, these concepts have been put into practice experimentally in different QIP architectures, namely atomic ensembles [41, 42], trapped ions [43, 44] and superconducting qubits [45].

Here, we apply these ideas to a quantum dot system and investigate a scheme for the deterministic generation of steady-state entanglement between the two spatially separated nuclear spin ensembles in an electrically defined double quantum dot (DQD), operated in the Pauli-blockade regime [3, 25]. We develop in detail the underlying theoretical framework, and discuss in great depth the coherent phenomena emerging from the hyperfine coupled electron and nuclear dynamics in a DQD in spin blockade regime. The analysis is based on the fact that the electron spins evolve rapidly on typical timescales of the nuclear spin dynamics. This allows us to derive a coarse-grained quantum master equation for the nuclear spins only, disclosing the nuclei as the quantum system coupled to an electronic environment with an exceptional degree of tunability; see Fig. 3.1 for a schematic illustration. This approach provides valuable insights by building up a straightforward analogy between mesoscopic solid-state physics and a generic setting in quantum optics (compare, for example, Ref. [41]): The nuclear spin ensemble can be identified with an atomic ensemble, with individual nuclear spins corresponding to the internal levels of a single atom and electrons playing the role of photons [46].

Our theoretical analysis goes beyond this simple analogy by incorporating nonlinear, feedback-driven effects resulting from a backaction of the effective magnetic field generated by the nuclei (Overhauser shift) on the electron energy levels. In accordance with previous theoretical [32, 34, 47–52] and experimental [12, 24, 53–56] observations, this feedback mechanism is shown to lead to a rich set of phe-

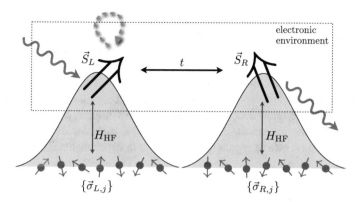

Fig. 3.1 Schematic illustration of the DQD system under study. Two nuclear spin ensembles are hyperfine-coupled to the electronic subsytem; due to various *fast* coherent (*double-arrow*) and incoherent processes (*arrows*) the latter settles to a quasisteady state on a timescale shorter than the nuclear dynamics

nomena such as multistability, criticality, and dynamic nuclear polarization (DNP). In our model, we study the nuclear dynamics in a systematic expansion of the master equation governing the evolution of the combined electron-nuclear system, which allows us efficiently trace out the electronic degrees of freedom yielding a compact dynamical equation for the nuclear system alone. This mathematical description can be understood in terms of the so-called slaving principle: The electronic subsystem settles to a *quasisteady* state on a timescale much faster than the nuclear dynamics, and creates an effective environment with tunable properties for the nuclear spins. Consequently, we analyze the nuclear dynamics subject to this artificial environment. Feedback effects kick in as the generated nuclear spin polarization acts back on the electronic subsystem via the Overhauser shift changing the electronic quasisteady state. We derive explicit expressions for the nuclear steady state which allows us to fully assess the nuclear properties in dependence on the external control parameters. In particular, we find that, depending on the parameter regime, the polarization of the nuclear ensemble can show two distinct behaviors: The nuclear spins either saturate in a dark state without any nuclear polarization or, upon surpassing a certain threshold gradient, turn self-polarizing and build up sizable Overhauser field differences. Notably, the high-polarization stationary states feature steady-state entanglement between the two nuclear spin ensembles, even though the electronic quasisteady state is separable, underlining the very robustness of our scheme against electronic noise.

To analyze the nuclear spin dynamics in detail, we employ different analytical approaches, namely a semiclassical calculation and a fully quantum mechanical treatment. This is based on a hierarchy of timescales: While the nuclear polarization process occurs on a typical timescale of $\tau_{\text{pol}} \gtrsim 1\,\text{s}$, the timescale for building up quantum correlations τ_{gap} is collectively [46] enhanced by a factor $N \sim 10^5 - 10^6$; i.e., $\tau_{\text{gap}} \approx (3-30)\,\mu\text{s}$. Since nuclear spins dephase due to internal dipole-dipole interactions on a timescale of $\tau_{dec} \approx (0.1-1)$ ms [4, 26, 57], our system exhibits the following separation of typical timescales: $\tau_{pol} \gg \tau_{dec} \gg \tau_{gap}$. While the first inequality allows us to study the (slow) dynamics of the macroscopic semiclassical part of the nuclear fields in a mean-field treatment (which essentially disregards quantum correlations) on long timescales, based on the second inequality we investigate the generation of (comparatively small) quantum correlations on a much faster timescale where we neglect decohering processes due to internal dynamics among the nuclei. Lastly, numerical results complement our analytical findings and we discuss in detail detrimental effects typically encountered in experiments.

3.2 Executive Summary: Reader's Guide

This chapter is organized as follows. In Sect. 3.3, we provide an intuitive picture of our basic ideas, allowing the reader to grasp our main results on a qualitative level. Section 3.4 introduces the master-equation-based theoretical framework. Based on a simplified model, in Sect. 3.5 we study the coupled electron nuclear dynamics. Using adiabatic elimination techniques, we can identify two different regimes as possible fixed points of the nuclear evolution which differ remarkably in their nuclear polarization and entanglement properties. Subsequently, in Sect. 3.6 the underlying multi-stability of the nuclear system is revealed within a semiclassical model. Based on a self-consistent Holstein-Primakoff approximation, in Sect. 3.7 we study in great detail the nuclear dynamics in the vicinity of a high-polarization fixed point. This analysis puts forward the main result of this chapter, the steady-state generation of entanglement between the two nuclear spin ensembles in a DQD. Within the framework of the Holstein-Primakoff analysis, Sect. 3.8 highlights the presence of a dissipative phase transition in the nuclear spin dynamics. Generalizations of our findings to inhomogeneous hyperfine coupling and other weak undesired effects are covered in Sect. 3.9. Finally, in Sect. 3.10 we draw conclusions and give an outlook on possible future directions of research.

3.3 Main Results

In this chapter, we propose a scheme for the dissipative preparation of steady-state entanglement between the two nuclear spin ensembles in a double quantum dot (DQD) in the Pauli-blockade regime [3]. The entanglement arises from an interference between different hyperfine-induced processes lifting the Pauli-blockade; compare Fig. 3.2 for a schematic illustration. This becomes possible by suitably engineering the effective electronic environment, which ensures a *collective* coupling of electrons and nuclei (i.e., each flip can happen either in the left or the right

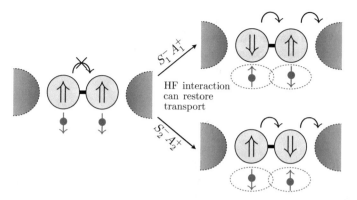

Fig. 3.2 Schematic illustration of nuclear entanglement generation via electron transport. Whenever the Pauli-blockade is lifted via the HF interaction with the nuclear spins, a nuclear flip can occur in *either* of the *two dots*. The local nature of the HF interaction is masked by the non-local character of the electronic level $|\lambda_2\rangle$ which predominantly couples to the Pauli-blocked triplets $|T_\pm\rangle$

QD and no which-way information is leaked), and that just two such processes with a common entangled stationary state are dominant. Engineering of the electronic system via external gate voltages facilitates the control of the desired steady-state properties. Exploiting the separation of electronic and nuclear time-scales allows to derive a quantum master equation in which the interference effect becomes apparent: It features non-local jump operators which drive the nuclear system into an entangled steady state of EPR-type [41]. Since the entanglement is actively stabilized by the dissipative dynamics, our approach is inherently robust against weak random perturbations [35–37, 41, 58]. The entanglement build-up is accompanied by a self-polarization of the nuclear system towards large Overhauser (OH) field gradients if a small initial gradient is provided. Upon surpassing a certain threshold value of this field the nuclear dynamics turn self-polarizing, and drive the system to even larger gradients. Entanglement is then generated in the quantum fluctuations around these macroscopic nuclear polarizations. Furthermore, feedback between electronic and nuclear dynamics leads to multi-stability and criticality in the steady-state solutions. Our scheme may provide a long-lived, solid-state entanglement resource and a new route for nuclear-spin-based information storage and manipulation.

3.4 The System

This section presents a detailed description of the system under study, a gate-defined double quantum dot (DQD) in the Pauli-blockade regime. To model the dynamics of this system, we employ a master equation formalism [46]. This allows us to study the irreversible dynamics of the DQD coupled to source and drain electron reservoirs. By tracing out the unobserved degrees of freedom of the leads, we show that—under appropriate conditions to be specified below—the dynamical evolution of the reduced density matrix of the system ρ can formally be written as

$$\dot{\rho} = \underbrace{-i\,[H_{el}, \rho] + \mathcal{V}\rho + \mathcal{L}_{\Gamma}\rho}_{\textcircled{1}} + \underbrace{\mathcal{L}_{\pm}\rho + \mathcal{L}_{deph}\rho,}_{\textcircled{2}} \tag{3.1}$$

Here, H_{el} describes the electronic degrees of freedom of the DQD in the relevant two-electron regime, $\mathcal{V}$ refer to the coherent hyperfine coupling between electronic and nuclear spins and $\mathcal{L}_{\Gamma}$ is a Liouvillian of Lindblad form describing electron transport in the spin-blockade regime. The last two terms labeled by $\textcircled{2}$ account for different physical mechanisms such as cotunneling, spin-exchange with the leads or spin-orbital coupling in terms of effective dissipative terms in the electronic subspace.

3.4.1 Microscopic Model

We consider an electrically defined DQD in the Pauli-blockade regime [3, 25]. Microscopically, our analysis is based on a two-site Anderson Hamiltonian: Due to strong confinement, both the left and right dot are assumed to support a single orbital level ε_i ($i = L, R$) only which can be Zeeman split in the presence of a magnetic field and occupied by up to two electrons forming a localized spin singlet. For now, excited states, forming on-site triplets that could lift spin-blockade, are disregarded, since they are energetically well separated by the singlet-triplet splitting $\Delta_{st} \gtrsim 400\,\mu eV$ [3]. Cotunneling effects due to energetically higher lying localized triplet states will be addressed separately below.

Formally, the Hamiltonian for the global system $\mathcal{H}$ can be decomposed as

$$\mathcal{H} = H_{DQD} + H_B + H_T, \tag{3.2}$$

where H_B refers to two independent reservoirs of non-interacting electrons, the left (L) and right (R) lead, respectively,

$$H_B = \sum_{i,k,\sigma} \varepsilon_{ik} c_{ik\sigma}^{\dagger} c_{ik\sigma}, \tag{3.3}$$

with $i = L, R$, $\sigma = \uparrow, \downarrow$ and H_T models the coupling of the DQD to the leads in terms of the tunnel Hamiltonian

$$H_T = \sum_{i,k,\sigma} T_i d_{i\sigma}^{\dagger} c_{ik\sigma} + \text{h.c.}. \tag{3.4}$$

The tunnel matrix element T_i, specifying the transfer coupling between the leads and the system, is assumed to be independent of momentum k and spin σ of the electron. The fermionic operator $c_{ik\sigma}^\dagger$ ($c_{ik\sigma}$) creates (annihilates) an electron in lead $i = L, R$ with wavevector k and spin $\sigma = \uparrow, \downarrow$. Similarly, $d_{i\sigma}^\dagger$ creates an electron with spin σ inside the dot in the orbital $i = L, R$. Accordingly, the localized electron spin operators are

$$\vec{S}_i = \frac{1}{2} \sum_{\sigma, \sigma'} d_{i\sigma}^\dagger \vec{\sigma}_{\sigma\sigma'} d_{i\sigma'}, \tag{3.5}$$

where $\vec{\sigma}$ refers to the vector of Pauli matrices. Lastly,

$$H_{DQD} = H_S + H_t + V_{HF} \tag{3.6}$$

describes the coherent electron-nuclear dynamics inside the DQD. In the following, H_S, H_t and V_{HF} are presented. First, H_S accounts for the bare electronic energy levels in the DQD and Coulomb interaction terms

$$H_S = \sum_{i\sigma} \varepsilon_{i\sigma} n_{i\sigma} + \sum_i U_i n_{i\uparrow} n_{i\downarrow} + U_{LR} n_L n_R, \tag{3.7}$$

where U_i and U_{LR} refer to the on-site and interdot Coulomb repulsion; $n_{i\sigma} = d_{i\sigma}^\dagger d_{i\sigma}$ and $n_i = n_{i\uparrow} + n_{i\downarrow}$ are the spin-resolved and total electron number operators, respectively. Typical values are $U_i \approx 1-4$ meV and $U_{LR} \approx 200\,\mu$eV [3, 25, 59]. Coherent, spin-preserving interdot tunneling is described by

$$H_t = t \sum_\sigma d_{L\sigma}^\dagger d_{R\sigma} + \text{h.c.} \tag{3.8}$$

Spin-blockade regime—By appropriately tuning the chemical potentials of the leads μ_i, one can ensure that at maximum two conduction electrons reside in the DQD [3, 37]. Moreover, for $\varepsilon_{R\sigma} < \mu_R$ the right dot always stays occupied. In what follows, we consider a transport setting where an applied bias between the two dots approximately compensates the Coulomb energy of two electrons occupying the right dot, that is $\varepsilon_L \approx \varepsilon_R + U_R - U_{LR}$. Then, a source drain bias across the DQD device induces electron transport via the cycle $(0, 1) \rightarrow (1, 1) \rightarrow (0, 2)$. Here, (m, n) refers to a configuration with m (n) electrons in the left (right) dot, respectively. In our Anderson model, the only energetically accessible $(0, 2)$ state is the localized singlet, referred to as $|S_{02}\rangle = d_{R\uparrow}^\dagger d_{R\downarrow}^\dagger |0\rangle$. As a result of the Pauli principle, the interdot charge transition $(1, 1) \rightarrow (0, 2)$ is allowed only for the $(1, 1)$ spin singlet $|S_{11}\rangle = (|\uparrow\downarrow\rangle - |\downarrow\uparrow\rangle)/\sqrt{2}$, while the spin triplets $|T_\pm\rangle$ and $|T_0\rangle = (|\uparrow\downarrow\rangle + |\downarrow\uparrow\rangle)/\sqrt{2}$ are Pauli blocked. Here, $|T_+\rangle = |\uparrow\uparrow\rangle$, $|T_-\rangle = |\downarrow\downarrow\rangle$, and $|\sigma\sigma'\rangle = d_{L\sigma}^\dagger d_{R\sigma'}^\dagger |0\rangle$. For further details on how to realize this regime we refer to Appendix 3.11.1.

Hyperfine interaction.—The electronic spins $\vec{S}_i$ confined in either of the two dots $(i = L, R)$ interact with two different sets of nuclear spins $\left\{\sigma_{i,j}^\alpha\right\}$ in the semiconductor host environment via hyperfine (HF) interaction. It is dominated by the isotropic Fermi contact term [13] given by

$$H_{\mathrm{HF}} = \frac{a_{\mathrm{hf}}}{2} \sum_{i=L,R} \left(S_i^+ A_i^- + S_i^- A_i^+\right) + a_{\mathrm{hf}} \sum_{i=L,R} S_i^z A_i^z. \qquad (3.9)$$

Here, S_i^α and $A_i^\alpha = \sum_j a_{i,j}\sigma_{i,j}^\alpha$ for $\alpha = \pm, z$ denote electron and collective nuclear spin operators. The coupling coefficients $a_{i,j}$ are proportional to the weight of the electron wavefunction at the jth lattice site and define the individual unitless HF coupling constant between the electron spin in dot i and the jth nucleus. They are normalized such that $\sum_{j=1}^{N_i} a_{i,j} = N$, where $N = (N_L + N_R)/2 \sim 10^6$; a_{hf} is related to the total HF coupling strength $A_{\mathrm{HF}} \approx 100\,\mu\mathrm{eV}$ via $a_{\mathrm{hf}} = A_{\mathrm{HF}}/N$ and $g_{\mathrm{hf}} = A_{\mathrm{HF}}/\sqrt{N} \approx 0.1\,\mu\mathrm{eV}$ quantifies the typical HF interaction strength. The individual nuclear spin operators $\sigma_{i,j}^\alpha$ are assumed to be spin-$\frac{1}{2}$ for simplicity. We neglect the nuclear Zeeman and dipole-dipole terms which will be slow compared to the system's dynamics [13]; these simplifications will be addressed in more detail in Sect. 3.9.

The effect of the hyperfine interaction can be split up into a perpendicular component

$$H_{\mathrm{ff}} = \frac{a_{\mathrm{hf}}}{2} \sum_{i=L,R} \left(S_i^+ A_i^- + S_i^- A_i^+\right), \qquad (3.10)$$

which exchanges excitations between the electronic and nuclear spins, and a parallel component, referred to as Overhauser (OH) field,

$$H_{\mathrm{OH}} = a_{\mathrm{hf}} \sum_{i=L,R} S_i^z A_i^z. \qquad (3.11)$$

The latter can be recast into the following form

$$H_{\mathrm{OH}} = H_{\mathrm{sc}} + H_{zz}, \qquad (3.12)$$

where

$$H_{\mathrm{sc}} = \bar{\omega}_{\mathrm{OH}} \left(S_L^z + S_R^z\right) + \Delta_{\mathrm{OH}} \left(S_R^z - S_L^z\right) \qquad (3.13)$$

describes a (time-dependent) semiclassical OH field which comprises a homogeneous $\bar{\omega}_{\mathrm{OH}}$ and inhomogeneous Δ_{OH} component, respectively,

$$\bar{\omega}_{\mathrm{OH}} = \frac{a_{\mathrm{hf}}}{2} \left(\langle A_L^z\rangle_t + \langle A_R^z\rangle_t\right), \qquad (3.14)$$

$$\Delta_{\mathrm{OH}} = \frac{a_{\mathrm{hf}}}{2} \left(\langle A_R^z\rangle_t - \langle A_L^z\rangle_t\right), \qquad (3.15)$$

and

$$H_{zz} = a_{hf} \sum_{i=L,R} S_i^z \delta A_i^z, \qquad (3.16)$$

with $\delta A_i^z = A_i^z - \langle A_i^z \rangle_t$, refers to residual quantum fluctuations due to deviations of the Overhauser field from its expectation value [46]. The semiclassical part H_{sc} only acts on the electronic degrees of freedom and can therefore be absorbed into H_S. Then, the coupling between electronic and nuclear degrees of freedom is governed by the operator

$$V_{HF} = H_{ff} + H_{zz}. \qquad (3.17)$$

3.4.2 Master Equation

To model the dynamical evolution of the DQD system, we use a master equation approach. Starting from the full von Neumann equation for the global density matrix ϱ

$$\dot{\varrho} = -i [\mathcal{H}, \varrho], \qquad (3.18)$$

we employ a Born-Markov treatment, trace out the reservoir degrees of freedom, apply the so-called approximation of independent rates of variation [60], and assume fast recharging of the DQD which allows us to eliminate the single-electron levels [61, 62]; for details, see Appendix 3.11.2. Then, we arrive at the following master equation for the system's density matrix $\rho = \text{Tr}_B [\varrho]$

$$\dot{\rho} = -i [H_{el}, \rho] + \mathcal{L}_\Gamma \rho + \mathcal{V}\rho, \qquad (3.19)$$

where $\text{Tr}_B [\dots]$ denotes the trace over the bath degrees of freedom in the leads. In the following, the Hamiltonian H_{el} and the superoperators $\mathcal{L}_\Gamma$, $\mathcal{V}$ will be discussed in detail [cf. Eqs. (3.20), (3.22) and (3.24), respectively].

Electronic Hamiltonian.—In Eq. (3.19), H_{el} describes the electronic degrees of freedom of the DQD within the relevant two-electron subspace. It can be written as ($\hbar = 1$)

$$H_{el} = \omega_0 \left(S_L^z + S_R^z \right) + \Delta \left(S_R^z - S_L^z \right) - \varepsilon \left| S_{02} \right\rangle \left\langle S_{02} \right|$$
$$+ t \left(\left| \Uparrow \Downarrow \right\rangle \left\langle S_{02} \right| - \left| \Downarrow \Uparrow \right\rangle \left\langle S_{02} \right| + \text{h.c.} \right), \qquad (3.20)$$

where the nuclear-polarization-dependent 'mean-field' quantities $\bar{\omega}_{OH}$ and Δ_{OH} have been absorbed into the definitions of ω_0 and Δ as $\omega_0 = \omega_{ext} + \bar{\omega}_{OH}$ and $\Delta = \Delta_{ext} + \Delta_{OH}$, respectively. In previous theoretical work, this feedback of the Overhauser shift on the electronic energy levels has been identified as a means for controlling the nuclear spins via instabilities towards self-polarization; compare for example Ref. [32]. Apart from the OH contributions, ω_{ext} and Δ_{ext} denote the Zee-

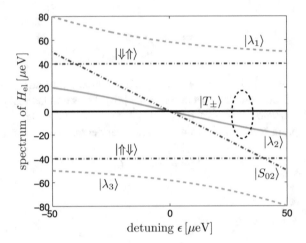

Fig. 3.3 Spectrum of H_{el} in the relevant two-electron regime for $\Delta = 40\,\mu\text{eV}$ and $t = 30\,\mu\text{eV}$; shown here as a function of the interdot detuning parameter ε. The three hybridized electronic eigenstates $|\lambda_k\rangle$ within the $S_{\text{tot}}^z = 0$ subspace are displayed in *red*, while the bare electronic states are shown in *blue (dash-dotted lines)*. The homogeneous Zeeman splitting ω_0 has been set to zero, so that the Pauli-blocked triplets $|T_{\pm}\rangle$ are degenerate. In this setting, the levels $|\lambda_{1,3}\rangle$ are far detuned from $|T_{\pm}\rangle$. Therefore, the spin-blockade is lifted pre-dominantly via the non-local electronic level $|\lambda_2\rangle$. The *black dashed ellipse* refers to a potential operational area of our scheme

man splitting due to the homogeneous and inhomogeneous component of a potential external magnetic field, respectively. Furthermore, ε refers to the relative interdot energy detuning between the left and right dot. The interdot tunneling with coupling strength t occurs exclusively in the singlet subspace due to Pauli spin-blockade. It is instructive to diagonalize the effective five-dimensional electronic Hamiltonian H_{el}. The eigenstates of H_{el} within the $S_{\text{tot}}^z = S_L^z + S_R^z = 0$ subspace can be expressed as

$$|\lambda_k\rangle = \mu_k\,|\Uparrow\Downarrow\rangle + \nu_k\,|\Downarrow\Uparrow\rangle + \kappa_k\,|S_{02}\rangle\,, \qquad (3.21)$$

for $k = 1, 2, 3$ with corresponding eigenenergies ε_k; compare Fig. 3.3.[1] Note that, throughout this work, the hybridized level $|\lambda_2\rangle$ plays a crucial role for the dynamics of the DQD system: Since the levels $|\lambda_{1,3}\rangle$ are energetically separated from all other electronic levels (for $t \gg \omega_0, g_{\text{hf}}$), $|\lambda_2\rangle$ represents the dominant channel for lifting of the Pauli-blockade; compare Fig. 3.3.

Electron transport.—After tracing out the reservoir degrees of freedom, electron transport induces dissipation in the electronic subspace: The Liouvillian

$$\mathcal{L}_{\Gamma}\rho = \sum_{k,\nu=\pm} \Gamma_k \mathcal{D}\left[|T_{\nu}\rangle\,\langle\lambda_k|\right]\rho, \qquad (3.22)$$

---

[1]The analytic expressions for the amplitudes μ_k, ν_k and κ_k are not instructive and therefore not explicitly given.

with the short-hand notation for the Lindblad form $\mathcal{D}[c]\rho = c\rho c^\dagger - \frac{1}{2}\{c^\dagger c, \rho\}$, effectively models electron transport through the DQD; here, we have applied a rotating-wave approximation by neglecting terms rotating at a frequency of $\varepsilon_k - \varepsilon_l$ for $k \neq l$ (see Appendix 3.11.2 for details). Accordingly, the hybridized electronic levels $|\lambda_k\rangle$ ($k = 1, 2, 3$) acquire a finite lifetime [47] and decay with a rate

$$\Gamma_k = |\langle \lambda_k | S_{02}\rangle|^2 \Gamma = \kappa_k^2 \Gamma, \tag{3.23}$$

determined by their overlap with the localized singlet $|S_{02}\rangle$, back into the Pauli-blocked triplet subspace $\{|T_\pm\rangle\}$. Here, $\Gamma = \Gamma_R/2$, where Γ_R is the sequential tunneling rate to the right lead.

Hyperfine interaction.—After splitting off the semiclassical quantities $\bar{\omega}_{\text{OH}}$ and Δ_{OH}, the superoperator

$$\mathcal{V}\rho = -i[V_{\text{HF}}, \rho], \tag{3.24}$$

captures the remaining effects due to the HF coupling between electronic and nuclear spins. Within the eigenbasis of H_{el}, the hyperfine flip-flop dynamics H_{ff}, accounting for the exchange of excitations between the electronic and nuclear subsystem, takes on the form

$$H_{\text{ff}} = \frac{a_{\text{hf}}}{2} \sum_k \left[|\lambda_k\rangle \langle T_+| \otimes L_k + |\lambda_k\rangle \langle T_-| \otimes \mathbb{L}_k + \text{h.c.} \right], \tag{3.25}$$

where the *non-local* nuclear jump operators

$$L_k = \nu_k A_L^+ + \mu_k A_R^+, \tag{3.26}$$
$$\mathbb{L}_k = \mu_k A_L^- + \nu_k A_R^-, \tag{3.27}$$

are associated with lifting the spin-blockade from $|T_+\rangle$ and $|T_-\rangle$ via $|\lambda_k\rangle$, respectively. These operators characterize the effective coupling between the nuclear system and its electronic environment; they can be controlled externally via gate voltages as the parameters t and ε define the amplitudes μ_k and ν_k. Since generically $\mu_k \neq \nu_k$, the non-uniform electron spin density of the hybridized eigenstates $|\lambda_k\rangle$ introduces an asymmetry to flip a nuclear spin on the first or second dot [47].

Electronic spin-blockade lifting.—Apart from the hyperfine mechanism described above, the Pauli blockade may also be lifted by other, purely electronic processes such as (i) cotunneling, (ii) spin-exchange with the leads, or (iii) spin-orbit coupling [63]. Although they do not exchange excitations with the nuclear spin bath, these processes have previously been shown to be essential to describe the nuclear spin dynamics in the Pauli blockade regime [32, 47, 64]. In our analysis, it is crucial to include them as they affect the average electronic quasisteady state seen by the nuclei, while the exact, microscopic nature of the electronic decoherence processes does not play an important role for our proposal. Therefore, for concreteness, here we only describe *exemplarily* virtual tunneling processes via the doubly occupied triplet

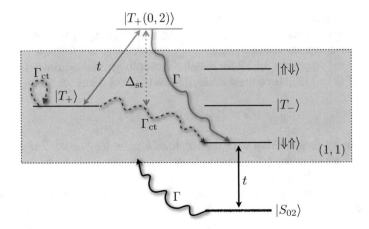

Fig. 3.4 Scheme for the phenomenological cotunneling analysis. The spin-blocked triplet $|T_+\rangle$ is tunnel-coupled to the (virtually occupied) triplet $|T_+ (0, 2)\rangle$, localized on the right dot. Due to Pauli exclusion, this level is energetically well separated by the singlet-triplet splitting $\Delta_{st} \gtrsim 400\,\mu$eV. It has a finite lifetime Γ^{-1} and may decay back (via a singly occupied level on the right dot) to $|T_+\rangle$ or via a series of fast coherent and incoherent intermediate processes end up in any level within the $(1, 1)$ charge sector (*shaded box*), since $|S_{02}\rangle$ decays with a rate Γ to all four $(1, 1)$ states. The overall effectiveness of the process is set by the effective rate $\Gamma_{ct} \approx (t/\Delta_{st})^2 \Gamma$, depicted by *dashed arrows*

state labeled as $|T_+ (0, 2)\rangle$, while spin-exchange with the leads or spin-orbital effects are discussed in detail in Appendix 3.11.4. Cotunneling via $|T_- (0, 2)\rangle$ or $|T_0 (0, 2)\rangle$ can be analyzed along the same lines. As schematically depicted in Fig. 3.4, the triplet $|T_+\rangle$ with $(1, 1)$ charge configuration is coherently coupled to $|T_+ (0, 2)\rangle$ by the interdot tunnel-coupling t. This transition is strongly detuned by the singlet-triplet splitting Δ_{st}. Once, the energetically high lying level $|T_+ (0, 2)\rangle$ is populated, it quickly decays with rate Γ either back to $|T_+\rangle$ giving rise to a pure dephasing process within the low-energy subspace or to $\{|T_-\rangle, |\lambda_k\rangle\}$ via some fast intermediate steps, mediated by fast discharging and recharging of the DQD with the rate Γ.[2] In our theoretical model (see below), the former is captured by the pure dephasing rate Γ_{deph}, while the latter can be absorbed into the dissipative mixing rate $\Gamma_\pm$; compare Fig. 3.6 for a schematic illustration of $\Gamma_\pm$ and Γ_{deph}, respectively. Since the singlet-triplet splitting is the largest energy scale in this process ($t, \Gamma \ll \Delta_{st}$), the effective rate for this virtual cotunneling mechanism can be estimated as

$$\Gamma_{ct} \approx (t/\Delta_{st})^2 \Gamma. \tag{3.28}$$

---

[2]Note that, in the spirit of the approximation of independent rates of variation [60], the relevant (in-)coherent couplings are shown in Fig. 3.4 in terms of the bare basis $\{|\sigma, \sigma'\rangle, |S_{02}\rangle\}$ for $\sigma = \Uparrow, \Downarrow$. After a basis transformation to the dressed basis $\{|T_\pm\rangle, |\lambda_k\rangle\}$ one obtains the dissipative terms given in Eq. (3.30).

Equation (3.28) describes a virtually assisted process by which t couples $|T_+\rangle$ to a virtual level, which can then escape via sequential tunneling $\sim\Gamma$; thus, it can be made relatively fast compared to typical nuclear timescales by working in a regime of efficient electron exchange with the leads $\sim\Gamma$.[3] For example, taking $t \approx 30\,\mu eV$, $\Delta_{st} \approx 400\,\mu eV$ and $\Gamma \approx 50\,\mu eV$, we estimate $\Gamma_{ct} \approx 0.3\,\mu eV$, which is fast compared to typical nuclear timescales. Note that, for more conventional, *slower* electronic parameters ($t \approx 5\,\mu eV$, $\Gamma \approx 0.5\,\mu eV$), indirect tunneling becomes negligibly small, $\Gamma_{ct} \approx 5 \times 10^{-5}\,\mu eV \approx 5 \times 10^4\,s^{-1}$, in agreement with values given in Ref. [47]. Our analysis, however, is restricted to the regime, where indirect tunneling is fast compared to the nuclear dynamics; this regime of motional averaging has previously been shown to be beneficial for e.g. nuclear spin squeezing [32, 34]. Alternatively, spin-blockade may be lifted via spin-exchange with the leads. The corresponding rate Γ_{se} scales as $\Gamma_{se} \sim \Gamma^2$, as compared to $\Gamma_{ct} \sim t^2\Gamma$. Moreover, Γ_{se} depends strongly on the detuning of the $(1, 1)$ levels from the Fermi levels of the leads. If this detuning is $\sim 500\,\mu eV$ and for $\Gamma \approx 100\,\mu eV$, we estimate $\Gamma_{se} \approx 0.25\,\mu eV$, which is commensurate with the desired motional averaging regime, whereas, for less efficient transport ($\Gamma \approx 1\,\mu eV$) and stronger detuning $\sim 1\,meV$, one obtains a negligibly small rate, $\Gamma_{se} \approx 6 \times 10^{-6}\,\mu eV \approx 6 \times 10^3\,s^{-1}$. Again, this is in line with Ref. [47]. As discussed in more detail in Appendix 3.11.4, these spin-exchange processes as well as spin-orbital effects can be treated on a similar footing as the interdot cotunneling processes discussed here. Therefore, to describe the net effect of various non-hyperfine mechanisms and to complete our theoretical description of electron transport in the spin-blockade regime, we add the following phenomenological Lindblad terms to our model

$$\mathcal{L}_{deph}\rho = \frac{\Gamma_{deph}}{2}\mathcal{D}\left[|T_+\rangle\langle T_+| - |T_-\rangle\langle T_-|\right]\rho, \tag{3.29}$$

$$\mathcal{L}_\pm\rho = \Gamma_\pm \sum_{\nu=\pm}\mathcal{D}\left[|T_{\bar{\nu}}\rangle\langle T_\nu|\right]\rho \tag{3.30}$$

$$+\Gamma_\pm\sum_{k,\nu}\mathcal{D}\left[|T_\nu\rangle\langle\lambda_k|\right]\rho + \mathcal{D}\left[|\lambda_k\rangle\langle T_\nu|\right]\rho.$$

Summary.—Before concluding the description of the system under study, let us quickly reiterate the ingredients of the master equation as stated in Eq. (3.1): It accounts for (i) the unitary dynamics within the DQD governed by $-i\,[H_{el} + V_{HF}, \rho]$, (ii) electron-transport-mediated dissipation via $\mathcal{L}_\Gamma$ and (iii) dissipative mixing and dephasing processes described by $\mathcal{L}_\pm$ and $\mathcal{L}_{deph}$, respectively. Finally, the most important parameters of our model are summarized in Fig. 3.5.

---

[3]If the nuclear spins in the second dot are highly polarized, the doubly occupied triplet levels get Zeeman split by the corresponding local Overhauser field. In principle, this could lead to a small asymmetry in the incoherent rates describing decay from $|T_+\rangle$ to $|T_-\rangle$ and vice versa. However, in the Appendix it is shown that in this regime a similar incoherent decay process within the Pauli-blocked triplet space $\{|T_+\rangle, |T_-\rangle\}$ can be made much more efficient by working in a regime of fast spin-exchange with the reservoirs in the leads.

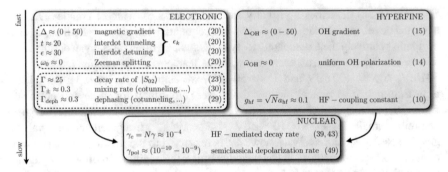

Fig. 3.5 Schematic overview of the most important parameters in our model, grouped into electronic, hyperfine and HF-mediated nuclear quantities. Within the electronic quantities, we can differentiate between coherent and incoherent processes (compare *dashed boxes*). Typical numbers are given in μeV, while the numbers in parentheses ($\cdot$) refer to the corresponding equations in the text

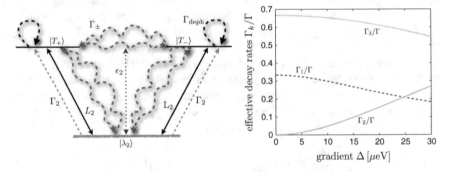

Fig. 3.6 *Left plot* Schematic illustration of coherent and incoherent processes within the effective three-level system $\{|T_\pm\rangle, |\lambda_2\rangle\}$: The level $|\lambda_2\rangle$ is detuned from $|T_\pm\rangle$ by ε_2 and decays according to its overlap with the localized singlet with the rate $\Gamma_2 = \kappa_2^2 \Gamma$. Moreover, it is coherently coupled to the triplets $|T_+\rangle$ and $|T_-\rangle$ via the *nonlocal* nuclear operators L_2 and $\mathbb{L}_2$, respectively. Purely electronic spin-blockade lifting mechanisms such as cotunneling or spin-orbital effects result in effective dephasing and dissipative mixing rates, labeled as Γ_{deph} and $\Gamma_\pm$, respectively. The latter do not affect the nuclei directly, but lead to an unbiased population transfer within the electronic three-level system. In particular, mixing between $|T_\pm\rangle$ can arise from virtual occupation of $|\lambda_{1,3}\rangle$ or spin-orbit coupling. *Right plot* Effective decay rates $\Gamma_k = \kappa_k^2 \Gamma$, shown here for $\varepsilon = t = 30\,\mu$eV. For small gradients, $|\lambda_2\rangle \approx |T_0\rangle$ and therefore it does not decay due to Pauli-blockade

3.5 Effective Nuclear Dynamics

In this section we develop the general theoretical framework of our analysis which is built upon the fact that, generically, the nuclear spins evolve slowly on typical electronic timescales. Due to this separation of electronic and nuclear timescales, the system is subject to the slaving principle [65] implying that the electronic subsystem settles to a quasisteady state on a timescale much shorter than the nuclear dynamics. This allows us to adiabatically eliminate the electronic coordinates yielding an effective master equation on a coarse-grained timescale. Furthermore, the electronic quasisteady state is shown to depend on the state of the nuclei resulting in feedback mechanisms between the electronic and nuclear degrees of freedom. Specifically, here we analyze the dynamics of the nuclei coupled to the electronic three-level subspace spanned by the levels $|T_\pm\rangle$ and $|\lambda_2\rangle$. This simplification is justified for $t \gg \omega_0, g_{hf}$, since in this parameter regime the electronic levels $|\lambda_{1,3}\rangle$ are strongly detuned from the manifold $\{|T_\pm\rangle, |\lambda_2\rangle\}$; compare Fig. 3.3. Effects due to the presence of $|\lambda_{1,3}\rangle$ will be discussed separately in Sects. 3.7 and 3.8. Here, due to their fast decay with a rate $\Gamma_{1,3}$, they have already been eliminated adiabatically from the dynamics, leading to a dissipative mixing between the blocked triplet states $|T_\pm\rangle$ with rate $\Gamma_\pm$; alternatively, this mixing could come from spin-orbit coupling (see Appendix 3.11.4 for details). Moreover, for simplicity, we assume $\omega_0 = 0$ and neglect nuclear fluctuations arising from H_{zz}. This approximation is in line with the semiclassical approach used below in order to study the nuclear polarization dynamics; for details we refer to Appendix 3.11.6. In summary, all relevant coherent and incoherent processes within the effective three-level system $\{|T_\pm\rangle, |\lambda_2\rangle\}$ are schematically depicted in Fig. 3.6.

Intuitive picture.—The main results of this section can be understood from the fact that the level $|\lambda_2\rangle$ decays according to its overlap with the localized singlet, that is with a rate

$$\Gamma_2 = |\langle\lambda_2|S_{02}\rangle|^2 \, \Gamma \xrightarrow{\Delta\to0} 0 \qquad (3.31)$$

which in the low-gradient regime $\Delta \approx 0$ tends to zero, since then $|\lambda_2\rangle$ approaches the triplet $|T_0\rangle$ which is dark with respect to tunneling and therefore does not allow for electron transport; see Fig. 3.6. In other words, in the limit $\Delta \to 0$, the electronic level $|\lambda_2\rangle \to |T_0\rangle$ gets stabilized by Pauli-blockade. In this regime, we expect the nuclear spins to undergo some form of random diffusion process since the dynamics lack any *directionality*: the operators L_2 ($\mathbb{L}_2$) and their respective adjoints $L_2^\dagger(\mathbb{L}_2^\dagger)$ act with equal strength on the nuclear system. In contrast, in the high-gradient regime, $|\lambda_2\rangle$ exhibits a significant singlet character and therefore gets depleted very quickly. Thus, $|\lambda_2\rangle$ can be eliminated adiabatically from the dynamics, the electronic subsystem settles to a maximally mixed state in the Pauli-blocked $|T_\pm\rangle$ subspace and the nuclear dynamics acquire a certain directionality in that now the nuclear spins experience dominantly the action of the non-local operators L_2 and $\mathbb{L}_2$, respectively. As will be shown below, this directionality features both the build-up of an Overhauser field gradient and entanglement generation between the two nuclear spin ensembles.

3.5.1 Adiabatic Elimination of Electronic Degrees of Freedom

Having separated the macroscopic semiclassical part of the nuclear Overhauser fields, the problem at hand features a hierarchy in the typical energy scales since the typical HF interaction strength is slow compared to all relevant electronic timescales. This allows for a perturbative approach to second order in $\mathcal{V}$ to derive an effective master equation for the nuclear subsystem [46, 66]. To stress the perturbative treatment, the full quantum master equation can formally be decomposed as

$$\dot{\rho} = [\mathcal{L}_0 + \mathcal{V}] \rho, \tag{3.32}$$

where the superoperator $\mathcal{L}_0$ acts on the electron degrees of freedom only and the HF interaction represents a perturbation. Thus, in zeroth order the electronic and nuclear dynamics are decoupled. In what follows, we will determine the effective nuclear evolution in the submanifold of the electronic quasisteady states of $\mathcal{L}_0$. The electronic Liouvillian $\mathcal{L}_0$ features a *unique* steady state [37], that is $\mathcal{L}_0 \rho_{ss}^{el} = 0$ for

$$\rho_{ss}^{el} = p \left(|T_+\rangle \langle T_+| + |T_-\rangle \langle T_-| \right) + (1 - 2p) \, |\lambda_2\rangle \langle \lambda_2| \,, \tag{3.33}$$

where

$$p = \frac{\Gamma_\pm + \Gamma_2}{3\Gamma_\pm + 2\Gamma_2}, \tag{3.34}$$

completely defines the electronic quasisteady state. It captures the competition between undirected population transfer within the the manifold $\{|T_\pm\rangle, |\lambda_2\rangle\}$ due to $\Gamma_\pm$ and a unidirectional, electron-transport-mediated decay of $|\lambda_2\rangle$. Moreover, it describes feedback between the electronic and nuclear degrees of freedom as the rate Γ_2 depends on the gradient Δ which incorporates the nuclear-polarization-dependent Overhauser gradient Δ_{OH}. We can immediately identify two important limits which will be analyzed in greater detail below: For $\Gamma_\pm \gg \Gamma_2$ we get $p = 1/3$, whereas $\Gamma_\pm \ll \Gamma_2$ results in $p = 1/2$, that is a maximally mixed state in the $|T_\pm\rangle$ subspace, since a fast decay rate Γ_2 leads to a complete depletion of $|\lambda_2\rangle$.

Since ρ_{ss}^{el} is unique, the projector $\mathcal{P}$ on the subspace of zero eigenvalues of $\mathcal{L}_0$, i.e., the zeroth order steady states, is given by

$$\mathcal{P}\rho = \mathrm{Tr}_{el} [\rho] \otimes \rho_{ss}^{el} = \sigma \otimes \rho_{ss}^{el}. \tag{3.35}$$

By definition, we have $\mathcal{P}\mathcal{L}_0 = \mathcal{L}_0 \mathcal{P} = 0$ and $\mathcal{P}^2 = \mathcal{P}$. The complement of $\mathcal{P}$ is $\mathcal{Q} = \mathbb{1} - \mathcal{P}$. Projection of the master equation on the $\mathcal{P}$ subspace gives in second-order perturbation theory

$$\frac{d}{dt}\mathcal{P}\rho = \left[\mathcal{P}\mathcal{V}\mathcal{P} - \mathcal{P}\mathcal{V}\mathcal{Q}\mathcal{L}_0^{-1}\mathcal{Q}\mathcal{V}\mathcal{P} \right] \rho, \tag{3.36}$$

from which we can deduce the required equation of motion $\dot{\sigma} = \mathcal{L}_{eff} [\sigma]$ for the reduced density operator of the nuclear subsystem $\sigma = \mathrm{Tr}_{el} [\mathcal{P}\rho]$ as

$$\dot{\sigma} = \mathrm{Tr}_{el} \left[\mathcal{P}\mathcal{V}\mathcal{P}\rho - \mathcal{P}\mathcal{V}\mathcal{Q}\mathcal{L}_0^{-1}\mathcal{Q}\mathcal{V}\mathcal{P}\rho \right]. \tag{3.37}$$

The subsequent, full calculation follows the general framework developed in Ref. [66] and is presented in detail in Appendices 3.11.5 and 3.11.9. We then arrive at the following effective master equation for nuclear spins

$$\dot{\sigma} = \gamma \{ p [\mathcal{D}[L_2]\sigma + \mathcal{D}[\mathbb{L}_2]\sigma] \qquad (3.38)$$
$$+ (1 - 2p) \left[\mathcal{D}\left[L_2^{\dagger} \right] + \mathcal{D}\left[\mathbb{L}_2^{\dagger} \right] \sigma \right] \}$$
$$+ i\delta \left\{ p \left(\left[L_2^{\dagger}L_2, \sigma \right] + \left[\mathbb{L}_2^{\dagger}\mathbb{L}_2, \sigma \right] \right) \right.$$
$$\left. - (1 - 2p) \left(\left[L_2 L_2^{\dagger}, \sigma \right] + \left[\mathbb{L}_2\mathbb{L}_2^{\dagger}, \sigma \right] \right) \right\}.$$

Here, we have introduced the effective quantities

$$\gamma = \frac{a_{\mathrm{hf}}^2 \tilde{\Gamma}}{2 \left[\tilde{\Gamma}^2 + \varepsilon_2^2 \right]}, \qquad (3.39)$$

$$\delta = \frac{a_{\mathrm{hf}}^2 \varepsilon_2}{4 \left[\tilde{\Gamma}^2 + \varepsilon_2^2 \right]}, \qquad (3.40)$$

and

$$\tilde{\Gamma} = \Gamma_2 + 2\Gamma_{\pm} + \frac{\Gamma_{\mathrm{deph}}}{4}. \qquad (3.41)$$

The master equation in Eq. (3.38) is our first main result. It is of Lindblad form and incorporates electron-transport-mediated jump terms as well as Stark shifts. The two main features of Eq. (3.38) are: (i) The dissipative nuclear jump terms are governed by the *nonlocal* jump operators L_2 and $\mathbb{L}_2$, respectively. (ii) The effective dissipative rates $\sim p\gamma$ incorporate intrinsic electron-nuclear feedback effects as they depend on the macroscopic state of the nuclei via the parameter p and the decay rate Γ_2. Because of this feedback mechanism, we can distinguish two very different fixed points for the coupled electron-nuclear evolution. This is discussed below.

3.5.2 Low-Gradient Regime: Random Nuclear Diffusion

As argued qualitatively above, in the low-gradient regime where $|\lambda_2\rangle \approx |T_0\rangle$, the nuclear master equation given in Eq. (3.38) lacks any directionality. Accordingly, the resulting dynamics may be viewed as a random nuclear diffusion process. Indeed, in the limit $\Gamma_2 \to 0$, it is easy to check that $p = 1/3$ and $\sigma_{\mathrm{ss}} \propto \mathbb{1}$ is a steady-state solution. Therefore, both the electronic and the nuclear subsystem settle into the fully mixed state with no preferred direction nor any peculiar polarization characteristics.

   This analytical argument is corroborated by exact numerical simulations (i.e., without having eliminated the electronic degrees of freedom) for the full five-level electronic system coupled to ten ($N_L = N_R = 5$) nuclear spins. Here, we assume homogeneous HF coupling (effects due to non-uniform HF couplings are discussed in Sect. 3.9): Then, the total spins J_i are conserved and it is convenient to describe the nuclear spin system in terms of Dicke states $|J_i, m_i\rangle$ with total spin quantum number J_i and spin projection $m_i = -J_i, \ldots, J_i$. Fixing the (conserved) total spin quantum numbers $J_i = N_i/2$, we write in short $|J_L, m_L\rangle \otimes |J_R, m_R\rangle = |m_L, m_R\rangle$. In order to realistically mimic the perturbative treatment of the HF coupling in an experimentally relevant situation where $N \approx 10^6$, here the HF coupling constant $g_{hf} = A_{HF}/\sqrt{N}$ is scaled down to a constant value of $g_{hf} = 0.1\,\mu$eV. Moreover, let us for the moment neglect the nuclear fluctuations due to H_{zz}, in order to restrict the following analysis to the semiclassical part of the nuclear dynamics; compare also previous theoretical studies [20, 32, 47]. In later sections, this part of the dynamics will be taken into account again. In particular, we compute the steady state and analyze its dependence on the gradient Δ: Experimentally, Δ could be induced intrinsically via a nuclear Overhauser gradient Δ_{OH} or extrinsically via a nano- or micro-magnet [61, 67]. The results are displayed in Fig. 3.7: Indeed, in the low-gradient regime the nuclear subsystem settles into the fully mixed state. However, outside of the low-gradient

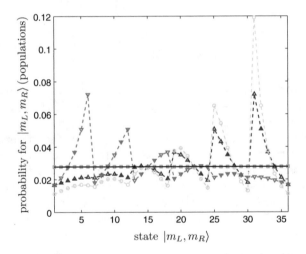

Fig. 3.7 Exact steady-state as a function of the inhomogeneous splitting Δ; results for 10 nuclear spins, five in each quantum *dot*. We plot the diagonal elements of the nuclear steady-state density matrix σ_{ss} (i.e. the nuclear populations); its dimension is $(2J_L + 1)(2J_R + 1) = 36$. For a small external gradient of the order of natural fluctuations of the Overhauser field (*red squares*) the nuclear system settles into the fully mixed state, as evidenced by the uniform populations of the nuclear levels. However, as we increase the gradient Δ, the nuclear steady state starts to display a structure different from the fully mixed state, showing a dominant peak in the occupation of the nuclear level with maximum gradient, that is $|-J_L, J_R\rangle$ and $|J_L, -J_R\rangle$ for $\Delta > 0$ and $\Delta < 0$, respectively. The *upward triangles, downward triangles* and *circles* refer to $\Delta = 5\,\mu$eV, $\Delta = -5\,\mu$eV and $\Delta = 10\,\mu$eV, respectively. Other numerical parameters are: $\Gamma = 10\,\mu$eV, $\Gamma_\pm = 0.3\,\mu$eV, $\Gamma_{deph} = 3\,\mu$eV, $\omega_0 = 0$, $t = 20\,\mu$eV and $\varepsilon = 30\,\mu$eV

regime, the nuclear subsystem is clearly driven away from the fully mixed state and shows a tendency towards the build-up of a nuclear Overhauser gradient. For $\Delta > 0$, we find numerically an increasing population (in descending order) of the levels $|-J_L, J_R\rangle$, $|-J_L + 1, J_R - 1\rangle$ etc., whereas for $\Delta < 0$ strong weights are found at $|J_L, -J_R\rangle$, $|J_L - 1, -J_R + 1\rangle$, ... which effectively increases Δ such that the nuclear spins actually tend to self-polarize. This trend towards self-polarization and the peculiar structure of the nuclear steady state σ_{ss} displayed in Fig. 3.7 is in very good agreement with the ideal nuclear two-mode squeezedlike steady-state that we are to construct analytically in the next subsection.

3.5.3 High-Gradient Regime: Entanglement Generation

In the high-gradient regime the electronic level $|\lambda_2\rangle$ overlaps significantly with the localized singlet $|S_{02}\rangle$. For $\Gamma_2 \gg \Gamma_\pm$ it decays sufficiently fast such that it can be eliminated adiabatically from the dynamics. As can be seen from Eqs. (3.33) and (3.34), on typical nuclear timescales, the electronic subsystem then quickly settles into the quasisteady state given by $\rho_{ss}^{el} = (|T_+\rangle \langle T_+| + |T_-\rangle \langle T_-|)/2$ and the effective master equation for the nuclear spin density matrix σ simplifies to

$$\dot{\sigma} = \frac{\gamma}{2} \left[\mathcal{D}[L_2]\sigma + \mathcal{D}[\mathbb{L}_2]\sigma \right] + i \frac{\delta}{2} \left(\left[L_2^\dagger L_2, \sigma \right] + \left[\mathbb{L}_2^\dagger \mathbb{L}_2, \sigma \right] \right). \tag{3.42}$$

For later reference, the typical timescale of this dissipative dynamics is set by the rate

$$\gamma_c = N\gamma = \frac{g_{hf}^2 \tilde{\Gamma}}{2 \left[\tilde{\Gamma}^2 + \varepsilon_2^2 \right]}, \tag{3.43}$$

which is *collectively* enhanced by a factor of $N \approx 10^6$ to account for the norm of the collective nuclear spin operators $A_i^\pm$. This results in the typical HF-mediated interaction strength of $g_{hf} = \sqrt{N} a_{hf}$ [46], and for typical parameter values we estimate $\gamma_c \approx 10^{-4} \mu eV$.

   This evolution gives rise to the desired, entangling nuclear squeezing dynamics: It is easy to check that all *pure* stationary solutions $|\xi_{ss}\rangle$ of this Lindblad evolution can be found via the dark-state condition $L_2 |\xi_{ss}\rangle = \mathbb{L}_2 |\xi_{ss}\rangle = 0$. Next, we explicitly construct $|\xi_{ss}\rangle$ in the limit of equal dot sizes ($N_L = N_R$) and uniform HF coupling $\left(a_{i,j} = N/N_i \right)$, and generalize our results later. In this regime, again it is convenient to describe the nuclear system in terms of Dicke states $|J_i, k_i\rangle$, where $k_i = 0, \ldots, 2J_i$. For the symmetric scenario $J_L = J_R = J$, one can readily verify that the dark state condition is satisfied by the (unnormalized) pure state

$$|\xi_{ss}\rangle = \sum_{k=0}^{2J} \xi^k |J, k\rangle_L \otimes |J, 2J - k\rangle_R. \tag{3.44}$$

This nuclear state may be viewed as an extension of the two-mode squeezed state familiar from quantum optics [41] to finite dimensional Hilbert spaces; for an explicit construction of $|\xi_{ss}\rangle$, we refer to Appendix 3.11.7. The parameter $\xi = -\nu_2/\mu_2$ quantifies the entanglement and polarization of the nuclear system. Note that unlike in the bosonic case (discussed in detail in Sect. 3.7), the modulus of ξ is unconfined. Both $|\xi| < 1$ and $|\xi| > 1$ are allowed and correspond to states of large positive (negative) OH field gradients, respectively, and the system is invariant under the corresponding symmetry transformation ($\mu_2 \leftrightarrow \nu_2$, $A^z_{L,R} \to -A^z_{L,R}$). As we discuss in detail in Sect. 3.6, this symmetry gives rise to a bistability in the steady state, as for every solution with positive OH field gradient ($\Delta_{OH} > 0$), we find a second one with negative gradient ($\Delta_{OH} < 0$). As a first indication for this bistability, also compare the green and blue curve in Fig. 3.7: For $\Delta \gg 0$, the dominant weight of the nuclear steady state is found in the level $|-J_L, J_R\rangle$, that is the Dicke state with maximum *positive* Overhauser gradient, whereas for $\Delta \ll 0$, the weight of the nuclear stationary state is peaked symmetrically at $|J_L, -J_R\rangle$, corresponding to the Dicke state with maximum *negative* Overhauser gradient.

In the asymmetric scenario $J_L \neq J_R$, one can readily show that a *pure* dark-state solution does not exist. Thus, we resort to exact numerical solutions for small system sizes $J_i \approx 3$ to compute the nuclear steady state-solution σ_{ss}. To verify the creation of steady-state entanglement between the two nuclear spin ensembles, we take the EPR uncertainty as a figure of merit. It is defined via

$$\Delta_{EPR} = \frac{\text{var}\left(I^x_L + I^x_R\right) + \text{var}\left(I^y_L + I^y_R\right)}{|\langle I^z_L\rangle| + |\langle I^z_R\rangle|}, \tag{3.45}$$

and measures the degree of nonlocal correlations. For an arbitrary state, $\Delta_{EPR} < 1$ implies the existence of such non-local correlations, whereas $\Delta_{EPR} \geq 1$ for separable states [41, 42]. The results are displayed in Fig. 3.8. First of all, the numerical solutions confirm the analytical result in the symmetric limit where the asymmetry parameter $\Delta_J = J_R - J_L$ is zero. In the asymmetric setting, where $J_L \neq J_R$, the steady state σ_{ss} is indeed found to be mixed, that is $\text{Tr}\left[\sigma^2_{ss}\right] < 1$. However, both the amount of generated entanglement as well as the purity of σ_{ss} tend to increase, as we increase the system size $J_L + J_R$ for a fixed value of Δ_J. For fixed J_i, we have also numerically verified that the steady-state solution is *unique*.

In practical experimental situations one deals with a mixture of different J_i subspaces. The width of the nuclear spin distribution is typically $\Delta_J \sim \sqrt{N}$, but may even be narrowed further actively; see for example Refs. [20, 32]. The numerical results displayed above suggest that the amount of entanglement and purity of the nuclear steady state increases for smaller absolute values of the relative asymmetry $\Delta_J/J = (J_R - J_L)/(J_L + J_R)$. In Fig. 3.8, $\Delta_{EPR} < 1$ is still observed even for $|\Delta_J|/J = 2.5/3.5 \approx 0.7$. Thus, experimentally one might still obtain entanglement in a mixture of different *large* J_i subspaces for which the relative width is comparatively small, $\Delta_J/J \approx \sqrt{N}/N \approx 10^{-3} \ll 1$. Intuitively, the idea is that for every pair $\{J_L, J_R\}$ with $J_L \approx J_R$ the system is driven towards a state similar to the ideal

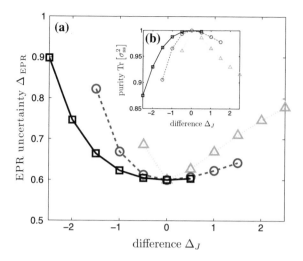

Fig. 3.8 EPR-uncertainty Δ_{EPR} and purity of the (exact) nuclear dark states fulfilling $\mathcal{D}[L_2]\sigma_{\mathrm{ss}} + \mathcal{D}[\mathbb{L}_2]\sigma_{\mathrm{ss}} = 0$ for small system sizes J_i; here, we have set $\delta = 0$ for simplicity. However, similar results (not shown) have been obtained for finite values of δ. We fix J_L to $J_L = 1$ (*triangles*), $J_L = 2$ (*circles*) and $J_L = 3$ (*squares*) and compute σ_{ss} for different values of Δ_J; J_R runs from 0.5 up to 3.5. In the symmetric scenario $\Delta_J = 0$, σ_{ss} is pure and given by the two-mode squeezed like state $\sigma_{\mathrm{ss}} = |\xi_{\mathrm{ss}}\rangle\langle\xi_{\mathrm{ss}}|$. For $\Delta_J \neq 0$, σ_{ss} is mixed; however, the purity $\mathrm{Tr}\left[\sigma_{\mathrm{ss}}^2\right]$ (*inset*) as well as Δ_{EPR} increase with the system size $J_L + J_R$. In all cases, σ_{ss} was found to be unique. Here, we have set $|\xi| = 0.25$

two-mode squeezedlike state given in Eq. (3.44). This will also be discussed in more detail in Sect. 3.7.

3.6 Dynamic Nuclear Polarization

In the previous section we have identified a low-gradient regime, where the nuclear spins settle into a fully mixed state, and a high-gradient regime, where the ideal nuclear steady state was found to be a highly polarized, entangled two-mode squeezedlike state. Now, we provide a thorough analysis which reveals the multistability of the nuclear subsystem and determines the connection between these two very different regimes. It is shown that, beyond a critical polarization, the nuclear spin system becomes self-polarizing and is driven towards a highly polarized OH gradient.

To this end, we analyze the nuclear spin evolution within a semiclassical approximation which neglects coherences among different nuclei. This approach has been well studied in the context of central spin systems (see for example Ref. [57] and references therein) and is appropriate on timescales longer than nuclear dephasing times [68]. This approximation will be justified self-consistently. The analysis is

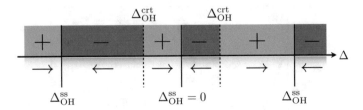

Fig. 3.9 Schematic representation of the multistability of the nuclear dynamics. For initial nuclear gradients smaller than Δ_{OH}^{crt} the nuclear system is attracted towards the trivial zero-polarization solution ($\Delta_{OH}^{ss} = 0$). Upon surpassing Δ_{OH}^{crt}, however, the system enters into an electron-nuclear feedback loop and the nuclear dynamics turn self-polarizing such that large OH gradients can be reached in the steady state. This is schematically denoted by $\pm$ referring to the sign of $\dot{\Delta}_{Iz}$ which determines the stable fixed point the nuclear system is attracted to in the steady state (see *arrows*)

based on the effective QME given in Eq. (3.38). First, assuming homogeneous HF coupling and equal dot sizes ($N_L = N_R = N$), we construct dynamical equations for the expectation values of the collective nuclear spins $\langle I_i^z \rangle_t$, $i = L, R$, where $I_i^\nu = \sum_j \sigma_{i,j}^\nu$ for $\nu = \pm, z$. To close the corresponding differential equations we use a semiclassical factorization scheme resulting in two equations of motion for the two nuclear dynamical variables $\langle I_L^z \rangle_t$ and $\langle I_R^z \rangle_t$, respectively. This extends previous works on spin dynamics in double quantum dots, where a single dynamical variable for the nuclear polarization was used to explain the feedback mechanism in this system; see for example Refs. [32, 49]. The corresponding nonlinear differential equations are then shown to yield nonlinear equations for the equilibrium polarizations. Generically, the nuclear polarization is found to be multi-stable (compare also Refs. [47, 51]) and, depending on the system's parameters, we find up to three stable steady state solutions for the OH gradient Δ_{OH}^{ss}, two of which are highly polarized in opposite directions and one is unpolarized; compare Fig. 3.9 for a schematic illustration.

At this point, some short remarks are in order: First, the analytical results obtained within the semiclassical approach are confirmed by exact numerical results for small sets of nuclei; see Appendix 3.11.8. Second, by virtue of the semiclassical decoupling scheme used here, our results can be generalized to the case of inhomogeneous HF coupling in a straightforward way with the conclusions remaining essentially unchanged. Third, for simplicity here we assume the symmetric scenario of vanishing external fields $\omega_{ext} = \Delta_{ext} = 0$; therefore, $\Delta = \Delta_{OH}$. However, as shown in Sect. 3.9 and Appendix 3.11.8 one may generalize our results to finite external fields: This opens up another experimental knob to tune the desired steady-state properties of the nuclei.

Intuitive picture.—Before going through the calculation, let us sketch an intuitive picture that can explain the instability of the nuclear spins towards self-polarization and the corresponding build-up of a macroscopic nuclear OH gradient: In the high-gradient regime, the nuclear spins predominantly experience the action of the nonlocal jump operators $L_2 = \nu_2 A_L^+ + \mu_2 A_R^+$ and $\mathbb{L}_2 = \mu_2 A_L^- + \nu_2 A_R^-$, respectively, both of them acting with the same rate γ on the nuclear spin ensembles. For example, for

$\Delta > 0$ and $\varepsilon > 0$, where $\mu_2 > \nu_2$, the first nuclear ensemble gets exposed more strongly to the action of the collective lowering operator A_L^-, whereas the second ensemble preferentially experiences the action of the raising operator A_R^+; therefore, the two nuclear ensembles are driven towards polarizations of opposite sign. The second steady solution featuring a large OH gradient with opposite sign is found along the same lines for $\mu_2 < \nu_2$. Therefore, our scheme provides a good dynamic nuclear polarization (DNP) protocol for $\mu_2 \gg \nu_2$ ($|\xi| \ll 1$), or vice-versa for $\mu_2 \ll \nu_2$ ($|\xi| \gg 1$).

Semiclassical analysis.—Using the usual angular momentum commutation relations $[I^z, I^\pm] = \pm I^\pm$ and $[I^+, I^-] = 2I^z$, Eq. (3.38) readily yields two rate equations for the nuclear polarizations $\langle I_i^z \rangle_t$, $i = L, R$. We then employ a semiclassical approach by neglecting correlations among different nuclear spins, that is

$$\langle \sigma_i^+ \sigma_j^- \rangle = \begin{cases} 0 & , i \neq j \\ \langle \sigma_i^z \rangle + \frac{1}{2} & , i = j \end{cases} \tag{3.46}$$

which allows us to close the equations of motion for the nuclear polarizations $\langle I_i^z \rangle$. This leads to the two following nonlinear equations of motion,

$$\frac{d}{dt} \langle I_L^z \rangle_t = -\gamma_{\text{pol}} \left[\langle I_L^z \rangle_t + \frac{N}{2} \frac{\chi}{\gamma_{\text{pol}}} \right], \tag{3.47}$$

$$\frac{d}{dt} \langle I_R^z \rangle_t = -\gamma_{\text{pol}} \left[\langle I_R^z \rangle_t - \frac{N}{2} \frac{\chi}{\gamma_{\text{pol}}} \right], \tag{3.48}$$

where we have introduced the effective HF-mediated depolarization rate γ_{pol} and pumping rate χ as

$$\gamma_{\text{pol}} = \gamma \left(\mu_2^2 + \nu_2^2 \right) (1 - p), \tag{3.49}$$

$$\chi = \gamma \left(\mu_2^2 - \nu_2^2 \right) (3p - 1), \tag{3.50}$$

with the rate γ given in Eq. (3.39). Clearly, Eqs. (3.47) and (3.48) already suggest that the two nuclear ensembles are driven towards opposite polarizations. The nonlinearity is due to the fact that both χ and γ_{pol} depend on the gradient Δ which itself depends on the nuclear polarizations $\langle I_i^z \rangle_t$; at this stage of the analysis, however, Δ simply enters as a parameter of the underlying effective Hamiltonian. Equivalently, the macroscopic dynamical evolution of the nuclear system may be expressed in terms of the total net polarization $P(t) = \langle I_L^z \rangle_t + \langle I_R^z \rangle_t$ and the polarization gradient $\Delta_{I^z} = \langle I_R^z \rangle_t - \langle I_L^z \rangle_t$ as

$$\dot{P}(t) = -\gamma_{\text{pol}} P(t), \tag{3.51}$$

$$\frac{d}{dt} \Delta_{I^z} = -\gamma_{\text{pol}} \left[\Delta_{I^z} - N \frac{\chi}{\gamma_{\text{pol}}} \right]. \tag{3.52}$$

Fixed-point analysis.—In what follows, we examine the fixed points of the semi-classical equations derived above. First of all, since $\gamma_{\text{pol}} > 0 \,\forall P, \Delta_{I^z}$, Eq. (3.51) simply predicts that in our system no homogeneous nuclear net polarization P will be produced. In contrast, any potential initial net polarization is exponentially damped to zero in the long-time limit, since in the steady state $\lim_{t\to\infty} P(t) = 0$. This finding is in agreement with previous theoretical results showing that, due to angular momentum conservation, a net nuclear polarization cannot be pumped in a system where the HF-mediated relaxation rate for the blocked triplet levels $|T_+\rangle$ and $|T_-\rangle$, respectively, is the same; see, e.g., Ref. [32] and references therein.

The dynamical equation for Δ_{I^z}, however, is more involved: The effective rates $\gamma_{\text{pol}} = \gamma_{\text{pol}}(\Delta)$ and $\chi = \chi(\Delta)$ in Eq. (3.52) depend on the nuclear-polarization dependent parameter Δ. This nonlinearity opens up the possibility for multiple steady-state solutions. From Eqs. (3.47) and (3.48) we can immediately identify the fixed points $\langle I_i^z\rangle_{\text{ss}}$ of the nuclear polarization dynamics as $\pm (N/2)\chi/\gamma_{\text{pol}}$. Consequently, the two nuclear ensembles tend to be polarized along opposite directions, that is $\langle I_L^z\rangle_{\text{ss}} = -\langle I_R^z\rangle_{\text{ss}}$. The corresponding steady-state nuclear polarization gradient $\Delta_{I^z}^{\text{ss}}$, scaled in terms of its maximum value N, is given by

$$\frac{\Delta_{I^z}^{\text{ss}}}{N} = \mathcal{R}(\Delta) = \Lambda \frac{3p - 1}{1 - p}. \tag{3.53}$$

Here, we have introduced the nonlinear function $\mathcal{R}(\Delta)$ which depends on the purely electronic quantity

$$\Lambda = \Lambda(\Delta) = \frac{\mu_2^2 - \nu_2^2}{\mu_2^2 + \nu_2^2} = \frac{1 - \xi^2}{1 + \xi^2}. \tag{3.54}$$

According to Eq. (3.53), the function $\mathcal{R}(\Delta)$ determines the nuclear steady-state polarization. While the functional dependence of Λ on the gradient Δ can give rise to two highly polarized steady-state solutions with opposite nuclear spin polarization, for $|\mu_2| \gg |\nu_2|$ and $|\mu_2| \ll |\nu_2|$, respectively, the second factor in Eq. (3.53) may prevent the system from reaching these highly polarized fixed points. Based on Eq. (3.53), we can identify the two important limits discussed previously: For $\Gamma_2 \ll \Gamma_\pm$, the electronic subsystem settles into the steady-state solution $p = 1/3$ and the nuclear system is unpolarized, as the second factor in Eq. (3.53) vanishes. This is what we identified above as the nuclear diffusion regime in which the nuclear subsystem settles into the unpolarized fully mixed state. In the opposite limit, where $\Gamma_2 \gg \Gamma_\pm$, the electronic subsystem settles into $p \approx 1/2$. In this limit, the second factor in Eq. (3.53) becomes 1 and the functional dependence of $\Lambda(\Delta)$ dominates the behavior of $\mathcal{R}(\Delta)$ such that large nuclear OH gradients can be achieved in the steady state. The electron-nuclear feedback loop can then be closed self-consistently via $\Delta_{\text{OH}}^{\text{ss}}/\Delta_{\text{OH}}^{\text{max}} = \mathcal{R}(\Delta_{\text{OH}}^{\text{ss}})$, where, in analogy to Eq. (3.53), $\Delta_{\text{OH}}^{\text{ss}}$ has been scaled in units of its maximum value $\Delta_{\text{OH}}^{\text{max}} = A_{\text{HF}}/2$. Points fulfilling this condition can be found at intersections of $\mathcal{R}(\Delta)$ with $\Delta_{\text{OH}}^{\text{ss}}/\Delta_{\text{OH}}^{\text{max}}$. This is elaborated below.

To gain further insights into the nuclear polarization dynamics, we evaluate $\dot{\Delta}_{I^z}$ as given in Eq. (3.52). The results are displayed in Fig. 3.10. *Stable* fixed points

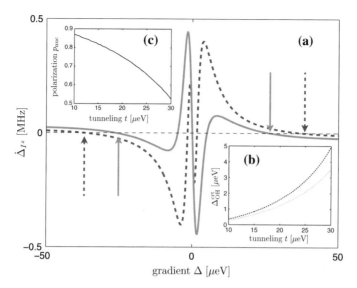

Fig. 3.10 Semiclassical solution to the nuclear polarization dynamics: tristability of the nuclear steady state. **a** Instantaneous nuclear polarization rate $\dot{\Delta}_{Iz}$ for $t = 20\,\mu\text{eV}$ (*dashed*) and $t = 30\,\mu\text{eV}$ (*solid*), respectively. Stable fixed points are found at $\dot{\Delta}_{Iz} = 0$ and $d\dot{\Delta}_{Iz}/d\Delta < 0$. The nuclear system is driven towards one of the highly polarized fixed points (indicated by *arrows*), if the initial gradient Δ exceeds a critical threshold $\left|\Delta_{\text{OH}}^{\text{crt}}\right|$, shown in (**b**) for $\Gamma_{\pm} = 0.1\,\mu\text{eV}$ (*dashed*) and $\Gamma_{\pm} = 0.05\,\mu\text{eV}$ (*solid*), respectively. **c** By tuning t, one can achieve $|\xi| \ll 1$ leading to a nuclear polarization of $\lesssim 90\,\%$. Other numerical parameters in μeV: $\Gamma = 25$, $\varepsilon = 30$, $\Gamma_{\pm} = 0.1$ (except for the *solid line* in (**b**) where $\Gamma_{\pm} = 0.05$) and $\Gamma_{\text{deph}} = 0.1$

of the dynamics are determined by $\dot{\Delta}_{Iz} = 0$ and $d\dot{\Delta}_{Iz}/d\Delta < 0$ as opposed to unstable fixed points where $d\dot{\Delta}_{Iz}/d\Delta > 0$. In this way it is ensured that fluctuations of Δ_{Iz} away from a stable fixed point are corrected by a restoring intrinsic pump effect [20, 51, 69]. We can identify parameter regimes in which the nuclear system features three stable fixed points. As schematically depicted in Fig. 3.9, they are interspersed by two unstable points referred to as $\Delta_{\text{OH}}^{\text{crt}}$. Therefore, in general, the nuclear steady-state polarization is found to be tri-stable: Two of the stable fixed points are high-polarization solutions of opposite sign, supporting a macroscopic OH gradient, while one is the trivial zero-polarization solution. The unstable points $\Delta_{\text{OH}}^{\text{crt}}$ represent critical values for the initial OH gradient marking the boundaries of a critical region. If the initial gradient lies outside of this critical region, the OH gradient runs into one of the highly polarized steady states. Otherwise, the nuclear system gets stuck in the zero-polarization steady state. Note that $\Delta_{\text{OH}}^{\text{crt}}$ is tunable: To surpass the critical region one needs $\Gamma_2 \gg \Gamma_{\pm}$; thus, the critical region can be destabilized by making $\Gamma_{\pm}$ smaller [compare Fig. 3.10b] which is lower bounded by $\Gamma_{\pm} \gg \gamma_c$ in order to justify the elimination of the electronic degrees of freedom. For typical parameters we thus estimate $\Delta_{\text{OH}}^{\text{crt}} \approx (3 - 5)\,\mu\text{eV}$ which sets the required initial Δ in order to kick-start the nuclear self-polarization process. Experimentally, this could be realized either via an initial nuclear polarization of $p_{nuc} \approx (5-10)\%$ or an on-chip nanomagnet [61, 67].

Timescales.—In order to reach a highly polarized steady state, approximately $\sim 10^5$ nuclear spin flips are required. We estimate $\dot{\Delta}_{Iz} \approx 0.1\,\text{MHz}$ and, thus, the total time for the polarization process is therefore approximately $\sim 10^5/0.1\,\text{MHz} \approx 1\,\text{s}$. This order of magnitude estimate is in very good agreement with typical timescales observed in nuclear polarization experiments [26]. Moreover, $\gamma_{pol}^{-1} \approx 1\,\text{s}$ is compatible with our semiclassical approach, since nuclear spins typically dephase at a rate of $\sim \text{kHz}$ [26, 57]. Finally, in any experimental situation, the nuclear spins are subject to relaxation and diffusion processes which prohibit complete polarization of the nuclear spins. Therefore, in order to capture other depolarizing processes that go beyond our current analysis, one could add an additional phenomenological nuclear depolarization rate γ_{dp} by simply making the replacement $\gamma_{pol}(\Delta) \rightarrow \gamma_{pol}(\Delta) + \gamma_{dp}$. Since typically $\gamma_{dp}^{-1} \approx 15\,\text{s}$ [51], however, these additional processes are slow in comparison to the intrinsic rate γ_{pol} and should not lead to any qualitative changes of our results.

3.7 Steady-State Entanglement Generation

In Sect. 3.5 we have identified a high-gradient regime which—after adiabatically eliminating all electronic coordinates—supports a rather simple description of the nuclear dynamics on a coarse-grained timescale. Now, we extend our previous analysis and provide a detailed analysis of the nuclear dynamics in the high-gradient regime. In particular, this includes perturbative effects due to the presence of the so far neglected levels $|\lambda_{1,3}\rangle$. To this end, we apply a self-consistent Holstein-Primakoff approximation, which reexpresses nuclear fluctuations around the semiclassical state in terms of bosonic modes. This enables us to approximately solve the nuclear dynamics analytically, to directly relate the ideal nuclear steady state to a two-mode squeezed state familiar from quantum optics and to efficiently compute several entanglement measures.

3.7.1 Extended Nuclear Master Equation in the High-Gradient Regime

In the high-gradient regime the electronic system settles to a quasisteady state $\rho_{ss}^{el} = \rho_{target}^{el} = (|T_+\rangle\langle T_+| + |T_-\rangle\langle T_-|)/2$ [compare Eqs. (3.33) and (3.34)] on a timescale short compared to the nuclear dynamics; deviations due to (small) populations of the hybridized levels are discussed in Appendix 3.11.11. We then follow the general adiabatic elimination procedure discussed in Sect. 3.5 to obtain an effective master equation for the nuclear spins in the submanifold of the electronic quasisteady state

$\rho_{\text{target}}^{\text{el}}$. The full calculation is presented in detail in Appendix 3.11.9. In summary, the generalized effective master equation reads

$$\dot{\sigma} = \sum_k \left[\frac{\gamma_k^+}{2} \mathcal{D}[L_k]\sigma + \frac{\gamma_k^-}{2} \mathcal{D}[\mathbb{L}_k]\sigma \right] + i\,[H_{\text{Stark}}, \sigma]$$
$$+ \gamma_{zz} \sum_{i,j} \left[\delta A_i^z \sigma \delta A_j^z - \frac{1}{2} \left\{ \delta A_j^z \delta A_i^z, \sigma \right\} \right]. \tag{3.55}$$

Here, we have introduced the effective HF-mediated decay rates

$$\gamma_k^+ = \frac{a_{\text{hf}}^2 \tilde{\Gamma}_k}{2\left[\Delta_k^2 + \tilde{\Gamma}_k^2\right]}, \tag{3.56}$$

$$\gamma_k^- = \frac{a_{\text{hf}}^2 \tilde{\Gamma}_k}{2\left[\delta_k^2 + \tilde{\Gamma}_k^2\right]}, \tag{3.57}$$

where $\tilde{\Gamma}_k = \Gamma_k + 3\Gamma_\pm + \Gamma_{\text{deph}}/4$ and the detuning parameters

$$\Delta_k = \varepsilon_k - \omega_0, \tag{3.58}$$
$$\delta_k = \varepsilon_k + \omega_0, \tag{3.59}$$

specify the splitting between the electronic eigenstate $|\lambda_k\rangle$ and the Pauli-blocked triplet states $|T_+\rangle$ and $|T_-\rangle$, respectively. The effective nuclear Hamiltonian

$$H_{\text{Stark}} = \sum_k \frac{\Delta_k^+}{2} L_k^\dagger L_k + \frac{\Delta_k^-}{2} \mathbb{L}_k^\dagger \mathbb{L}_k \tag{3.60}$$

is given in terms of the second-order Stark shifts

$$\Delta_k^+ = \frac{a_{\text{hf}}^2 \Delta_k}{4\left[\Delta_k^2 + \tilde{\Gamma}_k^2\right]}, \tag{3.61}$$

$$\Delta_k^- = \frac{a_{\text{hf}}^2 \delta_k}{4\left[\delta_k^2 + \tilde{\Gamma}_k^2\right]}. \tag{3.62}$$

Lastly, in Eq. (3.55) we have set $\gamma_{zz} = a_{\text{hf}}^2/(5\Gamma_\pm)$. For $\omega_0 = 0$, we have $\gamma_k^+ = \gamma_k^-$ and $\Delta_k^+ = \Delta_k^-$. When disregarding effects due to H_{zz} and neglecting the levels $|\lambda_{1,3}\rangle$, i.e., only keeping $k = 2$ in Eq. (3.55), indeed, we recover the result of the Sect. 3.5; see Eq. (3.42). As shown in Appendix 3.11.10, the nuclear HF-mediated jump terms in Eq. (3.55) can be brought into diagonal form which features a clear hierarchy due to the predominant coupling to $|\lambda_2\rangle$. To stress this hierarchy in the effective nuclear dynamics $\dot{\sigma} = \mathcal{L}_{\text{eff}}\sigma$, we write

$$\dot{\sigma} = \mathcal{L}_{\text{id}}\sigma + \mathcal{L}_{\text{nid}}\sigma, \tag{3.63}$$

where the first term captures the dominant coupling to the electronic level $|\lambda_2\rangle$ only and is given as

$$\mathcal{L}_{\text{id}}\sigma = \frac{\gamma_2^+}{2}\mathcal{D}[L_2]\sigma + \frac{\gamma_2^-}{2}\mathcal{D}[\mathbb{L}_2]\sigma$$
$$+ i\frac{\Delta_2^+}{2}\left[L_2^\dagger L_2, \sigma\right] + i\frac{\Delta_2^-}{2}\left[\mathbb{L}_2^\dagger\mathbb{L}_2, \sigma\right], \tag{3.64}$$

whereas the remaining non-ideal part $\mathcal{L}_{\text{nid}}$ captures all remaining effects due to the coupling to the far-detuned levels $|\lambda_{1,3}\rangle$ and the OH fluctuations described by H_{zz}.

3.7.2 Holstein-Primakoff Approximation and Bosonic Formalism

To obtain further insights into the nuclear spin dynamics in the high-gradient regime, we now restrict ourselves to uniform hyperfine coupling $(a_{i,j} = N/N_i)$ and apply a Holstein-Primakoff (HP) transformation to the collective nuclear spin operators $I_i^\alpha = \sum_j \sigma_{i,j}^\alpha$ for $\alpha = \pm, z$; generalizations to non-uniform coupling will be discussed separately below in Sect. 3.9. This treatment of the nuclear spins has proven valuable already in previous theoretical studies [70]. In the present case, it allows for a detailed study of the nuclear dynamics including perturbative effects arising from $\mathcal{L}_{\text{nid}}$.

The (exact) Holstein-Primakoff (HP) transformation expresses the truncation of the collective nuclear spin operators to a total spin J_i subspace in terms of a bosonic mode [70]. Note that for uniform HF coupling the total nuclear spin quantum numbers J_i are conserved quantities. Here, we consider two nuclear spin ensembles that are polarized in opposite directions of the quantization axis $\hat{z}$. Then, the HP transformation can explicitly be written as

$$I_L^- = \sqrt{2J_L}\sqrt{1 - \frac{b_L^\dagger b_L}{2J_L}}\, b_L, \tag{3.65}$$

$$I_L^z = b_L^\dagger b_L - J_L,$$

for the first ensemble, and similarly for the second ensemble

$$I_R^+ = \sqrt{2J_R}\sqrt{1 - \frac{b_R^\dagger b_R}{2J_R}}\, b_R, \tag{3.66}$$

$$I_R^z = J_R - b_R^\dagger b_R.$$

Here, b_i denotes the annihilation operator of the bosonic mode $i = L, R$. Next, we expand the operators of Eqs. (3.65) and (3.66) in orders of $\varepsilon_i = 1/\sqrt{J_i}$ which can be identified as a perturbative parameter [70]. This expansion can be justified self-consistently provided that the occupation numbers of the bosonic modes b_i are small compared to $2J_i$. Thus, here we consider the subspace with large collective spin quantum numbers, that is $J_i \sim \mathcal{O}(N/2)$. Accordingly, up to second order in $\varepsilon_L \approx \varepsilon_R$, the hyperfine Hamiltonian can be rewritten as

$$H_{HF} = H_{sc} + H_{ff} + H_{zz}, \tag{3.67}$$

where the semiclassical part H_{sc} reads

$$H_{sc} = a_R J_R S_R^z - a_L J_L S_L^z \tag{3.68}$$
$$= \bar{\omega}_{OH} \left(S_L^z + S_R^z\right) + \Delta_{OH} \left(S_R^z - S_L^z\right). \tag{3.69}$$

Here, we have introduced the individual HF coupling constants $a_i = A_{HF}/N_i$ and

$$\bar{\omega}_{OH} = \Delta_{OH}^{max} \left(p_R - p_L\right)/2, \tag{3.70}$$
$$\bar{\Delta}_{OH} = \Delta_{OH}^{max} \left(p_L + p_R\right)/2, \tag{3.71}$$

with $p_i = J_i/J_i^{max} = 2J_i/N_i$ denoting the degree of polarization in dot $i = L, R$ and $\Delta_{OH}^{max} = A_{HF}/2 \approx 50\,\mu eV$. Within the HP approximation, the hyperfine dynamics read

$$H_{ff} = \frac{a_L}{2}\sqrt{2J_L} S_L^+ b_L + \frac{a_R}{2}\sqrt{2J_R} S_R^+ b_R^\dagger + \text{h.c.}, \tag{3.72}$$

and

$$H_{zz} = a_L S_L^z b_L^\dagger b_L - a_R S_R^z b_R^\dagger b_R. \tag{3.73}$$

Note that, due to the different polarizations in the two dots, the collective nuclear operators I_i^- map onto bosonic annihilation (creation) operators in the left (right) dot, respectively. The expansion given above implies a clear hierarchy in the Liouvillian $\mathcal{L}_0 + \mathcal{V}$ allowing for a perturbative treatment of the leading orders and adiabatic elimination of the electron degrees of freedom whose evolution is governed by the fastest timescale of the problem: while the semiclassical part $H_{sc}/J_i \sim \mathcal{O}(1)$, the HF interaction terms scales as $H_{ff}/J_i \sim \mathcal{O}(\varepsilon)$ and $H_{zz}/J_i \sim \mathcal{O}(\varepsilon^2)$; also compare Ref. [70]. To make connection with the analysis of the previous subsection, we give the following explicit mapping

$$A_L^+ \approx \eta_L b_L^\dagger, \quad \delta A_L^z = \zeta_L b_L^\dagger b_L, \tag{3.74}$$
$$A_R^+ \approx \eta_R b_R, \quad \delta A_R^z = -\zeta_R b_R^\dagger b_R.$$

Here, the parameters $\zeta_i = N/N_i$ and $\eta_i = \zeta_i \sqrt{2J_i}$ capture imperfections due to either different dot sizes ($N_L \neq N_R$) and/or different total spin manifolds ($J_L \neq J_R$).

Moreover, within the HP treatment $\mathcal{V}$ can be split up into a first ($\mathcal{L}_{\mathrm{ff}}$) and a second-order effect ($\mathcal{L}_{\mathrm{zz}}$); therefore, in second-order perturbation theory, the effective nuclear dynamics simplify to [compare Eq. (3.37)]

$$\dot{\sigma} = \mathrm{Tr}_{\mathrm{el}}\left[\mathcal{P}\mathcal{L}_{\mathrm{ff}}\mathcal{P}\rho + \mathcal{P}\mathcal{L}_{\mathrm{zz}}\mathcal{P}\rho - \mathcal{P}\mathcal{L}_{\mathrm{ff}}\mathcal{Q}\mathcal{L}_{0}^{-1}\mathcal{Q}\mathcal{L}_{\mathrm{ff}}\mathcal{P}\rho\right], \tag{3.75}$$

since higher-order effects due to $\mathcal{L}_{\mathrm{zz}}$ can be neglected self-consistently to second order.

Ideal nuclear target state.—Within the HP approximation and for the symmetric setting $\eta_1 = \eta_2 = \eta$, the dominant nuclear jump operators L_2 and $\mathbb{L}_2$, describing the lifting of the spin blockade via the electronic level $|\lambda_2\rangle$, can be expressed in terms of *nonlocal* bosonic modes as

$$L_2 = \eta\sqrt{\mu_2^2 - \nu_2^2}\, a, \tag{3.76}$$

$$\mathbb{L}_2 = \eta\sqrt{\mu_2^2 - \nu_2^2}\, \tilde{a}, \tag{3.77}$$

where $a = \nu b_L^\dagger + \mu b_R$ and $\tilde{a} = \mu b_L + \nu b_R^\dagger$. Here, $\mu = \mu_2/\sqrt{\mu_2^2 - \nu_2^2}$ and $\nu = \nu_2/\sqrt{\mu_2^2 - \nu_2^2}$, such that $\mu^2 - \nu^2 = 1$. Therefore, due to $[a, a^\dagger] = 1 = [\tilde{a}, \tilde{a}^\dagger]$ and $[a, \tilde{a}^\dagger] = 0 = [a, \tilde{a}]$, the operators a and $\tilde{a}$ refer to two independent, properly normalized nonlocal bosonic modes. In this picture, the (unique) ideal nuclear steady state belonging to the dissipative evolution $\mathcal{L}_{\mathrm{id}}\sigma$ in Eq. (3.64) is well known to be a two-mode squeezed state

$$|\Psi_{\mathrm{TMS}}\rangle = \mu^{-1}\sum_n \xi^n |n\rangle_L \otimes |n\rangle_R \tag{3.78}$$

with $\xi = -\nu/\mu$ [41]: $|\Psi_{\mathrm{TMS}}\rangle$ is the common vacuum of the non-local bosonic modes a and $\tilde{a}$, $a|\Psi_{\mathrm{TMS}}\rangle = \tilde{a}|\Psi_{\mathrm{TMS}}\rangle = 0$. It features entanglement between the number of excitations n in the first and second dot. Going back to collective nuclear spins, this translates to perfect correlations between the degree of polarization in the two nuclear ensembles. Note that $|\Psi_{\mathrm{TMS}}\rangle$ represents the dark state $|\xi_{\mathrm{ss}}\rangle$ given in Eq. (3.44) in the zeroth-order HP limit where the truncation of the collective spins to J_i subspaces becomes irrelevant.

Bosonic steady-state solution.—Within the HP approximation, the nuclear dynamics generated by the full effective Liouvillian $\dot{\sigma} = \mathcal{L}_{\mathrm{eff}}\sigma$ are quadratic in the bosonic creation $b_i^\dagger$ and annihilation operators b_i. Therefore, the nuclear dynamics are purely Gaussian and an exact solution is feasible. Based on Eqs. (3.55) and (3.74), one readily derives a closed dynamical equation for the second-order moments

$$\frac{d}{dt}\gamma = \mathcal{M}\gamma + \boldsymbol{C}, \tag{3.79}$$

where γ is a vector comprising the second-order moments, that is $\gamma = \left(\left\langle b_i^\dagger b_j \right\rangle_t ,$ $\left\langle b_i^\dagger b_j^\dagger \right\rangle_t , \dots \right)^\top$ and C is a constant vector. The solution to Eq. (3.79) is given by

$$\gamma(t) = e^{\mathcal{M}t} c_0 - \mathcal{M}^{-1} C, \qquad (3.80)$$

where c_0 is an integration constant. Accordingly, provided that the dynamics generated by $\mathcal{M}$ is contractive (see Sect. 3.8 for more details), the steady-state solution is found to be

$$\gamma_{ss} = -\mathcal{M}^{-1} C. \qquad (3.81)$$

Based on γ_{ss}, one can construct the steady-state covariance matrix (CM), defined as $\Gamma_{ij}^{CM} = \left\langle \{R_i, R_j\} \right\rangle - 2 \left\langle R_i \right\rangle \left\langle R_j \right\rangle$, where $\{R_i, i = 1, \dots, 4\} = \{X_L, P_L, X_R, P_R\}$; here, $X_i = (b_i + b_i^\dagger)/\sqrt{2}$ and $P_i = i(b_i^\dagger - b_i)/\sqrt{2}$ refer to the quadrature operators related to the bosonic modes b_i. By definition, Gaussian states are fully characterized by the first and second moments of the field operators R_i. Here, the first order moments can be shown to vanish. The entries of the CM are real numbers: since they constitute the variances and covariances of quantum operators, they can be detected experimentally via nuclear spin variance and correlation measurements [71].

We now turn to the central question of whether the steady-state entanglement inherent to the ideal target state $|\Psi_{TMS}\rangle$ is still present in the presence of the undesired terms described by $\mathcal{L}_{nid}$. In our setting, this is conveniently done via the CM, which encodes all information about the entanglement properties [72]: It allows us to compute certain entanglement measures efficiently in order to make qualitative and quantitative statements about the degree of entanglement [72]. Here, we will consider the following quantities: For symmetric states, the entanglement of formation E_F can be computed easily [73, 74]. It measures the minimum number of singlets required to prepare the state through local operations and classical communication. For symmetric states, this quantification of entanglement is fully equivalent to the one provided by the logarithmic negativity $E_\mathcal{N}$; the latter is determined by the smallest symplectic eigenvalues of the CM of the partially transposed density matrix [75]. Lastly, in the HP picture the EPR uncertainty defined in Eq. (3.45) translates to $\Delta_{EPR} = [\text{var}(X_L + X_R) + \text{var}(P_L - P_R)]/2$. For the ideal target state $|\Psi_{TMS}\rangle$, we find $\Delta_{EPR}^{id} = (\mu - \nu)^2 = (1 - |\xi|)/(1 + |\xi|) < 1$. Finally, one can also compute the fidelity $\mathcal{F}(\sigma_{ss}, \sigma_{target})$ which measures the overlap between the steady state generated by the full dynamics $\dot{\sigma} = \mathcal{L}_{eff}\sigma$ and the ideal target state $\sigma_{target} = |\Psi_{TMS}\rangle \langle \Psi_{TMS}|$ [72].

As illustrated in Figs. 3.11 and 3.12, the generation of steady-state entanglement persists even in presence of the undesired noise terms described by $\mathcal{L}_{nid}$, asymmetric dot sizes ($N_L \neq N_R$) and classical uncertainty in total spins J_i: The maximum amount of entanglement that we find (in the symmetric scenario $N_L = N_R$) is approximately $E_\mathcal{N} \approx 1.5$, corresponding to an entanglement of formation $E_F \approx (1-2)$ebit and an EPR uncertainty of $\Delta_{EPR} \approx 0.4$. When tuning the interdot tunneling parameter from $t = 10\,\mu\text{eV}$ to $t = 35\,\mu\text{eV}$, the squeezing parameter $|\xi| = |\nu_2/\mu_2|$ increases from

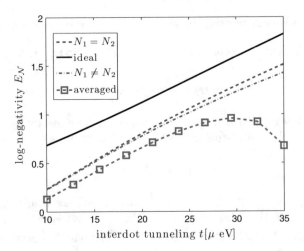

Fig. 3.11 Steady-state entanglement generation between the two nuclear spin ensembles. $E_{\mathcal{N}} > 0$ indicates the creation of entanglement. The *black solid curve* refers to the idealized, symmetric setting where the undesired HF-coupling to $|\lambda_{1,3}\rangle$ has been ignored and where $J_L = J_R = p J_{max}$; here, the nuclear polarization p = 0.8 and $N_L = N_R = 2 J_{max} = 10^6$. The *blue-dashed line* then also takes into account coupling to $|\lambda_{1,3}\rangle$, while the *red (dash-dotted) curve* in addition accounts for an asymmetric dot size: $N_R = 0.8 N_L = 8 \times 10^5$. Additionally, classical uncertainty (*red squares*) in the total spin J_i quantum numbers leads to a reduced amount of entanglement, but does not disrupt it completely; here, we have set the range of the (uniform) distribution to $\Delta_{J_i} = 50\sqrt{N_i}$. Other numerical parameters: $\omega_0 = 0$, $\Gamma = 25\,\mu eV$, $\varepsilon = 30\,\mu eV$, $3\Gamma_{\pm} + \Gamma_{deph}/4 = 0.5\,\mu eV$

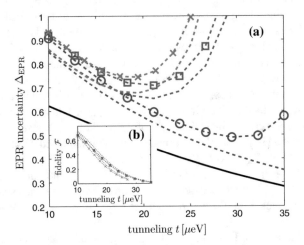

Fig. 3.12 Steady-state entanglement between the two nuclear spin ensembles quantified via **a** the EPR-uncertainty Δ_{EPR} and **b** fidelity $\mathcal{F}$ of the nuclear steady state with the two-mode squeezed target state. The *black solid curve* refers to the idealized setting where the undesired HF coupling to $|\lambda_{1,3}\rangle$ has been ignored and where $J_1 = J_2 = p J_{max}$, p = 0.8 and $N_1 = N_2 = 2 J_{max} = 10^6$, corresponding to $\Delta_{OH} = 40\,\mu eV$. The *blue-dashed line* then also takes into account coupling to $|\lambda_{1,3}\rangle$ while the *red-dashed curve* in addition accounts for an asymmetric dot size: $N_2 = 0.8 N_1 = 8 \times 10^5$. The amount of entanglement decreases for a smaller nuclear polarization: p = 0.7 (*green dashed curve*). For other numerical parameters see Fig. 3.11

~0.2 to ~0.6, respectively; this is because (for fixed Δ, $\varepsilon > 0$) and increasing t, ε_2 approaches 0 and the relative weight of ν_2 as compared to μ_2 increases. Ideally, this implies stronger squeezing of the steady state of $\mathcal{L}_{id}$ and therefore a greater amount of entanglement (compare the solid line in Fig. 3.11), but, at the same time, it renders the target state more susceptible to undesired noise terms. For $|\xi| \approx 0.2$, we obtain a relatively high fidelity $\mathcal{F}$ with the ideal two-mode squeezed state, close to 80 %. Stronger squeezing leads to a larger occupation of the bosonic HP modes (pictorially, the nuclear target state leaks farther into the Dicke ladder) and eventually to a break-down of the approximative HP description. The associated critical behavior in the nuclear spin dynamics can be understood in terms of a dynamical phase transition [70], which will be analyzed in greater detail in the next section.

3.8 Criticality

Based on the Holstein-Primakoff analysis outlined above, we now show that the nuclear spin dynamics exhibit a dynamical quantum phase transition which originates from the competition between dissipative terms and unitary dynamics. This rather generic phenomenon in open quantum systems results in nonanalytic behaviour in the spectrum of the nuclear spin Liouvillian, as is well known from the paradigm example of the Dicke model [70, 76–78].

The nuclear dynamics in the vicinity of the stationary state are described by the stability matrix $\mathcal{M}$. Resulting from a systematic expansion in the system size, the (complex) eigenvalues of $\mathcal{M}$ correspond exactly to the low-excitation spectrum of the full system Liouvillian given in Eq. (3.1) in the thermodynamic limit ($J \to \infty$). A non-analytic change of steady state properties (indicating a steady state phase transition) can only occur if the spectral gap of $\mathcal{M}$ closes [70, 79]. The relevant gap in this context is determined by the eigenvalue with the largest real part different from zero [from here on referred to as the asymptotic decay rate (ADR)]. The ADR determines the rate by which the steady state is approached in the long time limit.

As depicted in Fig. 3.13, the system reaches such a critical point at $t_{crt} \approx 37\,\mu eV$ where the ADR (red/blue dotted lines closest to zero) becomes zero. At this point, the dynamics generated by $\mathcal{M}$ become non-contractive [compare Eq. (3.80)] and the nuclear fluctuations diverge, violating the self-consistency condition of low occupation numbers in the bosonic modes b_i and thus leading to a break-down of the HP approximation. Consequently, the dynamics cannot further be described by the dynamical matrix $\mathcal{M}$ indicating a qualitative change in the system properties and a steady state phase transition.

To obtain further insights into the cross-over of the maximum real part of the eigenvalues $\lambda_{\mathcal{M}}$ of the matrix $\mathcal{M}$ from negative to positive values, we analyze the effect of the nuclear Stark shift terms [Eq. (3.60)] in more detail. In the HP regime, up to irrelevant constant terms, the Stark shift Hamiltonian H_{Stark} can be written as

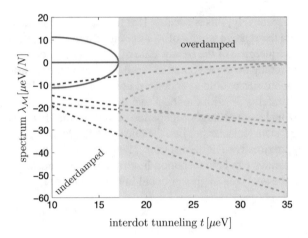

Fig. 3.13 Spectrum of $\mathcal{M}$. Real (*dashed*) and imaginary parts (*solid*) for $\Gamma = 25\,\mu\text{eV}$ (*blue*) and $\Gamma = 50\,\mu\text{eV}$ (*red*). A dynamical phase transition is found at the bifurcation separating an underdamped from an overdamped region (see *gray shading* for the spectrum displayed in *red*). The critical point $t_{\text{crt}} \approx 37\,\mu\text{eV}$ is reached where the smallest decay rate (ADR) becomes zero. Other numerical parameters as those for the dashed curve in Fig. 3.11

$$H_{\text{Stark}} = \varepsilon_L^{\text{st}} b_L^\dagger b_L + \varepsilon_R^{\text{st}} b_R^\dagger b_R + \varepsilon_{LR}^{\text{st}} \left[b_L b_R + b_L^\dagger b_R^\dagger \right], \qquad (3.82)$$

The relevant parameters $\varepsilon_\nu^{\text{st}}$ introduced above are readily obtained from Eqs. (3.60) and (3.74). In the symmetric setting $\eta_L = \eta_R$, it is instructive to re-express H_{Stark} in terms of the squeezed, non-local bosonic modes $a = \nu b_L^\dagger + \mu b_R$ and $\tilde{a} = \mu b_L + \nu b_R^\dagger$ [see Eqs. (3.76) and (3.77)] whose common vacuum is the ideal steady state of $\mathcal{L}_{\text{id}}$. Up to an irrelevant constant term, H_{Stark} takes on the form

$$H_{\text{Stark}} = \Delta_a a^\dagger a + \Delta_{\tilde{a}} \tilde{a}^\dagger \tilde{a} + g_{a\tilde{a}} \left(a\tilde{a} + a^\dagger \tilde{a}^\dagger \right). \qquad (3.83)$$

With respect to the entanglement dynamics, the first two terms do not play a role as the ideal steady state $|\Psi_{\text{TMS}}\rangle$ is an eigenstate thereof. However, the last term is an active squeezing term in the non-local bosonic modes: It does not preserve the excitation number in the modes $a, \tilde{a}$ and may therefore drive the nuclear system away from the vacuum by pumping excitations into the system. Numerically, we find that the relative strength of $g_{a\tilde{a}}$ increases compared to the desired entangling dissipative terms when tuning the interdot tunneling parameter t towards t_{crt}. We therefore are confronted with two competing effects while tuning the interdot coupling t. On the one hand, the dissipative dynamics tries to pump the system into the vacuum of the modes a and $\tilde{a}$ [see Eqs. (3.76) and (3.77)], which become increasingly squeezed as we increase t. On the other hand, an increase in t leads to enhanced coherent dynamics (originating from the nuclear Stark shift H_{Stark}) which try to pump excitations in the system [Eq. (3.83)]. This competition between dissipative and coherent dynamics

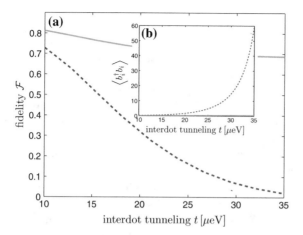

Fig. 3.14 **a** Fidelity $\mathcal{F}$ of the nuclear steady-state with the two-mode squeezed target state. The *blue-dashed line* accounts for the full nuclear Liouvillian $\mathcal{L}_{\mathrm{eff}}$ for the symmetric setting ($N_L = N_R$), while the *green solid line* refers to the same setting in the absence of any Stark-shift terms. Therefore, the decreasing fidelity $\mathcal{F}$ (*blue dashed line*) and a diverging number of HP bosons shown in (**b**) is due to undesired Stark shift terms included in $\mathcal{L}_{\mathrm{eff}}$. Here, $\eta_L = \eta_R$ and therefore $\langle b_L^\dagger b_L \rangle = \langle b_R^\dagger b_R \rangle$; asymmetric settings where $\eta_L \neq \eta_R$ entail small asymmetries in the number of HP bosons. For other numerical parameters compare the dashed curve in Fig. 3.11

is known to be at the origin of many dissipative phase transitions, and has been extensively studied, e.g., in the context of the Dicke phase transition [76, 77].

As shown in Fig. 3.14, the observed critical behaviour in the nuclear spin dynamics can indeed be traced back to the presence of the nuclear Stark shift terms H_{Stark}: here, when tuning the system towards the critical point t_{crt}, the diverging number of HP bosons is shown to be associated with the presence of H_{Stark}. Moreover, for relatively low values of the squeezing parameter $|\xi|$, we obtain a relatively high fidelity $\mathcal{F}$ with the ideal two-mode squeezed state, close to 80 %. For stronger squeezing, however, the target state becomes more susceptible to the undesired noise terms, first leading to a reduction of $\mathcal{F}$ and eventually to a break-down of the HP approximation.

Aside from this phase transition in the steady state, we find nonanalyticities at non-zero values of the nuclear ADR, indicating a change in the dynamical properties of the system which cannot be detected in steady-state observables [70]. Rather, the system displays anomalous behaviour approaching the stationary state: As shown Fig. 3.13, we can distinguish two *dynamical* phases [80–83], an underdamped and an overdamped one, respectively. The splitting of the real parts of $\mathcal{M}$ coincides with vanishing imaginary parts. Thus, in the overdamped regime, perturbing the system away from its steady state leads to an exponential, non-oscillating return to the stationary state. A similar underdamped region in direct vicinity of the phase transition can be found in the dissipative Dicke phase transition [76, 77].

3.9 Implementation

This section is devoted to the experimental realization of our proposal. First, we summarize the experimental requirements of our scheme. Thereafter, we address several effects that are typically encountered in realistic systems, but which have been neglected so far in our analysis. This includes non-uniform HF coupling, larger individual nuclear spins ($I > 1/2$), external magnetic fields, different nuclear species, internal nuclear dynamics and charge noise.

Experimental requirements.—Our proposal relies on the predominant spin-blockade lifting via the electronic level $|\lambda_2\rangle$ and the adiabatic elimination of the electronic degrees of freedom: First, the condition $t \gg \omega_0, g_{hf}$ ascertains a predominant lifting of the Pauli-blockade via the hybridized, nonlocal level $|\lambda_2\rangle$. To reach the regime in which the electronic subsystem settles into the desired quasisteady state $\rho_{ss}^{el} = (|T_+\rangle \langle T_+| + |T_-\rangle \langle T_-|)/2$ on a timescale much shorter than the nuclear dynamics, the condition $\Gamma_2 \gg \Gamma_\pm \gg \gamma_c$ must be fulfilled. Both, $t \gg \omega_0, g_{hf}$ and $\Gamma_\pm \gg \gamma_c$ can be reached thanks to the extreme, separate, in-situ tunability of the relevant, electronic parameters t, ε and Γ [3]. Moreover, to kick-start the nuclear self-polarization process towards a high-gradient stable fixed point, where the condition $\Gamma_2 \gg \Gamma_\pm$ is fulfilled, an initial gradient of approximately $\sim (3-5)$ μeV, corresponding to a nuclear polarization of $\sim (5-10)$ %, is required; as shown in Sect.3.6, this ensures $\kappa_2^2 \gg x_\pm$, where we estimate the suppression factor $x_\pm = \Gamma_\pm / \Gamma \approx 10^{-3}$. The required gradient could be provided via an on-probe nanomagnet [61, 67] or alternative dynamic polarization schemes [22, 26, 54, 57]; experimentally, nuclear spin polarizations of up to 50 % have been reported for electrically defined quantum dots [53, 61].

Inhomogeneous HF coupling.—Within the HP analysis presented in Sect. 3.7, we have restricted ourselves to uniform HF coupling. Physically, this approximation amounts to the assumption that the electron density is flat in the dots and zero outside [34]. In Ref. [84], it was shown that corrections to this idealized scenario are of the order of $1 - p_{nuc}$ for a high nuclear polarization p_{nuc}. Thus, the HP analysis for uniform HF coupling is correct to zeroth order in the small parameter $1 - p_{nuc}$. To make connection with a more realistic setting, where—according to the electronic s-type wavefunction—the HF coupling constants $a_{i,j}$ typically follow a Gaussian distribution, one may express them as $a_{i,j} = \bar{a} + \delta_{i,j}$. Then, the uniform contribution $\bar{a}$ enables an efficient description within fixed J_i subspaces, whereas the non-uniform contribution leads to a coupling between different J_i subspaces on a much longer timescale. As shown in Ref. [68], the latter is relevant in order to avoid low-polarization dark states and to reach highly polarized nuclear states. Let us stress that (for uniform HF coupling) we have found that the generation of nuclear steady-state entanglement persists in the presence of asymmetric ($N_L \neq N_R$) dot sizes which represents another source of inhomogeneity in our system.

In what follows, we show that our scheme works even in the case of non-uniform coupling, provided that the two dots are sufficiently similar. If the HF coupling constants are completely inhomogeneous, that is $a_{i,j} \neq a_{i,k}$ for all $j \neq k$, but the two dots are identical $(a_{1,j} = a_{2,j} \equiv a_j \ \forall j = 1, 2, \ldots, N_L \equiv N_R \equiv N)$, such that the nuclear spins can be grouped into pairs according to their HF coupling constants, the two dominant nuclear jump operators L_2 and $\mathbb{L}_2$ simplify to

$$L_2 = \sum_j a_j l_j, \qquad \mathbb{L}_2 = \sum_j a_j \mathbb{l}_j, \qquad (3.84)$$

where the nuclear operators $l_j = \nu_2 \sigma_{Lj}^+ + \mu_2 \sigma_{Rj}^+$ and $\mathbb{l}_j = \mu_2 \sigma_{Lj}^- + \nu_2 \sigma_{Rj}^-$ are nonlocal nuclear operators, comprising two nuclear spins that belong to different nuclear ensembles, but have the same HF coupling constant a_j. For one such pair of nuclear spins, the unique, common nuclear dark state fulfilling

$$l_j \, |\xi\rangle_j = \mathbb{l}_j \, |\xi\rangle_j = 0, \qquad (3.85)$$

is easily verified to be

$$|\xi\rangle_j = \mathcal{N}_\xi \left(|\downarrow_j, \uparrow_j\rangle + \xi \, |\uparrow_j, \downarrow_j\rangle \right), \qquad (3.86)$$

where $\mathcal{N}_\xi = 1/\sqrt{1 + \xi^2}$ for normalization. Therefore, in the absence of degeneracies in the HF coupling constants $(a_{i,j} \neq a_{i,k} \ \forall j \neq k)$, the *pure, entangled* ideal nuclear dark state fulfilling $L_2 \, |\xi_{ss}\rangle = \mathbb{L}_2 \, |\xi_{ss}\rangle = 0$ can be constructed as a tensor product of entangled pairs of nuclear spins,

$$|\xi_{ss}\rangle = \otimes_{j=1}^N |\xi\rangle_j. \qquad (3.87)$$

Again, the parameter $\xi = -\nu_2/\mu_2$ fully quantifies polarization and entanglement properties of the nuclear stationary state; compare Eq. (3.44): First, for small values of the parameter $|\xi|$ the ideal nuclear dark state $|\xi_{ss}\rangle$ features an arbitrarily high polarization gradient

$$\Delta_{I^z} = \langle I_R^z \rangle_{ss} - \langle I_L^z \rangle_{ss} = N \frac{1 - \xi^2}{1 + \xi^2}, \qquad (3.88)$$

whereas the homogeneous net polarization $P = \langle I_L^z \rangle_{ss} + \langle I_R^z \rangle_{ss}$ vanishes. The stationary solution for the nuclear gradient Δ_{I^z} is bistable as it is positive (negative) for $|\xi| < 1$ ($|\xi| > 1$), respectively. Second, the amount of entanglement inherent to the stationary solution $|\xi_{ss}\rangle$ can be quantified via the EPR uncertainty ($\Delta_{EPR} < 1$ indicates entanglement) and is given by $\Delta_{EPR} = (1 - |\xi|)^2 / |1 - \xi^2|$.

Our analytical findings are verified by exact diagonalization results for small sets of inhomogeneously coupled nuclei. Here, we compute the exact (possibly mixed) solutions σ_{ss} to the dark state equation $\mathcal{D}[L_2] \sigma_{ss} + \mathcal{D}[\mathbb{L}_2] \sigma_{ss} = 0$; compare Fig. 3.8 for the special case of uniform HF coupling. As shown in Fig. 3.15, our numerical evidence indicates that small deviations from the perfect symmetry (that is for $a_{Lj} \approx a_{Rj}$) between the QDs still yield a (mixed) *unique* entangled steady state close to

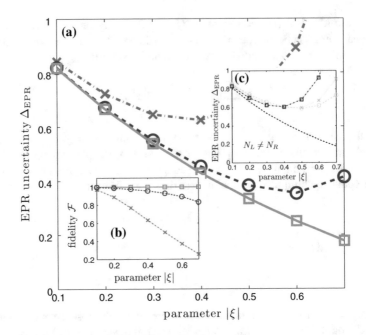

Fig. 3.15 EPR uncertainty (**a**) and fidelity $\mathcal{F}$ with the ideal nuclear target state $|\xi_{ss}\rangle$ given in Eq.(3.87) (**b**) as a function of the squeezing-like parameter $|\xi|$ for $N_L = N_R = 3$ *inhomogeneously* coupled nuclei. The *blue curve* (*squares*) refers to a symmetric setting where $\vec{a}_L = \vec{a}_R = (1.11, 1.67, 0.22)$, whereas the *green* (*circles*) and *red* (*crosses*) solutions incorporate asymmetries: $\vec{a}_L = (1.18, 1.61, 0.21)$, $\vec{a}_R = (1.11, 1.67, 0.22)$ and $\vec{a}_L = (1.0, 1.5, 0.5)$, $\vec{a}_R = (1.24, 1.55, 0.21)$, respectively. **c** Exact results for the asymmetric scenario $N_L = 2 \neq 3 = N_R$. Here, $\vec{a}_L = (1.0, 1.5)$ was held fixed while the *green* (*circles*), *orange* (*crosses*) and *dark blue* (*squares*) curves refer to $\vec{a}_R = (0.98, 1.47, 0.05)$, $\vec{a}_R = (0.93, 1.39, 0.18)$ and $\vec{a}_R = (0.76, 1.14, 0.60)$, respectively; as a benchmark, the *black dashed curve* refers to the ideal results in the symmetric setting. Due to the absence of degeneracies, the steady state solution σ_{ss} is unique in all cases considered here

$|\xi_{ss}\rangle$. In the ideal case $a_{Lj} = a_{Rj}$, we recover the pure steady state given in Eq. (3.87). Moreover, we find that the generation of steady-state entanglement even persists for asymmetric dot sizes, i.e. for $N_L \neq N_R$. Exact solutions for $N_L = 2 \neq 3 = N_R$ are displayed in Fig. 3.15. Here, we still find strong traces of the ideal dark state $|\xi_{ss}\rangle$, provided that one can approximately group the nuclear spins into pairs of similar HF coupling strength. The interdot correlations $\left\langle \sigma_{Lj}^+ \sigma_{Rj}^- \right\rangle$ are found to be close to the ideal value of $\xi / \left(1 + \xi^2\right)$ for nuclear spins with a similar HF constant, but practically zero otherwise. In line with this reasoning, the highest amount of entanglement in Fig. 3.15 is observed in the case where one of the nuclear spins belonging to the bigger second ensemble is practically uncoupled. Lastly, we note that one can 'continuously' go from the case of non-degenerate HF coupling constants (the case considered in detail here) to the limit of uniform HF coupling [compare Eq. (3.44)] by grouping spins with the same HF coupling constants to 'shells', which form collective nuclear spins. For degenerate couplings, however, there are additional conserved quantities, namely the respective total spin quantum numbers, and therefore multiple stationary states of the above form. As argued in Sect. 3.5, a mixture of different J-subspaces should still be entangled provided that the range of J-subspaces involved in this mixture is small compared to the average J value.

Larger nuclear spins.—All natural isotopes of Ga and As carry a nuclear spin $I = 3/2$ [13], whereas we have considered $I = 1/2$ for the sake of simplicity. For our purposes, however, this effect can easily be incorporated as an individual nuclear spin with $I = 3/2$ maps onto 3 homogeneously coupled nuclear spins with individual $I = 1/2$ which are already in the fully symmetric Dicke subspace $J = 3/2$.

External magnetic fields.—For simplicity, our previous analysis has focused on a symmetric setting of vanishing external fields, $\Delta_{\text{ext}} = \omega_{\text{ext}} = 0$. Non-vanishing external fields, however, may be used as further experimental knobs to tune the desired nuclear steady-state properties: First, as mentioned above, a non-zero external gradient Δ_{ext} is beneficial for our proposal as it can provide an efficient way to destabilize the zero-polarization solution $\left(\Delta_{\text{OH}}^{\text{ss}} = 0\right)$ by initiating the nuclear self-polarization process. Second, non-vanishing $\omega_{\text{ext}} \neq 0$ gives rise to another electron-nuclear feedback-driven experimental knob for controlling the nuclear stationary state. In the framework of Sect. 3.6, for $\omega_{\text{ext}} \neq 0$ the semiclassical dynamical equations can be generalized to

$$\frac{d}{dt} \langle I_L^z \rangle_t = \alpha_+ N_{L\downarrow} - \beta_- N_{L\uparrow}, \tag{3.89}$$

$$\frac{d}{dt} \langle I_R^z \rangle_t = \beta_+ N_{R\downarrow} - \alpha_- N_{R\uparrow}, \tag{3.90}$$

where we have introduced the number of nuclear spin-up and spin-down spins as $N_{i\uparrow} = N_i/2 + \langle I_i^z \rangle$ and $N_{i\downarrow} = N_i/2 - \langle I_i^z \rangle$, respectively, and the generalized polarization rates

$$\alpha_\pm = p\gamma^\pm \nu_2^2 + (1 - 2p)\gamma^\mp \mu_2^2, \tag{3.91}$$

$$\beta_\pm = p\gamma^\pm \mu_2^2 + (1 - 2p)\gamma^\mp \nu_2^2. \tag{3.92}$$

They depend on the generalized HF-mediated decay rate

$$\gamma^\pm = \frac{a_{\text{hf}}^2 \tilde{\Gamma}}{2\left[(\varepsilon_2 \mp \omega_0)^2 + \tilde{\Gamma}^2\right]}, \tag{3.93}$$

which accounts for different detunings for $\omega_0 \neq 0$; compare Eq. (3.39). As shown in Fig. 3.16, in the presence of an external magnetic splitting ω_{ext}, the nuclear spins build up a homogeneous Overhauser field $\bar{\omega}_{\text{OH}}$ in the steady state to partially compensate the external component. The steady state solution then locally fulfills a detailed-balance principle, namely $\alpha_+ N_{L\downarrow} = \beta_- N_{L\uparrow}$ and $\beta_+ N_{R\downarrow} = \alpha_- N_{R\uparrow}$, which

Fig. 3.16 Buildup of a homogeneous nuclear Overhauser field component $\bar{\omega}_{OH}^{ss}$ which partially compensates an applied external magnetic field, shown here for $t = 10\,\mu eV$ (*red solid*) and $t = 20\,\mu eV$ (*blue dashed*). Other numerical parameters: $\Gamma = 25\,\mu eV$, $\varepsilon = 30\,\mu eV$, $\Gamma_{\pm} = \Gamma_{deph} = 0.1\,\mu eV$

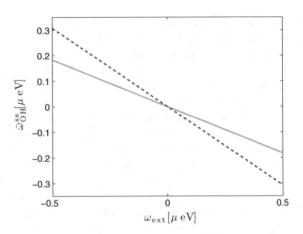

is determined by effective nuclear flip rates and the number of spins available for a spin flip. Intuitively, this finding can be understood as follows: For $\omega_{ext} \neq 0$, the degeneracy between $|T_+\rangle$ and $|T_-\rangle$ is lifted with one of them being less detuned from $|\lambda_2\rangle$ than the other. This favors the build-up of a nuclear net polarization P which, however, counteracts the splitting ω_{ext}; for $\omega_{ext} = 0$, this mechanism stabilizes $\bar{\omega}_{OH} = P = 0$ in the stationary state. This result has also been confirmed by numerical results presented in Appendix 3.11.8.

Species inhomogeneity.—Nonzero external magnetic fields, however, induce nuclear Zeeman splittings, with the nuclear magnetic moment being about three orders of magnitude smaller than the Bohr magneton for typical quantum dots [3, 13]. Most QDs consist of a few (in GaAs three) different species of nuclei with strongly varying g factors. In principle, this species-inhomogeneity can cause dephasing between the nuclear spins. However, for a uniform external magnetic field this dephasing mechanism only applies to nuclei belonging to different species. In a rotating wave approximation, this leads to few mutually decohered subsystems (in GaAs three) each of which being driven towards a two-mode squeezedlike steady state: note that, because of the opposite polarizations in the two dots, the nuclear target state $|\xi\rangle_{ss}$ is invariant under the application of a homogeneous magnetic field. This argument, however, does not hold for an inhomogeneous magnetic field which causes dephasing of $|\xi\rangle_{ss}$ as the nuclear states $|m, -m\rangle$ (m is the nuclear spin projection) pick up a phase $\exp\left[2im\Delta_{ext}^{nuc}t\right]$, where $\Delta_{ext}^{nuc} \approx 10^{-3}\Delta_{ext}$. If one uses an external magnetic gradient to incite the nuclear self-polarization process, after successful polarization one should therefore switch off the gradient[4] to support the generation of entanglement between the two ensembles.

---

[4]We have numerically checked that the entanglement generation persists for a small external gradient Δ_{ext} provided that the corresponding induced nuclear Zeeman splitting is small compared to the dissipative gap ADR $= \tau_{gap}^{-1}$. Thus, to support the generation of steady-state entanglement, we estimate that the magnetic gradient should be smaller than ~ 1 mT.

Weak nuclear interactions.—We have neglected nuclear dipole-dipole interactions among the nuclear spins. The strength of the effective magnetic dipole-dipole interaction between *neighboring* nuclei in GaAs is about $g_{dd} \sim (100\,\mu s)^{-1}$ [3, 13]. Spin-nonconserving terms and flip-flop terms between different species can be suppressed efficiently by applying an external magnetic field of $B_{ext} \gtrsim 10\,mT$ [85]. As discussed above, the corresponding (small) electron Zeeman splitting $\omega_{ext} \approx 0.25\,\mu eV$ does not hamper our protocol. Then, it is sufficient to consider so-called homonuclear flip-flop terms between nuclei of the same species only and phase changing zz-terms. First, nuclear spin diffusion processes—governing the dynamics of the spatial profile of the nuclear polarization by changing A_i^z—have basically no effect within an (almost) completely symmetric Dicke subspace. With typical timescales of $\gtrsim 10\,s$, they are known to be very slow and therefore always negligible on the timescale considered here [63, 85, 86]. Second, the interactions $\propto \sigma_i^z \sigma_j^z$ lead to dephasing similar to the nuclear Zeeman terms discussed above: In a mean-field treatment one can estimate the effective Zeeman splitting of a single nuclear spin in the field of its surrounding neighbors to be a few times g_{dd} [68]. This mean field is different only for different species and thus does not cause any homonuclear dephasing. Still, the variance of this effective field may dephase spins of the same species, but for a high nuclear polarization p_{nuc} this effect is further suppressed by a factor $\sim (1 - p_{nuc}^2)$ as the nuclei experience a *sharp* field for a sufficiently high nuclear polarization p_{nuc}. Lastly, we refer to recently measured nuclear decoherence times of $\sim 1\,ms$ in vertical double quantum dots [26]. Since this is slow compared to the dissipative gap of the nuclear dynamics $\tau_{gap} \approx (3 - 30)\,\mu s$ for $N \approx 10^5 - 10^6$, we conclude that it should be possible to create entanglement between the two nuclear spin ensembles faster than it gets disrupted due to dipole-dipole interactions among the nuclear spins or other competing mechanisms [34]. Moreover, since strain is largely absent in electrically defined QDs [4], nuclear quadrupolar interactions have been neglected as well. For a detailed analysis of the internal nuclear dynamics within a HP treatment, we refer to Ref. [72].

Charge noise.—Nearly all solid-state qubits suffer from some kind of charge noise [87]. In a DQD device background charge fluctuations and noise in the gate voltages may cause undesired dephasing processes. In a recent experimental study [87], voltage fluctuations in ε have been identified as the source of the observed dephasing in a singlet-triplet qubit. In our setting, however, the electronic subsystem quickly settles into the quasisteady state ρ_{ss}^{el} which lives solely in the (1, 1) triplet subspace spanned by $\{|T_\pm\rangle\}$ and is thus relatively robust against charge noise. Still, voltage fluctuations in ε lead to fluctuations in the parameter ξ characterizing the nuclear two-mode target state given in Eq. (3.44). For typical parameters ($t = 20\,\mu eV$, $\varepsilon = 30\,\mu eV$, $\Delta = 40\,\mu eV$), however, ξ turns out to be rather insensitive to fluctuations in ε, that is $|d\xi/d\varepsilon| \approx 10^{-2}/\mu eV$. Note that the system can be made even more robust (while keeping ξ constant) by increasing both ε and t: For $t = 50\,\mu eV$, $\varepsilon = 90\,\mu eV$, the charge noise sensitivity is further reduced to $|d\xi/d\varepsilon| \approx 3 \times 10^{-3}/\mu eV$. We can then estimate the sensitivity of the generated steady-state entanglement via $|d\Delta_{EPR}/d\varepsilon| = |(d\xi/d\varepsilon)(d\Delta_{EPR}/d\xi)| \lesssim 2 \times 10^{-2}/\mu eV$, where we have used $|d\Delta_{EPR}/d\xi| = 2/(1 + \xi)^2 < 2$. Typical fluctuations in ε of the order of $\sim (1-3)\,\mu eV$ as reported in Ref. [59] may then cause a reduction of entanglement in the nuclear steady state of approximately $\sim 5\,\%$ as compared to the optimal value of ε. If the typical timescale associated with charge noise τ_{noise} is fast compared to the dissipative gap of the nuclear dynamics, i.e., $\tau_{noise} \ll \tau_{gap}$, the nuclear spins effectively only experience the averaged value of ξ, coarse-grained over its fast fluctuations.

Detection.—Finally, entanglement could be detected by measuring the OH shift in each dot separately [3]; in combination with NMR techniques to rotate the nuclear spins [4] we can obtain all spin components and their variances which are sufficient to verify the presence of entanglement (similar to the proposal [34]).

3.10 Conclusion and Outlook

In summary, we have developed a theoretical master-equation-based framework for a DQD in the Pauli-blockade regime which features coupled dynamics of electron and nuclear spins as a result of the hyperfine interaction. Our analysis is based on the typical separation of timescales between (fast) electron spin evolution and (slow) nuclear spin dynamics, yielding a coarse-grained quantum master equation for the nuclear spins. This reverses the standard perspective in which the nuclei are considered as an environment for the electronic spins, but rather views the nuclear spins as the quantum system coupled to an electronic environment with an exceptional degree of tunability. Here, we have focused on a regime favorable for the generation of entanglement in the nuclear steady state, whereas the electrons are driven to an unpolarized, classically correlated separable state. Therefore, in this setting, electron dephasing turns out to be an asset rather than a liability. Our central master equation directly incorporates nonlinear feedback mechanisms resulting from the back-action of the Overhauser field on the electron energy levels and thus explains the nuclear multi-stability in a very transparent way. The associated instability of the nuclei towards self-polarization can be used as a means for controlling the nuclear spin distribution [32]. For example, as a prominent application, we predict the deterministic generation of entanglement between two (spatially separated) mesoscopic spin ensembles, induced by electron transport and the common, collective coupling of the nuclei to the electronic degrees of freedom of the DQD. The nuclear entangled state is of EPR type, which is known to play a key role in continuous variable quantum information processing [88, 89], quantum sensing [90] and metrology [91–93]. Since the entanglement generation does not rely on coherent evolution, but is rather stabilized by the dissipative dynamics, the proposed scheme is inherently robust against weak random perturbations. Moreover, as two large spin ensembles with $N \sim 10^6$ get entangled, the nuclear system has the potential to generate large amounts of entanglement, i.e., many ebits. Lastly, the apparent relatively large robustness of the nuclear steady state against charge noise shows that, when viewed as (for example) a platform for spin-based quantum memories, nuclear spin ensembles have certain, intrinsic advantages with respect to their electronic cousins.

Our results provide a clear picture of the feedback-driven polarization dynamics in a generic electron transport setting and, therefore, should serve as a useful guideline for future experiments aiming at an enhanced, dynamical control of the nuclear spins: While DNP experiments in double quantum dots, for example, have revealed an instability towards large Overhauser gradients, consistent with our results, the question of whether or not this instability results from dot asymmetry or some other mechanism is still unsettled [22, 54, 56]. Here, we study a generic DC setting, where the buildup of a large OH gradient straightforwardly emerges even in the presence of a completely symmetric coherent hyperfine interaction. From a more fundamental, conceptual point of view, our theory gives valuable insights into the complex, non-equilibrium many-body dynamics of localized electronic spins interacting with a mesoscopic number of nuclear spins. Understanding the quantum dynamics of this central spin model marks an important goal in the field of mesoscopic physics, as a notable number of unexpected and intriguing phenomena such as multi-stability, switching, hysteresis and long timescale oscillations have been observed in this system [12, 24, 51, 63].

On the one hand, reversing again our approach, our scheme may lead to a better quantum control over the nuclear spin bath and therefore improved schemes to coherently control electron spin qubits, by reducing the Overhauser field fluctuations and/or exploiting the gradient for electron spin manipulation (as demonstrated experimentally already for example in Ref. [22]). On the other hand, with nuclear spin coherence times ranging from hundreds of microseconds to a millisecond [4, 26], our work could be extended towards nuclear spin-based information storage and manipulation protocols. The nuclear spin ensembles could serve as a long-lived entanglement resource providing the basic building block for an on-chip (solid-state) quantum network. The nodes of this quantum network could be interconnected with electrons playing the role of photons in more conventional atomic, molecular, and optical (AMO) based approaches [94]. To wire up the system, coherent transport of electron spins over *long* distances (potentially tens of microns in state-of-the-art experimental setups) could be realized via QD arrays [95, 96], quantum Hall edge channels [97–101] or surface acoustic waves [102–105]. Building upon this analogy to quantum optics, the localized nuclei might also be used as a source to generate a current of *many* entangled electrons [106]. Using the aforementioned tunability of the electronic degrees of freedom, one could also engineer different electronic quasisteady states, possibly resulting in nuclear stationary states with on-demand properties. On a more fundamental level, our work could also be extended towards deeper studies of dissipative phase transitions in this rather generic transport setting. When combined with driving—realized via, for example, a magnetic field B_x perpendicular to the polarization direction—a variety of strong-correlation effects, nonequilibrium, and dissipative phase transitions can be expected [70, 107, 108] and could now be studied in a mesoscopic solid-state system, complementing other approaches to dissipative phase transitions in quantum dots [109–112].

3.11   Appendix to Chap. 3

3.11.1   Spin-Blockade Regime

In this Appendix, for completeness we explicitly derive inequalities involving the chemical potentials $\mu_{L(R)}$ of the left and right lead, respectively, as well as the Coulomb energies introduced in Eq. (3.7) that need to be satisfied in order to tune the DQD into the desired Pauli-blockade regime in which at maximum two electrons reside on the DQD. For simplicity, Zeeman splittings are neglected for the moment as they typically constitute a much smaller energy scale compared to the Coulomb energies. Still, an extension to include them is straight-forward. Then, the bare energies $E_{(m,n)}$ for a state with (m, n) charge configuration can easily be read off from the Anderson Hamiltonian H_S. In particular, we obtain

$$E_{(1,1)} = \varepsilon_L + \varepsilon_R + U_{LR}, \tag{3.94}$$

$$E_{(2,1)} = 2\varepsilon_L + \varepsilon_R + U_L + 2U_{LR}, \tag{3.95}$$

$$E_{(1,2)} = \varepsilon_L + 2\varepsilon_R + U_R + 2U_{LR}, \tag{3.96}$$

$$E_{(0,2)} = 2\varepsilon_R + U_R, \tag{3.97}$$

$$E_{(2,0)} = 2\varepsilon_L + U_L. \tag{3.98}$$

In order to exclude the occupation of $(2, 1)$ and $(1, 2)$ states if the DQD is in a $(1, 1)$ charge configuration the left chemical potential must fulfill the inequality $\mu_L < E_{(2,1)} - E_{(1,1)} = \varepsilon_L + U_L + U_{LR}$. An analog condition needs to be satisfied for the right chemical potential μ_R so that we can write in total

$$\mu_i < \varepsilon_i + U_i + U_{LR}. \tag{3.99}$$

The same requirement should hold if the DQD is in a $(0, 2)$ or $(2, 0)$ charge configuration which leads to

$$\mu_i < \varepsilon_i + 2U_{LR}. \tag{3.100}$$

At the same time, the chemical potentials μ_i are tuned sufficiently high so that an electron is added to the DQD from the leads whenever only a single electron resides in the DQD. For example, this results in $\mu_L > E_{(1,1)} - \varepsilon_R = \varepsilon_L + U_{LR}$. An analog condition needs to hold for the right lead which gives

$$\mu_i > \varepsilon_i + U_{LR}. \tag{3.101}$$

In particular this inequality guarantees that the right dot is always occupied, since $\mu_R > \varepsilon_R$. Moreover, localized singlet states cannot populated directly if $\mu_i < \varepsilon_i + U_i$ holds. Since $U_{LR} < U_i$, the conditions to realize the desired two-electron regime can be summarized as

$$\varepsilon_i + U_{LR} < \mu_i < \varepsilon_i + 2U_{LR}. \tag{3.102}$$

By applying a large bias that approximately compensates the charging energy of the two electrons residing on the right dot, that is $\varepsilon_L \approx \varepsilon_R + U_R - U_{LR}$, the occupation of a localized singlet with charge configuration $(2, 0)$ can typically be neglected [111, 113]. In this regime, only states with the charge configurations $(0, 1)$, $(1, 0)$, $(1, 1)$ and $(0, 2)$ are relevant. Also, due to the large bias, admixing within the one-electron manifold is strongly suppressed—for typical parameters we estimate $t / (\varepsilon_L - \varepsilon_R) \approx 10^{-2}$—such that the relevant single electron states that participate in the transport cycle in the spin-blockade regime are the two lowest ones $|0, \sigma\rangle = d_{R\sigma}^\dagger |0\rangle$ with $(0, 1)$ charge configuration [62].

3.11.2   Quantum Master Equation in Spin-Blockade Regime

Following the essential steps presented in Ref. [46], we now derive an effective master equation for the DQD system which experiences irreversible dynamics via the electron's coupling to the reservoirs in the leads. We start out from the von Neumann equation for the global density matrix given in Eq. (3.18). It turns out to be convenient to decompose $\mathcal{H}$ as

$$\mathcal{H} = H_0 + H_1 + H_T, \tag{3.103}$$

with $H_0 = H_S + H_B$ and $H_1 = V_{HF} + H_t$. We define the superoperator P as

$$P \varrho = \text{Tr}_B [\varrho] \otimes \rho_B^0. \tag{3.104}$$

It acts on the total system's density matrix ϱ and projects the environment onto their respective thermal equilibrium states, labeled as ρ_B^0. The map P satisfies $P^2 = P$ and is therefore called a projector. By deriving a closed equation for the projection $P \varrho$ and tracing out the unobserved reservoir degrees of freedom, we arrive at the Nakajima-Zwanzig master equation for the system's density matrix

$$\dot{\rho} = [\mathcal{L}_S + \mathcal{L}_1] \rho \tag{3.105}$$
$$+ \int_0^t d\tau \text{Tr}_B \left[\mathcal{L}_T e^{(\mathcal{L}_0 + \mathcal{L}_T + \mathcal{L}_1)\tau} \mathcal{L}_T \rho (t - \tau) \otimes \rho_B^0 \right].$$

where the Liouville superoperators are defined as usual via $\mathcal{L}_\alpha \cdot = -i [H_\alpha, \cdot]$. Next, we introduce two approximations: First, in the weak coupling limit, we neglect all orders higher than two in $\mathcal{L}_T$. This is well known as the Born approximation. Accordingly, we neglect $\mathcal{L}_T$ in the exponential of the integrand. Second, we apply the approximation of independent rates of variations [60] which can be justified self-consistently, if the bath correlation time τ_c is short compared to the typical timescales associated with the system's internal interactions, that is $g_{hf} \tau_c \ll 1$ and $t \tau_c \ll 1$, and if H_1 can be treated as a perturbation with respect to H_0. In our system, the latter is justified as H_0 incorporates the large Coulomb energy scales which energetically separate the manifold with two electrons on the DQD from the lower manifold with only one electron residing in the DQD, whereas H_1 induces couplings within these manifolds only. In this limit, the master equation then reduces to

$$\dot{\rho} = [\mathcal{L}_S + \mathcal{L}_1]\rho \tag{3.106}$$
$$+ \int_0^t d\tau \mathrm{Tr}_B \left[\mathcal{L}_T e^{\mathcal{L}_0 \tau} \mathcal{L}_T \rho(t-\tau) \otimes \rho_B^0 \right].$$

In the next step, we write out the tunnel Hamiltonian H_T in terms of the relevant spin-eigenstates. Here, we single out one term explicitly, but all others follow along the lines. We get

$$\dot{\rho} = \cdots + \sum_{\sigma} \int_0^t d\tau \mathcal{C}(\tau) |0, \sigma\rangle \langle S_{02}| \tag{3.107}$$
$$\left[e^{-iH_0\tau} \rho(t-\tau) e^{iH_0\tau} \right] |S_{02}\rangle \langle 0, \sigma|,$$

where

$$\mathcal{C}(\tau) = \int_0^\infty d\varepsilon J(\varepsilon) e^{i(\Delta E - \varepsilon)\tau}, \tag{3.108}$$

and $J(\varepsilon) = |T_R|^2 n_R(\varepsilon) [1 - f_R(\varepsilon)]$ is the spectral density of the right lead, with $n_R(\varepsilon)$ being the density of states per spin of the right lead; $f_\alpha(\varepsilon)$ denotes the Fermi function of lead $\alpha = L, R$ and ΔE is the energy splitting between the two levels involved, i.e., for the term explicitly shown above $\Delta E = \varepsilon_R + U_R$. The correlation time of the bath τ_c is determined by the decay of the memory-kernel $\mathcal{C}(\tau)$. The Markov approximation is valid if the spectral density $J(\varepsilon)$ is flat on the scale of all the effects that we have neglected in the previous steps. Typically, the effective density of states $D(\varepsilon) = |T_R|^2 n_R(\varepsilon)$ is weakly energy dependent so that this argument is mainly concerned with the Fermi functions of the left (right) lead $f_{L(R)}(\varepsilon)$, respectively. Therefore, if $f_i(\varepsilon)$ is flat on the scale of $\sim t$, $\sim g_{\mathrm{hf}}$ and the dissipative decay rates $\sim \Gamma$, it can be evaluated at ΔE and a Markovian treatment is valid [46]. In summary, this results in

$$\dot{\rho} = \cdots + \Gamma_R \sum_{\sigma} \mathcal{D}[|0, \sigma\rangle \langle S_{02}|]\rho, \tag{3.109}$$

where Γ_R is the typical sequential tunneling rate $\Gamma_R = 2\pi |T_R|^2 n_R(\Delta E)$ $[1 - f_R(\Delta E)]$ describing direct hopping at leading order in the dot-lead coupling [46, 64].

Pauli blockade.—The derivation above allows for a clear understanding of the Pauli-spin blockade in which only the level $|S_{02}\rangle$ can decay into the right lead whereas all two electron states with $(1, 1)$ charge configuration are stable. If the $|S_{02}\rangle$ level decays, an energy of $\Delta E_2 = E_{(0,2)} - \varepsilon_R = \varepsilon_R + U_R$ is released on the DQD which has to be absorbed by the right reservoir due to energy conservation arguments. On the contrary, if one of the $(1, 1)$ levels were to decay to the right lead, an energy of $\Delta E_1 = E_{(1,1)} - \varepsilon_L = \varepsilon_R + U_{LR}$ would dissipate into the continuum. Therefore, the DQD is operated in the Pauli blockade regime if $f_R(\Delta E_2) = 0$ and $f_R(\Delta E_1) = 1$ is satisfied. Experimentally, this can be realized easily as ΔE_2 scales with the on-site Coulomb energy $\Delta E_2 \sim U_R$, whereas ΔE_1 scales only with the interdot Coulomb energy $\Delta E_1 \sim U_{LR}$.

Taking into account all relevant dissipative processes within the Pauli-blockade regime and assuming the Fermi function of the left lead $f_L(\varepsilon)$ to be sufficiently flat, the full quantum master equation for the DQD reads

$$\dot{\rho} = -i [H_S + H_1, \rho] + \Gamma_R \sum_\sigma \mathcal{D} [|0, \sigma\rangle \langle S_{02}|] \rho$$
$$+ \Gamma_L \left\{ \mathcal{D} [|T_+\rangle \langle 0, \uparrow\uparrow|] \rho + \mathcal{D} [|\downarrow\uparrow\rangle \langle 0, \uparrow\uparrow|] \rho \right\}$$
$$+ \Gamma_L \left\{ \mathcal{D} [|T_-\rangle \langle 0, \downarrow\downarrow|] \rho + \mathcal{D} [|\uparrow\downarrow\rangle \langle 0, \downarrow\downarrow|] \rho \right\}, \tag{3.110}$$

where the rate $\Gamma_R \sim [1 - f_R(\Delta E_2)]$ describes the decay of the localized singlet $|S_{02}\rangle$ into the right lead, while the second and third line represent subsequent recharging of the DQD with the corresponding rate $\Gamma_L \propto |T_L|^2$.[5]

We can obtain a simplified description for the regime in which on relevant timescales the DQD is always populated by two electrons. This holds for sufficiently strong recharging of the DQD which can be implemented experimentally by making the left tunnel barrier T_L more transparent than the right one T_R [46, 61, 62]. In this limit, we can eliminate the intermediate stage in the sequential tunneling process $(0, 2) \rightarrow (0, 1) \rightarrow (1, 1)$ and parametrize $H_S + H_1$ in the two-electron regime as $H_{el} + H_{ff} + H_{zz}$. Then, we arrive at the effective master equation

$$\dot{\rho} = -i [H_{el}, \rho] + \mathcal{K}_\Gamma \rho + \mathcal{V}\rho, \tag{3.111}$$

where the dissipator

$$\mathcal{K}_\Gamma \rho = \Gamma \sum_{x \in (1,1)} \mathcal{D} [|x\rangle \langle S_{02}|] \rho \tag{3.112}$$

models electron transport through the DQD; the sum runs over all four electronic bare levels with $(1, 1)$ charge configuration, i.e., $|\sigma, \sigma'\rangle$ for $\sigma, \sigma' = \uparrow, \downarrow$: Thus, in the limit of interest, the $(1, 1)$ charge states are reloaded with an effective rate $\Gamma = \Gamma_R/2$ via the decay of the localized singlet $|S_{02}\rangle$ [61, 62].

Transport dissipator in eigenbasis of H_{el}.—The electronic transport dissipator $\mathcal{K}_\Gamma$ as stated in Eq. (3.112) describes electron transport in the bare basis of the two-orbital Anderson Hamiltonian which does not correspond to the eigenbasis of H_{el} due

---

[5]In deriving Eq. (3.110), we have ignored level shifts arising from the coupling to the environment; as usual, they can be absorbed into renormalized energy levels. Moreover, we have applied the so-called secular approximation which is mathematically correct in the weak coupling limit and ensures the positivity of the dynamics [114].

to the presence of the interdot tunnel coupling H_t; in deriving Eq. (3.112) admixing due to H_t has been neglected based on the approximation of independent rates of variation [60]. It is valid if $t\tau_c \ll 1$ where $\tau_c \approx 10^{-15}$ s specifies the bath correlation time [46]. Performing a basis transformation $\tilde{\rho} = V^\dagger \rho V$ which diagonalizes the electronic Hamiltonian $\tilde{H}_{el} = V^\dagger H_{el} V = \text{diag}(\omega_0, -\omega_0, \varepsilon_1, \varepsilon_2, \varepsilon_3)$ and neglecting terms rotating at a frequency of $\varepsilon_l - \varepsilon_k$ for $k \neq l$, the electronic transport dissipator takes on the form[6]

$$\mathcal{K}_\Gamma \tilde{\rho} = \sum_{k,\nu=\pm} \Gamma_k \mathcal{D}\left[|T_\nu\rangle\langle\lambda_k|\right]\tilde{\rho} \qquad (3.113)$$
$$+ \sum_{k,j} \Gamma_{k\to j} \mathcal{D}\left[|\lambda_j\rangle\langle\lambda_k|\right]\tilde{\rho},$$

where $\Gamma_k = \kappa_k^2 \Gamma$ and $\Gamma_{k\to j} = \Gamma_k[1 - |\kappa_j|^2]$. Since only $(1, 1)$ states can be refilled, the rate at which the level $|\lambda_j\rangle$ is populated is proportional to $\sim [1 - |\kappa_j|^2]$; compare Ref. [61]. While the first line in Eq. (3.113) models the decay from the dressed energy eigenstates $|\lambda_k\rangle$ back to the Pauli-blocked triplet subspace $|T_\nu\rangle$ $(\nu = \pm)$ with an effective rate according to their overlap with the localized singlet, the second line refers to decay and dephasing processes acting entirely within the 'fast' subspace spanned by $\{|\lambda_k\rangle\}$. Intuitively, they should not affect the nuclear dynamics that take place on a much longer timescale. This intuitive picture is corroborated by exact diagonalization results: Leaving the HF interaction $\mathcal{V}$ aside for the moment, we compare the dynamics $\dot{\rho} = \mathcal{K}_0 \rho$ generated by the full electronic Liouvillian

$$\mathcal{K}_0 \rho = -i\,[H_{el}, \rho] + \mathcal{K}_\Gamma \rho \qquad (3.114)$$
$$+ \mathcal{K}_\pm \rho + \mathcal{L}_{deph}\rho,$$
$$\mathcal{K}_\pm \rho = \Gamma_\pm \sum_{\nu=\pm} \mathcal{D}\left[|T_{\bar{\nu}}\rangle\langle T_\nu|\right]\rho \qquad (3.115)$$
$$+ \Gamma_\pm \sum_{\nu=\pm} \left[\mathcal{D}\left[|T_\nu\rangle\langle T_0|\right]\rho + \mathcal{D}\left[|T_0\rangle\langle T_\nu|\right]\rho\right]$$

formulated in terms of the five undressed, bare levels $\left\{|\sigma, \sigma'\rangle, |S_{02}\rangle\right\}$ to the following Liouvillian

$$\mathcal{L}_0 \tilde{\rho} = -i\left[\tilde{H}_{el}, \tilde{\rho}\right] + \mathcal{L}_\Gamma \tilde{\rho}$$
$$+ \mathcal{L}_\pm \tilde{\rho} + \mathcal{L}_{deph}\tilde{\rho}, \qquad (3.116)$$

---

[6]Here, the tilde symbol explicitly refers to the dressed electronic basis $\{|T_\nu\rangle, |\lambda_k\rangle\}$. For notational convenience, it is dropped in the main text.

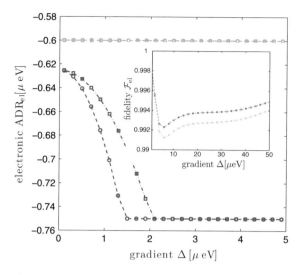

Fig. 3.17 Electronic asymptotic decay rate ADR$_{el}$ and fidelity $\mathcal{F}_{el}$ for the purely electronic Lindblad dynamics: The results obtained for the full dissipator given in Eq. (3.114) (*circles*) are in good agreement with the results we get for the simplified description as stated in Eq. (3.116) (*squares*). The *blue* and *red curves* correspond to $\Gamma = 25\,\mu eV$, $\Gamma_{\pm} = 0.25\,\mu eV$, $\Gamma_{deph} = 0.5\,\mu eV$ and $\Gamma = 25\,\mu eV$, $\Gamma_{\pm} = 0.3\,\mu eV$, $\Gamma_{deph} = 0$, respectively. *Inset* The fidelity $\mathcal{F}_{el}$ as a figure of merit for the similarity between the quasi-steady-state solutions ρ_{ss}^{el} and $\tilde{\rho}_{ss}^{el}$, respectively. Other numerical parameters are: $t = 20\,\mu eV$, $\varepsilon = 30\,\mu eV$ and $\omega_0 = 0$

which is based on the simplified form as stated in Eq. (3.113).[7] Here, we have also disregarded all dissipative processes acting entirely within the fast subspace, that is all terms of the form $\mathcal{D}\left[\left|\lambda_j\right\rangle\left\langle\lambda_k\right|\right]$; see the second line in Eq. (3.113). First, as shown in Fig. 3.17, we have checked numerically that both $\mathcal{K}_0$ and $\mathcal{L}_0$ feature very similar electronic quasisteady states, fulfilling $\mathcal{K}_0\left[\rho_{ss}^{el}\right] = 0$ and $\mathcal{L}_0\left[\tilde{\rho}_{ss}^{el}\right] = 0$, respectively, with a Uhlmann fidelity [115] $\mathcal{F}_{el}\left(\rho_{ss}^{el}, \tilde{\rho}_{ss}^{el}\right) = \left\|\sqrt{\rho_{ss}^{el}}\sqrt{\tilde{\rho}_{ss}^{el}}\right\|_{tr}$ exceeding 99 %; here, $\|\cdot\|_{tr}$ is the trace norm, the sum of the singular values. Second, we examine the electronic asymptotic decay rate ADR$_{el}$, corresponding to the eigenvalue with the largest real part different from zero, which quantifies the typical timescale on which the electronic subsystem reaches its quasi-steady state [70]. In other words, the ADR$_{el}$ gives the spectral gap of the electronic Liouvillian $\mathcal{K}_0$ ($\mathcal{L}_0$) setting the inverse relaxation time towards the steady state and therefore characterizes the long-time behaviour of the electronic system. The two models produce very similar results: Depending on the

---

[7]For simplicity, in the definition of $\mathcal{K}_\pm$ we have included dissipative cotunneling-mediated transitions in the bare triplet subspace only $\{|T_\pm\rangle, |T_0\rangle\}$. To make the comparison with the Liouvillian in the dressed basis $\mathcal{L}_0\tilde{\rho}$, the corresponding mixing terms $\mathcal{D}\left[|T_\nu\rangle\langle\lambda_k|\right]$ and $\mathcal{D}\left[|\lambda_k\rangle\langle T_\nu|\right]$ in $\mathcal{L}_\pm\tilde{\rho}$ appear with a rate $\Gamma_\pm|\langle\lambda_k|T_0\rangle|^2$. In particular, in the low-gradient regime where $|\lambda_2\rangle \approx |T_0\rangle$ this captures well the dissipative mixing between $|T_\pm\rangle$ and $|\lambda_2\rangle$ which is the most adverse process to our scheme. We have also verified that simply replacing $\Gamma_\pm|\langle\lambda_k|T_0\rangle|^2 \rightarrow \Gamma_\pm$ as it is stated in Eq. (3.30) does not change $\mathcal{F}_{el}$ nor ADR$_{el}$ severely.

particular choice of parameters, the electronic ADR_{el} is set either by the eigenvectors $|\lambda_2\rangle \langle T_\pm|$, $|T_+\rangle \langle T_-|$ and $|T_+\rangle \langle T_+| - |T_-\rangle \langle T_-|$ which explains the kinks observed in Fig. 3.17 as changes of the eigenvectors determining the ADR_{el}. In summary, both the electronic quasisteady state $\left(\rho_{ss}^{el} \approx \tilde{\rho}_{ss}^{el}\right)$ and the electronic asymptotic decay rate ADR_{el} are well captured by the approximative Liouvillian given in Eq. (3.116). Further arguments justifying this approximation are provided in Appendix 3.11.3.

3.11.3 Transport-Mediated Transitions in Fast Electronic Subspace

In this Appendix, we provide analytical arguments why one can drop the second line in Eq. (3.113) and keep only the first one to account for a description of electron transport in the eigenbasis of H_{el}. The second line, given by

$$\mathcal{L}_{fast}\rho = \sum_{k,j} \Gamma_{k \to j} \mathcal{D}\left[|\lambda_j\rangle \langle \lambda_k|\right]\rho, \qquad (3.117)$$

describes transport-mediated transitions in the fast subspace $\{|\lambda_k\rangle\}$. The transition rate $\Gamma_{k \to j} = \kappa_k^2 \left[1 - \kappa_j^2\right]\Gamma$ refers to a transport-mediated decay process from $|\lambda_k\rangle$ to $|\lambda_j\rangle$). Here, we show that $\mathcal{L}_{fast}$ simply amounts to an effective dephasing mechanism which can be absorbed into a redefinition of the effective transport rate Γ.

The only way our model is affected by $\mathcal{L}_{fast}$ is that it adds another dephasing channel for the coherences $|\lambda_k\rangle \langle T_\pm|$ which are created by the hyperfine flip-flop dynamics; see Appendix 3.11.9. In fact, we have

$$\mathcal{L}_{fast}\left[|\lambda_k\rangle \langle T_\pm|\right] = -\Gamma_{fast,k} |\lambda_k\rangle \langle T_\pm|, \qquad (3.118)$$

$$\Gamma_{fast,k} = \frac{1}{2}\sum_j \Gamma_{k \to j}. \qquad (3.119)$$

Due to the normalization condition $\sum_j \kappa_j^2 = 1$, the new effective dephasing rate $\Gamma_{fast,k}$ is readily found to coincide with the effective transport rate Γ_k, that is $\Gamma_{fast,k} = \Gamma_k = \kappa_k^2\Gamma$. This equality is readily understood since all four $(1, 1)$ levels are populated equally. While Γ_k describes the decay to the two Pauli-blocked triplet levels, $\Gamma_{fast,k}$ accounts for the remaining transitions within the $(1, 1)$ sector. Therefore, when accounting for $\mathcal{L}_{fast}$, the total effective dephasing rates $\tilde{\Gamma}_k$ needs to be modified as $\tilde{\Gamma}_k \to \tilde{\Gamma}_k + \Gamma_k = 2\Gamma_k + 3\Gamma_\pm + \Gamma_{deph}/4$. The factor of 2 is readily absorbed into our model by a simple redefinition of the overall transport rate $\Gamma \to 2\Gamma$.

3.11.4 Electronic Lifting of Pauli-Blockade

This Appendix provides a detailed analysis of purely electronic mechanisms which can lift the Pauli-blockade without affecting directly the nuclear spins. Apart from cotunneling processes discussed in the main text, here we analyze virtual spin exchange processes and spin-orbital effects [32, 47]. It is shown, that these mechanisms, though microscopically distinct, phenomenologically amount to effective incoherent mixing and pure dephasing processes within the $(1, 1)$ subspace which, for the sake of theoretical generality, are subsumed under the term $\textcircled{2}$ in Eq. (3.1).

Let us also note that electron spin resonance (ESR) techniques in combination with dephasing could be treated on a similar footing. As recently shown in Ref. [37], in the presence of a gradient Δ, ESR techniques can be used to drive the electronic system into the entangled steady state $|-\rangle = (|T_+\rangle - |T_-\rangle)/\sqrt{2}$. Magnetic noise may then be employed to engineer the desired electronic quasisteady state.

Spin Exchange with the Leads
In the Pauli-blockade regime the $(1, 1)$ triplet states $|T_\pm\rangle$ do not decay directly, but—apart from the cotunneling processes described in the main text—they may exchange electrons with the reservoirs in the leads via higher-order virtual processes [32, 47]. We now turn to these virtual, spin-exchange processes which can be analyzed along the lines of the interdot cotunneling effects. Again, for concreteness we fix the initial state of the DQD to be $|T_+\rangle$ and, based on the approximation of independent rates of variation [60], explain the physics in terms of the electronic bare states. The spin-blocked level $|T_+\rangle$ can virtually exchange an electron spin with the left lead yielding an incoherent coupling with the state $|\Downarrow\Uparrow\rangle$; this process is mediated by the intermediate singly occupied DQD level $|0, \Uparrow\rangle$ where no electron resides on the left dot. Then, from $|\Downarrow\Uparrow\rangle$ the system may decay back to the $(1, 1)$ subspace via the localized singlet $|S_{02}\rangle$. Therefore, for this analysis, in Fig. 3.4 we simply have to replace $|T_+(0, 2)\rangle$ and Γ_{ct} by $|0, \Uparrow\rangle$ and Γ_{se}, respectively. Along the lines of our previous analysis of cotunneling within the DQD, the bottleneck of the overall process is set by the first step, labeled as Γ_{se}. The main purpose of this Appendix is an estimate for the rate Γ_{se}.

The effective spin-exchange rate can be calculated in a "golden rule" approach in which transitions for different initial and final reservoir states are weighted according to the respective Fermi distribution functions and added incoherently [116]; for more details, see Refs. [117, 118]. Up to second order in H_T, the cotunneling rate Γ_{se} for the process $|T_+\rangle \rightsquigarrow |\Downarrow\Uparrow\rangle$ is then found to be

$$
\Gamma_{se} = 2\pi n_L^2 \, |T_L|^4 \int_{\mu_L}^{\mu_L+\Delta} d\varepsilon \frac{1}{(\varepsilon - \delta_+)^2}
$$

$$
\approx \frac{\Gamma_L^2}{2\pi} \frac{\Delta}{(\mu_L - \delta_+)^2}. \tag{3.120}
$$

Here, n_L is the left lead density of states at the Fermi energy, μ_L is the chemical potential of the left lead, $\Delta = E_{T_+} - E_{\Downarrow\Uparrow}$ is the energy released on the DQD (which gets absorbed by the reservoir) and $\delta_+ = E_{T_+} - E_{0\Uparrow} = \varepsilon_{L\uparrow} + U_{LR}$ refers to the energy difference between a doubly and singly occupied DQD in the intermediate virtual state. Moreover, Γ_L refers to the first-order sequential tunneling rates $\Gamma_L = 2\pi n_L |T_L|^2$ for the left (L) lead. Note that in the limit $T \to 0$ the DQD cannot be excited; accordingly, for $\Delta > 0$, the transition $|T_\pm\rangle \rightsquigarrow |\Uparrow\Downarrow\rangle$ is forbidden due to energy conservation [64]. As expected, Γ_{se} is proportional to $\sim |T_L|^4$, but suppressed by the energy penalty $\Delta_{se}^+ = \mu_L - \delta_+$ which characterizes the violation of the two-electron condition in Eq. (3.102) in the virtual intermediate step. Notably, this can easily be tuned electrostatically via the chemical potential μ_L. Comparing the parameter dependence $\Gamma_{se} \sim |T_L|^4$ to $\Gamma_{ct} \sim t^2 |T_L|^2$ shows that, in contrast to the cotunneling processes Γ_{ct}, Γ_{se} is independent of the interdot tunneling parameter t. Moreover, it can be made efficient by tuning properly the energy penalty Δ_{se}^+ and the tunnel coupling to the reservoir T_L. A similar analysis can be carried out for example for the effective decay process $|T_-\rangle \rightsquigarrow |\Downarrow\Uparrow\rangle$ by spin-exchange with the right reservoir. The corresponding rates are the same if $\Gamma_L/\Delta_{se}^+ = \Gamma_R/\Delta_{se}^-$, where $\Delta_{se}^- = \mu_R - (\varepsilon_{R\downarrow} + U_{LR})$, is satisfied. Taking the energy penalty as $\Delta_{se} \approx \Delta_{st}$, a comparison of Γ_{se} to interdot cotunneling transitions (as discussed in the main text) gives $\Gamma_{ct}/\Gamma_{se} \approx 2\pi t^2/(\Gamma\Delta)$. Thus, for $\Gamma \approx 2\pi t$ and $t \approx \Delta$ (as considered in this work), we get approximately $\Gamma_{ct} \approx \Gamma_{se}$.

The effective spin-exchange rate Γ_{se} can be made very efficient in the high gradient regime. For example, to obtain $\Gamma_{se} \approx 1\,\mu eV$ when $\Delta \approx 40\,\mu eV$, we estimate the required characteristic energy penalty to be $\Delta_{se} \approx 200\,\mu eV$. As stated in the main text, for an energy penalty of $\sim 500\,\mu eV$ and for $\Gamma_L \approx 100\,\mu eV$, we estimate $\Gamma_{se} \approx 0.25\,\mu eV$, making Γ_{se} fast compared to typical nuclear timescales; note that for less transparent barriers with $\Gamma_L \approx 1\,\mu eV$, Γ_{se} is four orders of magnitude smaller, in agreement with values given in Ref. [47]. Moreover, as apparent from Eq. (3.120), in the low gradient regime $\Gamma_{se} \sim \Delta$ is suppressed due to a vanishing phase space of reservoir electrons that can contribute to this process without violating energy conservation. To remedy this, one can lower the energy penalty Δ_{se}; however, if Δ_{se} becomes comparable to Γ, this leads to a violation of the Markov approximation and tunes the system away from the sequential tunneling regime. Note that the factor Δ appears in Eq. (3.120) as we consider explicitly the inelastic transition $|T_+\rangle \rightsquigarrow |\Downarrow\Uparrow\rangle$. In a more general analysis, Δ should be replaced by the energy separation ΔE (which is released by the DQD into the reservoir) for the particular transition at hand [64].

Here, we have considered spin-exchange via singly-occupied levels in the virtual intermediate stage only; they are detuned by the characteristic energy penalty $\delta = |\mu_i - (\varepsilon_i + U_{LR})|$ for $i = L, R$. In principle, spin exchange with the leads can also occur via electronic levels with $(1, 2)$ or $(2, 1)$ charge configuration. However, here the characteristic energy penalty can be estimated as $\delta = |\varepsilon_i + U_i + U_{LR} - \mu_i|$ which can be significantly bigger due to the appearance of the on-site Coulomb energies U_i in this expression. Therefore, they have been disregarded in the analysis above.

Spin Orbit Interaction

For the triplet states $|T_{\pm}\rangle$ interdot tunneling is suppressed due to Pauli spin blockade, but—apart from HF interaction with the nuclear spins—it can be mediated by spin-orbit interaction which does not conserve the electronic spin. In contrast to hyperfine mediated lifting of the spin blockade, spin-orbital effects provide another purely electronic alternative to escape the spin blockade, i.e., without affecting the nuclear spins. They describe interdot hopping accompanied by a spin rotation thereby coupling the triplet states $|T_{\pm}\rangle$ with single occupation of each dot to the singlet state $|S_{02}\rangle$ with double occupation of the right dot. Therefore, following Refs. [62, 111, 113, 119, 120], spin-orbital effects can be described phenomenologically in terms of the Hamiltonian

$$H_{so} = t_{so} \left(|T_+\rangle \langle S_{02}| + |T_-\rangle \langle S_{02}| + \text{h.c.} \right), \tag{3.121}$$

where the coupling parameter t_{so} in general depends on the orientation of the the DQD with respect to the crystallographic axes. Typical values of t_{so} can be estimated as $t_{so} \approx (d/l_{so})\, t$, where t is the usual spin-conserving tunnel coupling, d the interdot distance and l_{so} the material-specific spin-orbit length ($l_{so} \approx 1{-}10\,\mu\text{m}$ for GaAs); this estimate is in good agreement with the exact equation given in Ref. [113] and yields $t_{so} \approx (0.01{-}0.1)\, t$.

In Eq. (3.121) we have disregarded the spin-orbit coupling for the triplet $|T_0\rangle = \frac{1}{\sqrt{2}}(|\Uparrow\Downarrow\rangle + |\Downarrow\Uparrow\rangle)$. It may be taken into account by introducing the modified interdot tunneling Hamiltonian $H_t \rightarrow H_t'$ with $H_t' = t_{\uparrow\downarrow} |\Uparrow\Downarrow\rangle \langle S_{02}| - t_{\downarrow\uparrow} |\Downarrow\Uparrow\rangle \langle S_{02}| + \text{h.c.}$, where the tunneling parameters $t_{\uparrow\downarrow}$ and $t_{\downarrow\uparrow}$ are approximately given by $t_{\uparrow\downarrow(\downarrow\uparrow)} = t \pm t_{so}/\sqrt{2} \approx t$, since the second term marks only a small modification of the order of 5 %. While $|T_0\rangle$ is dark under tunneling in the singlet subspace, that is $H_t |T_0\rangle = 0$, similarly the slightly modified (unnormalized) state $|T_0'\rangle = t_{\downarrow\uparrow} |\Uparrow\Downarrow\rangle + t_{\uparrow\downarrow} |\Downarrow\Uparrow\rangle$ is dark under H_t'. Since this effect does not lead to any qualitative changes, it is disregarded.

Phenomenological treatment.—In the following, we first focus on the effects generated by H_{so} within the three-level subspace $\{|T_{\pm}\rangle, |S_{02}\rangle\}$. Within this reduced level scheme, the dynamics $\dot\rho = \mathcal{L}_{rd}\rho$ are governed by the Liouvillian

$$\mathcal{L}_{rd}\rho = -i\,[\mathcal{H}_{rd}, \rho] + \Gamma \sum_{\nu=\pm} \mathcal{D}\left[|T_\nu\rangle \langle S_{02}|\right]\rho \tag{3.122}$$

where the relevant Hamiltonian within this subspace is

$$\mathcal{H}_{rd} = \omega_0 \left(|T_+\rangle \langle T_+| - |T_-\rangle \langle T_-| \right) - \varepsilon\, |S_{02}\rangle \langle S_{02}| + H_{so}. \tag{3.123}$$

This situation is schematized in Fig. 3.18. The external Zeeman splitting ω_0 is assumed to be small compared to the interdot detuning ε yielding approximately equal detunings between the triplet states $|T_{\pm}\rangle$ and $|S_{02}\rangle$. In particular, we consider the regime $t_{so} \ll \varepsilon, \Gamma$, with the corresponding separation of timescales allow-

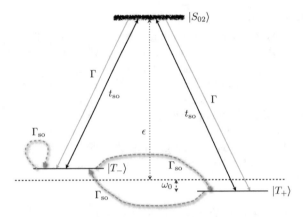

Fig. 3.18 Phenomenological treatment of spin-orbital effects in the spin-blockade regime. Scheme of the simplified electronic system: The triplet states $|T_\pm\rangle$ are coherently coupled to the local singlet $|S_{02}\rangle$ by spin-orbit interaction. Via coupling to the leads, the DQD is discharged and recharged again with an effective rate Γ. The triplet states may experience a Zeeman splitting ω_0. The parameter ε specifies the interdot energy offset. Since $\Gamma, \varepsilon \gg t_{so}$, the local singlet $|S_{02}\rangle$ can be eliminated adiabatically yielding effective dissipative processes of strength Γ_{so} (*green dashed arrows*)

ing for an alternative, effective description of spin-orbital effects. Since the short-lived singlet state $|S_{02}\rangle$ is populated negligibly throughout the dynamics, it can be eliminated adiabatically using standard techniques. The symmetric superposition $|-\rangle = (|T_+\rangle - |T_-\rangle)/\sqrt{2}$ is a dark state with respect to the spin-orbit Hamiltonian H_{so}. Therefore, it is instructive to formulate the resulting effective master equation in terms of the symmetric superposition states $|\pm\rangle = (|T_+\rangle \pm |T_-\rangle)/\sqrt{2}$. Within the two-dimensional subspace spanned by the symmetric superpositions $|\pm\rangle$, the effective dynamics is given by

$$\dot{\rho} = +i\omega_0\left[|-\rangle\langle+| + |+\rangle\langle-|, \rho\right] \tag{3.124}$$
$$-i\Omega_{so}\left[|+\rangle\langle+| - |-\rangle\langle-|, \rho\right]$$
$$+2\Gamma_{so}\mathcal{D}\left[|-\rangle\langle+|\right]\rho$$
$$+\frac{\Gamma_{so}}{2}\mathcal{D}\left[|+\rangle\langle+| - |-\rangle\langle-|\right]\rho,$$

where the effective rate

$$\Gamma_{so} = \frac{t_{so}^2}{\varepsilon^2 + \Gamma^2}\Gamma \tag{3.125}$$

governs decay as well as pure dephasing processes within the triplet subspace. We estimate $\Gamma_{so} \approx (0.2 - 0.3)\,\mu\text{eV}$ which is still fast compared to typical nuclear timescales. In Eq. (3.124) we have also introduced the quantity $\Omega_{so} = (\varepsilon/\Gamma)\Gamma_{so}$. As we are particularly concerned with the nuclear dynamics in the limit where one can eliminate the electronic degrees of freedom, Eq. (3.124) provides an alternative way of accounting for spin-orbit effects: In Eq. (3.124) we encounter a decay term—see the third line in Eq. (3.124)—which pumps the electronic subsystem towards the dark state of the spin-orbit Hamiltonian, namely the state $|-\rangle$. This state is also dark under the Stark shift and pure dephasing terms in the second and last line of Eq. (3.124), respectively. However, by applying an external magnetic field, the state $|-\rangle$ dephases due to the induced Zeeman splitting ω_0. This becomes apparent when examining the electronic quasisteady state corresponding to the evolution given in Eq. (3.124). In the basis $\{|T_+\rangle, |T_-\rangle\}$, it is found to be

$$\rho_{ss}^{el} = \begin{pmatrix} \frac{1}{2}\left[1 + \frac{\omega_0\Omega_{so}}{\omega_0^2+\Gamma_{so}^2+\Omega_{so}^2}\right] & -\frac{\Gamma_{so}^2+\Omega_{so}^2+i\Gamma_{so}\omega_0}{2\left(\omega_0^2+\Gamma_{so}^2+\Omega_{so}^2\right)} \\ -\frac{\Gamma_{so}^2+\Omega_{so}^2-i\Gamma_{so}\omega_0}{2\left(\omega_0^2+\Gamma_{so}^2+\Omega_{so}^2\right)} & \frac{1}{2}\left[1 - \frac{\omega_0\Omega_{so}}{\omega_0^2+\Gamma_{so}^2+\Omega_{so}^2}\right] \end{pmatrix}, \tag{3.126}$$

which in leading orders of ω_0^{-1} reduces to

$$\rho_{ss}^{el} \approx \begin{pmatrix} \frac{1}{2} + \frac{\Omega_{so}}{2\omega_0} & -i\frac{\Gamma_{so}}{2\omega_0} \\ i\frac{\Gamma_{so}}{2\omega_0} & \frac{1}{2} - \frac{\Omega_{so}}{2\omega_0} \end{pmatrix}. \tag{3.127}$$

Accordingly, for sufficiently large Zeeman splitting $\omega_0 \gg \Omega_{so}, \Gamma_{so}$, the electronic subsystem is driven towards the desired equal mixture of blocked triplet states $|T_+\rangle$ and $|T_-\rangle$. Alternatively, the off-diagonal elements of $|-\rangle\langle-|$ are damped out in the presence of dephasing processes either mediated intrinsically via cotunneling processes or extrinsically via engineered magnetic noise yielding approximately the equal mixture $\rho_{target}^{el} = (|T_+\rangle\langle T_+| + |T_-\rangle\langle T_-|)/2$ in the quasisteady state.

Numerical analysis.—To complement the perturbative, analytical study, we carry out a numerical evaluation of the electronic quasisteady state in the presence of spin-orbit coupling. In the two-electron subspace, the corresponding master equation (including spin-orbital effects) under consideration reads

$$\dot{\rho} = \tilde{\mathcal{K}}_0\rho = -i[H_{el} + H_{so}, \rho] + \mathcal{K}_\Gamma\rho + \mathcal{L}_{deph}\rho. \tag{3.128}$$

We evaluate the exact electronic quasisteady state ρ_{ss}^{el} fulfilling $\tilde{\mathcal{K}}_0\rho_{ss}^{el} = 0$. As a figure of merit, we compute the Uhlmann fidelity [115]

$$\mathcal{F}_{so} = \text{tr}\left[\left(\sqrt{\rho_{ss}^{el}}\,\rho_{target}^{el}\,\sqrt{\rho_{ss}^{el}}\right)^{1/2}\right]^2 \tag{3.129}$$

which measures how similar ρ_{ss}^{el} and ρ_{target}^{el} are. The results are illustrated in Fig. 3.19: For $\Gamma_{deph} = 0$ the electronic system settles into the pure dark state $|-\rangle\langle-|$. However, in the presence of dephasing, the coherences are efficiently damped out. In the low-gradient regime ρ_{ss}^{el} has a significant overlap with the triplet $|T_0\rangle$, whereas in the high-gradient regime it is indeed approximately given by the desired mixed target state

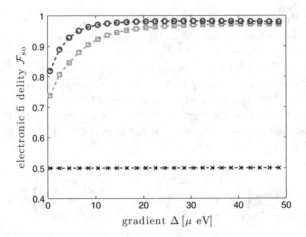

Fig. 3.19 Electronic quasi-steady-state fidelities in the presence of spin-orbit coupling for the dynamics generated by $\tilde{\mathcal{K}}_0$ as a function of the gradient Δ. As expected, in the absence of dephasing [$\Gamma_{\mathrm{deph}} = 0$ (*black curve*)], the system settles into the dark state $|-\rangle$. For $\Gamma_{\mathrm{deph}} = 1\,\mu\mathrm{eV}$ (*blue* and *red curves*), the off-diagonal elements of $|-\rangle$ are strongly suppressed, leading to a high fidelity $\mathcal{F}_{\mathrm{so}} \gtrsim 0.9$ with the desired mixed state $\rho^{\mathrm{el}}_{\mathrm{target}}$ in the high gradient regime: the *blue* and *red curve* refer to $t = 20\,\mu\mathrm{eV}$ and $t = 30\,\mu\mathrm{eV}$, respectively. Other numerical parameters are: $t_{\mathrm{so}} = 0.1t$, $\omega_0 = 0$, $\Gamma = 25\,\mu\mathrm{eV}$ and $\varepsilon = 30\,\mu\mathrm{eV}$

$\rho^{\mathrm{el}}_{\mathrm{target}}$. Lastly, we have checked that in the high-gradient regime the corresponding asymptotic decay rate can be approximated very well by $\mathrm{ADR}_{\mathrm{el}} \approx -2\Gamma_{\mathrm{so}}$.

3.11.5   Effective Nuclear Master Equation

In this Appendix, we present a detailed derivation of the effective nuclear dynamics presented in Sect. 3.5. We use standard adiabatic elimination techniques to derive an effective simplified description of the dynamics. To do so, we assume that electronic coherences decay quickly on typical nuclear timescales. Conservatively, i.e. not taking into account the detuning of the HF-mediated transitions, this holds for $2\Gamma_\pm + \Gamma_{\mathrm{deph}}/4 \gg g_{\mathrm{hf}}$, where g_{hf} quantifies the typical HF interaction strength. Alternatively, one may use a projection-operator based technique [46, 70]; this is done in detail in Appendix 3.11.9 for the high-gradient regime where $\rho^{\mathrm{el}}_{\mathrm{ss}} = (|T_+\rangle\langle T_+| + |T_-\rangle\langle T_-|)/2$, but a generalization for the electronic quasisteady state in Eq. (3.33) is straightforward.

Throughout this Appendix, for convenience we adopt the following notation: $|a\rangle = |T_+\rangle$, $|b\rangle = |\lambda_2\rangle$, $|c\rangle = |T_-\rangle$, $L = L_2$, $\mathbb{L} = \mathbb{L}_2$ and $\mathcal{D}[c]\rho = \mathcal{D}_c\rho$. Within this simplified three-level model system, the flip-flop Hamiltonian H_{ff} reads

$$H_{\mathrm{ff}} = \frac{a_{\mathrm{hf}}}{2}\left[L\,|b\rangle\langle a| + \mathbb{L}\,|b\rangle\langle c| + \mathrm{h.c.}\right]. \tag{3.130}$$

For simplicity, we assume $\omega_0 = 0$ and neglect nuclear fluctuations arising from H_{zz}. This approximation is in line with the semiclassical approximation for studying the nuclear polarization dynamics; for more details also see Appendix 3.11.6. Within this reduced scheme, the dynamics are then described by the Master equation

$$\dot{\rho} = -i\,[H_{\text{ff}}, \rho] - i\varepsilon_2\,[|b\rangle\,\langle b|\,, \rho] + \frac{\Gamma_{\text{deph}}}{2}\,\mathcal{D}_{|a\rangle\langle a|-|c\rangle\langle c|}\rho$$
$$+ \Gamma_{\pm}\left[\mathcal{D}_{|c\rangle\langle a|}\rho + \mathcal{D}_{|a\rangle\langle c|}\rho + \mathcal{D}_{|b\rangle\langle a|}\rho + \mathcal{D}_{|b\rangle\langle c|}\rho\right]$$
$$+ (\Gamma_{\pm} + \Gamma_2)\left[\mathcal{D}_{|a\rangle\langle b|}\rho + \mathcal{D}_{|c\rangle\langle b|}\rho\right]. \tag{3.131}$$

After adiabatic elimination of the electronic coherences $\rho_{ab} = \langle a|\rho|b\rangle$, ρ_{cb} and ρ_{ac} we obtain effective equations of motion for the system's density matrix projected onto the electronic levels $|a\rangle$, $|b\rangle$ and $|c\rangle$ as follows

$$\dot{\rho}_{aa} = \Gamma_{\pm}\,(\rho_{cc} - \rho_{aa}) + \Gamma_{\pm}\,(\rho_{bb} - \rho_{aa}) + \Gamma_2\rho_{bb} \tag{3.132}$$
$$+ \gamma\left[L^{\dagger}\rho_{bb}L - \frac{1}{2}\left\{L^{\dagger}L, \rho_{aa}\right\}\right]$$
$$+ i\delta\left[L^{\dagger}L, \rho_{aa}\right],$$
$$\dot{\rho}_{cc} = \Gamma_{\pm}\,(\rho_{aa} - \rho_{cc}) + \Gamma_{\pm}\,(\rho_{bb} - \rho_{cc}) + \Gamma_2\rho_{bb} \tag{3.133}$$
$$+ \gamma\left[\mathbb{L}^{\dagger}\rho_{bb}\mathbb{L} - \frac{1}{2}\left\{\mathbb{L}^{\dagger}\mathbb{L}, \rho_{cc}\right\}\right]$$
$$+ i\delta\left[\mathbb{L}^{\dagger}\mathbb{L}, \rho_{cc}\right],$$

and

$$\dot{\rho}_{bb} = -2\Gamma_2\rho_{bb} + \gamma\left[L\rho_{aa}L^{\dagger} - \frac{1}{2}\left\{LL^{\dagger}, \rho_{bb}\right\}\right] \tag{3.134}$$
$$- i\delta\left[LL^{\dagger}, \rho_{bb}\right]$$
$$+ \gamma\left[\mathbb{L}\rho_{cc}\mathbb{L}^{\dagger} - \frac{1}{2}\left\{\mathbb{L}\mathbb{L}^{\dagger}, \rho_{bb}\right\}\right] - i\delta\left[\mathbb{L}\mathbb{L}^{\dagger}, \rho_{bb}\right].$$
$$+ \Gamma_{\pm}\,(\rho_{aa} + \rho_{cc} - 2\rho_{bb}).$$

Since this set of equations is entirely expressed in terms of ρ_{aa}, ρ_{bb} and ρ_{cc}, the full density matrix of the system obeys a simple block structure, given by

$$\rho = \rho_{aa}\,|a\rangle\,\langle a| + \rho_{bb}\,|b\rangle\,\langle b| + \rho_{cc}\,|c\rangle\,\langle c|. \tag{3.135}$$

Therefore, the electronic decoherence is fast enough to prevent the entanglement between electronic and nuclear degrees of freedom and the total density matrix of the system ρ factorizes into a tensor product for the electronic and nuclear subsystem

[33], respectively, that is $\rho = \rho_{\mathrm{el}} \otimes \sigma$, where $\sigma = \mathsf{Tr}_{\mathrm{el}}[\rho]$ refers to the density matrix of the nuclear subsystem. This ansatz agrees with the projection operator approach where $\mathcal{P}\rho = \sigma \otimes \rho_{\mathrm{el}}$ and readily yields $\rho_{aa} = p_a\sigma$, where we have introduced the electronic populations

$$p_a = \langle a|\rho_{\mathrm{el}}|a\rangle = \mathsf{Tr}_{\mathrm{n}}[\rho_{aa}], \qquad (3.136)$$

and accordingly for p_b and p_c; here, $\mathsf{Tr}_{\mathrm{n}}[\dots]$ denotes the trace over the nuclear degrees of freedom. With these definitions, Eqs. (3.132), (3.133) and (3.134) can be rewritten as

$$
\begin{aligned}
\dot{p}_a &= \Gamma_\pm (p_c - p_a) + \Gamma_2 p_b + \gamma \left[p_b \langle LL^\dagger \rangle - p_a \langle L^\dagger L \rangle \right] \\
&\quad + \Gamma_\pm (p_b - p_a), \\
\dot{p}_c &= \Gamma_\pm (p_a - p_c) + \Gamma_2 p_b + \gamma \left[p_b \langle \mathbb{L}\mathbb{L}^\dagger \rangle - p_c \langle \mathbb{L}^\dagger \mathbb{L} \rangle \right] \\
&\quad + \Gamma_\pm (p_b - p_c), \\
\dot{p}_b &= -2\Gamma_2 p_b + \Gamma_\pm (p_a + p_c - 2p_b) \\
&\quad + \gamma \left[p_a \langle L^\dagger L \rangle - p_b \langle LL^\dagger \rangle + p_c \langle \mathbb{L}^\dagger \mathbb{L} \rangle - p_b \langle \mathbb{L}\mathbb{L}^\dagger \rangle \right].
\end{aligned} \qquad (3.137)
$$

Similarly, the effective Master equation for the nuclear density matrix $\sigma = \mathsf{Tr}_{\mathrm{el}}[\rho]$ is obtained from $\dot{\sigma} = \mathsf{Tr}_{\mathrm{el}}[\dot{\rho}] = \dot{\rho}_{aa} + \dot{\rho}_{bb} + \dot{\rho}_{cc}$, leading to

$$
\begin{aligned}
\dot{\sigma} &= \gamma \{ p_b \mathcal{D}_{L^\dagger}[\sigma] + p_b \mathcal{D}_{\mathbb{L}^\dagger}[\sigma] + p_a \mathcal{D}_L[\sigma] + p_c \mathcal{D}_{\mathbb{L}}[\sigma] \} \\
&\quad + i\delta \{ p_a \left[L^\dagger L, \sigma \right] + p_c \left[\mathbb{L}^\dagger \mathbb{L}, \sigma \right] \\
&\quad - p_b \left[LL^\dagger, \sigma \right] - p_b \left[\mathbb{L}\mathbb{L}^\dagger, \sigma \right] \}.
\end{aligned} \qquad (3.138)
$$

Equation (3.138) along with Eq. (3.137) describe the coupled electron-nuclear dynamics on a coarse-grained timescale that is long compared to electronic coherence timescales. Due to the normalization condition $p_a + p_b + p_c = 1$, this set of dynamical equations comprises three coupled equations. Differences in the populations of the levels $|a\rangle$ and $|c\rangle$ decay very quickly on timescales relevant for the nuclear evolution; that is,

$$
\begin{aligned}
\dot{p}_a - \dot{p}_c &= -3\Gamma_\pm (p_a - p_c) + \gamma \left[p_b \left(\langle LL^\dagger \rangle - \langle \mathbb{L}\mathbb{L}^\dagger \rangle \right) \right. \\
&\quad \left. - p_a \langle L^\dagger L \rangle + p_c \langle \mathbb{L}^\dagger \mathbb{L} \rangle \right]
\end{aligned} \qquad (3.139)
$$

Due to a separation of timescales, as $\Gamma_\pm \gg \gamma_c = N\gamma \approx 10^{-4}\,\mu\mathrm{eV}$, in a perturbative treatment the effect of the second term can be neglected and the electronic subsystem approximately settles into $p_a = p_c$. This leaves us with a single dynamical variable, namely p_a, entirely describing the electronic subsystem on relevant timescales. Thus, using $p_c = p_a$ and $p_b = 1 - 2p_a$, the electronic quasi steady state is uniquely defined by the parameter p_a and the nuclear evolution simplifies to

$$\dot{\sigma} = \gamma \{ p_a \left[\mathcal{D}_L \left[\sigma \right] + \mathcal{D}_{\mathbb{L}} \left[\sigma \right] \right] \tag{3.140}$$
$$+ (1 - 2 p_a) \left[\mathcal{D}_{L^\dagger} \left[\sigma \right] + \mathcal{D}_{\mathbb{L}^\dagger} \left[\sigma \right] \right] \}$$
$$+ i \delta \{ p_a \left(\left[L^\dagger L, \sigma \right] + \left[\mathbb{L}^\dagger \mathbb{L}, \sigma \right] \right)$$
$$- (1 - 2 p_a) \left(\left[L L^\dagger, \sigma \right] + \left[\mathbb{L} \mathbb{L}^\dagger, \sigma \right] \right) \},$$

with p_a obeying the dynamical equation

$$\dot{p}_a = \Gamma_\pm (1 - 3 p_a) + \Gamma_2 (1 - 2 p_a)$$
$$- \gamma \left[p_a \langle L^\dagger L \rangle + (1 - 2 p_a) \langle L L^\dagger \rangle \right].$$

Neglecting the HF terms in the second line, we recover the projection-operator-based result for the quasisteady state, $p_a \approx (\Gamma_\pm + \Gamma_2) / (3\Gamma_\pm + 2\Gamma_2)$ as stated in Eq. (3.34).

3.11.6 Effective Nuclear Dynamics: Overhauser Fluctuations

In Sect. 3.5 we have disregarded the effect of Overhauser fluctuations, described by $\dot{\rho} = -i [H_{zz}, \rho] = -i a_{hf} \sum_i [S_i^z \delta A_i^z, \rho]$. In the following analysis, this simplification is discussed in greater detail.

First of all, we note that this term cannot induce couplings within the effective electronic three level system, $\{ |T_\pm\rangle, |\lambda_2\rangle \}$, since $|T_\pm\rangle$ are eigenstates of S_i^z, that is explicitly $S_i^z |T_\pm\rangle = \pm \frac{1}{2} |T_\pm\rangle$, which leads to

$$\langle T_\pm | S_i^z | \lambda_2 \rangle = 0. \tag{3.141}$$

In other words, different S_{tot}^z subspaces are not coupled by the action of H_{zz}; this is in stark contrast to the flip flop dynamics H_{ff}.

When also accounting for Overhauser fluctuations, the dynamical equations for the coherences read

$$\dot{\rho}_{ab} = \left(i \varepsilon_2 - \tilde{\Gamma} \right) \rho_{ab} - i \left[L^\dagger \rho_{bb} - \rho_{aa} L^\dagger \right] \tag{3.142}$$
$$- i a_{hf} \sum_i \left[\langle S_i^z \rangle_a \delta A_i^z \rho_{ab} - \langle S_i^z \rangle_b \rho_{ab} \delta A_i^z \right],$$

where $\langle S_i^z \rangle_a = \langle a | S_i^z | a \rangle$; an analog equation holds for $\dot{\rho}_{cb}$. Typically, the second line is small compared to the fast electronic quantities ε_2, $\tilde{\Gamma}$ in the first line. Therefore, it will be neglected. In Eqs. (3.132), (3.133) and (3.134), the Overhauser fluctuations lead to the following additional terms

$$\dot{\rho}_{aa} = \cdots - \frac{i}{2} a_{hf} \sum_i \left[\delta A_i^z, \rho_{aa} \right], \qquad (3.143)$$

$$\dot{\rho}_{cc} = \cdots + \frac{i}{2} a_{hf} \sum_i \left[\delta A_i^z, \rho_{cc} \right], \qquad (3.144)$$

$$\dot{\rho}_{bb} = \cdots - i a_{hf} \sum_i \langle S_i^z \rangle_b \left[\delta A_i^z, \rho_{bb} \right]. \qquad (3.145)$$

First, this leaves the electronic populations $p_a = \text{Tr}_n [\rho_{aa}]$ untouched; H_{zz} does not induce any couplings between them. Second, the dynamical equation for the nuclear density matrix $\sigma = \text{Tr}_{el} [\rho]$ is modified as

$$\dot{\sigma} = \cdots - i a_{hf} \sum_i \left[\frac{1}{2} (p_a - p_c) + p_b \langle S_i^z \rangle_b \right] \left[\delta A_i^z, \sigma \right],$$

$$\approx \cdots - i (1 - 2p_a) a_{hf} \sum_i \langle S_i^z \rangle_b \left[\delta A_i^z, \sigma \right]. \qquad (3.146)$$

In the second step, we have used again that differences in p_a and p_c are quickly damped to zero with a rate of $3\Gamma_\pm$. Now, let us examine the effect of Eq. (3.146) for different important regimes: In the high gradient regime, where p_b is fully depleted, it does not give any contribution since the electronic quasi steady state does not have any magnetization $\left[\langle S_i^z \rangle_b = \langle S_i^z \rangle_{ss} = 0 \right]$ and $p_a = 1/2$. In the low gradient regime, $|b\rangle$ approaches the triplet $|T_0\rangle$ and again (since $\langle S_i^z \rangle_b = 0$) this term vanishes. Finally, the intermediate regime has been studied within a semiclassical approximation (see Sect. 3.6): Note that Eq. (3.146), however, leaves the dynamical equation for the nuclear polarizations I_i^z unchanged, since they commute with H_{zz}.

3.11.7 Ideal Nuclear Steady State

In this Appendix we analytically construct the ideal (pure) nuclear steady-state $|\xi_{ss}\rangle$, fulfilling $L_2 |\xi_{ss}\rangle = \mathbb{L}_2 |\xi_{ss}\rangle = 0$, in the limit of identical dots ($a_{1j} = a_{2j} \forall j = 1, \ldots, N_1 \equiv N_2 = N$) for uniform HF-coupling where $a_{ij} = N/N_i$. In this limit, the non-local nuclear jump operators simplify to

$$L_2 = \nu I_1^+ + \mu I_2^+, \qquad (3.147)$$

$$\mathbb{L}_2 = \mu I_1^- + \nu I_2^-. \qquad (3.148)$$

Here, to simplify the notation, we have replaced μ_2 and ν_2 by μ and ν, respectively. The common proportionality factor is irrelevant for this analysis and therefore has been dropped. The collective nuclear spin operators $I^{\alpha}_{1,2}$ form a spin algebra and the so-called Dicke states $|J_1, k_1\rangle \otimes |J_2, k_2\rangle \equiv |J_1, k_1; J_2, k_2\rangle$, where the total spin quantum numbers J_i are conserved and the spin projection quantum number $k_i = 0, 1, \ldots, 2J_i$, allow for an efficient description. Here, we restrict ourselves to the symmetric case where $J_1 = J_2 = J$; analytic and numerical evidence for small $J_i \approx 3$ shows, that for $J_1 \neq J_2$ one obtains a mixed nuclear steady state. The total spin quantum numbers $J_i = J$ are conserved and we set $|J, k_1; J, k_2\rangle = |k_1, k_2\rangle$. Using standard angular momentum relations, one obtains

$$L_2 |k_1, k_2\rangle = \nu j_{k_1} |k_1 + 1, k_2\rangle + \mu j_{k_2} |k_1, k_2 + 1\rangle, \tag{3.149}$$
$$\mathbb{L}_2 |k_1, k_2\rangle = \mu g_{k_1} |k_1 - 1, k_2\rangle + \nu g_{k_2} |k_1, k_2 - 1\rangle. \tag{3.150}$$

Here, we have introduced the matrix elements

$$j_k = \sqrt{J(J+1) - (k-J)(k-J+1)}, \tag{3.151}$$
$$g_k = \sqrt{J(J+1) - (k-J)(k-J-1)}. \tag{3.152}$$

Note that $j_{2J} = 0$ and $g_0 = 0$. Moreover, the matrix elements obey the symmetry

$$j_k = j_{2J-k-1}, \tag{3.153}$$
$$g_{k+1} = g_{2J-k}. \tag{3.154}$$

Now, we show that $|\xi_{ss}\rangle$ fulfills $L_2 |\xi_{ss}\rangle = \mathbb{L}_2 |\xi_{ss}\rangle = 0$. First, using the relations above, we have

$$
\begin{aligned}
L_2 |\xi_{ss}\rangle &= \sum_{k=0}^{2J} \xi^k \left[\nu j_k |k+1, 2J-k\rangle \right. \\
&\quad \left. + \mu j_{2J-k} |k, 2J-k+1\rangle \right] \\
&= \sum_{k=0}^{2J-1} \xi^k \left[\nu j_k |k+1, 2J-k\rangle \right. \\
&\quad \left. + \xi \mu j_{2J-k-1} |k+1, 2J-k\rangle \right] \\
&= \sum_{k=0}^{2J-1} \xi^k \nu \underbrace{\left[j_k - j_{2J-k-1} \right]}_{=0} |k+1, 2J-k\rangle.
\end{aligned}
$$

In the second step, since $j_{2J} = 0$, we have redefined the summation index as $k \to k + 1$. Along the same lines, one obtains

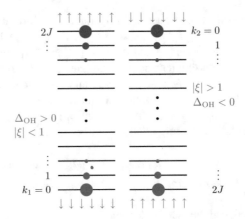

Fig. 3.20 Sketch of the ideal nuclear dark state for uniform HF coupling $|\xi_{ss}\rangle$. The Dicke states are labeled according to their spin projection $k_i = 0, 1, \ldots 2J$. Since $k_1 = k$ is strongly correlated with $k_2 = 2J - k$, the two Dicke ladders are arranged in opposite order. The bistability inherent to $|\xi_{ss}\rangle$ is schematized as well: The size of the spheres refers to $|\langle k_1, k_2 | \xi_{ss}\rangle|^2$ for $|\xi| < 1$ (*red*) and $|\xi| > 1$ (*blue*), respectively. As indicated by the arrows for individual nuclear spins, $|\xi| < 1$ ($|\xi| > 1$) corresponds to a nuclear OH gradient $\Delta_{OH} > 0$ ($\Delta_{OH} > 0$), respectively

$$
\mathbb{L}_2 |\xi_{ss}\rangle = \sum_{k=0}^{2J} \xi^k \left[\mu g_k |k-1, 2J-k\rangle \right.
$$
$$
\left. + \nu g_{2J-k} |k, 2J-k-1\rangle \right]
$$
$$
= \sum_{k=0}^{2J-1} \xi^k \left[\xi \mu g_{k+1} |k, 2J-k-1\rangle \right.
$$
$$
\left. + \nu g_{2J-k} |k, 2J-k-1\rangle \right]
$$
$$
= \sum_{k=0}^{2J-1} \xi^k \nu \underbrace{\left[g_{2J-k} - g_{k+1} \right]}_{=0} |k, 2J-k-1\rangle .
$$

This completes the proof. For illustration, the dark state $|\xi_{ss}\rangle$ is sketched in Fig. 3.20. In particular, the bistable polarization character inherent to $|\xi_{ss}\rangle$ is emphasized, as (in contrast to the bosonic case) the modulus of the parameter ξ is not confined to $|\xi| < 1$.

3.11.8   Numerical Results for DNP

In this Appendix the analytical findings of the semiclassical model are corroborated by exact numerical simulations for small sets of nuclear spins. This treatment complements our analytical DNP analysis in several aspects: First, we do not restrict ourselves to the effective three level system $\{|T_\pm\rangle, |\lambda_2\rangle\}$. Second, the electronic degrees of freedom are not eliminated adiabatically from the dynamics. Lastly, this approach does not involve the semiclassical decorrelation approximation stated in Eq. (3.46).

Technical details.—We consider the idealized case of homogeneous hyperfine coupling for which an exact numerical treatment is feasible even for a relatively large number of coupled nuclei as the system evolves within the totally symmetric low-dimensional subspace $\{|J, m\rangle, m = -J, \ldots, J\}$, referred to as *Dicke ladder*. We restrict ourselves to the fully symmetric subspace where $J_i = N_i/2 \approx 3$. Moreover, to mimic the separation of timescales in experiments where $N \approx 10^6$, the HF coupling is scaled down appropriately to the constant value $g_{hf} \approx 0.1\,\mu\text{eV}$; also compare the numerical results presented in Fig. 3.7.

Our first numerical approach is based on simulations of the time evolution. Starting out from nuclear states with different initial Overhauser gradient $\Delta_{OH}(t = 0)$, we make the following observations, depicted in Fig. 3.21: First of all, the tri-stability of the Overhauser gradient with respect to the initial nuclear polarization is confirmed. If the initial gradient $\Delta_{OH}(t = 0) + \Delta_{ext}$ exceeds a certain threshold value, the nuclear system runs into the highly-polarized steady state, otherwise it gets stuck in the trivial, zero-polarization solution. There are two symmetric high-polarization solutions that depend on the sign of $\Delta_{OH}(t = 0) + \Delta_{ext}$; also note that the Overhauser gradient Δ_{OH} may flip the sign as determined by the total initial gradient $\Delta_{OH}(t = 0) + \Delta_{ext}$. Second, in the absence of an external Zeeman splitting ω_{ext}, a potential initial homogeneous Overhauser polarizations $\bar{\omega}_{OH}$ is damped to zero in the steady state. For finite $\omega_{ext} \neq 0$, a homogeneous Overhauser polarization $\bar{\omega}_{OH}$ builds up which partially compensates ω_{ext}. Lastly, the high-polarization solutions $\Delta_{OH}^{ss} \approx 2\,\mu\text{eV}$ are far away from full polarization. This is an artifact of the small system sizes $J_i \approx 3$: As we deal with very short Dicke ladders, even the ideal, nuclear two-mode squeezedlike target state $|\xi\rangle_{ss}$ given in Eq. (3.8) does not feature a very high polarization. Pictorially, it leaks with a non-vanishing factor $\sim \xi^m$ into the low-polarization Dicke states. This argument is supported by the fact that (for the same set of parameters) we observe tendency towards higher polarization for an increasing number of nuclei N (which features a larger Dicke ladder) and confirmed by our second numerical approach to be discussed below.

Our second numerical approach is based on exact diagonalization: As we tune the parameter Δ, we compute the steady state for the full electronic-nuclear system directly giving the corresponding steady-state nuclear polarizations $\langle I_i^z \rangle_{ss}$. We see a clear instability towards the buildup of an Overhauser gradient Δ_{OH}^{ss} (Fig. 3.22): Inside the small-gradient region $(|\Delta| < |\Delta_{OH}^{crt}|)$ we observe negative feedback $\text{sgn}(\Delta_{OH}^{ss}) = -\text{sgn}(\Delta)$, whereas outside of it $(|\Delta| > |\Delta_{OH}^{crt}|)$ the nuclear system experiences positive feedback $\text{sgn}(\Delta_{OH}^{ss}) = \text{sgn}(\Delta)$. The latter leads to the build-up of large OH gradients, in agreement with our semiclassical analysis.

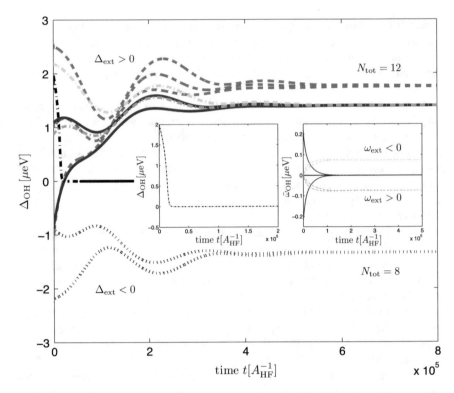

Fig. 3.21 Exact time evolution for $N = 8$ and $N = 12$ (*red dashed curves*) nuclear spins, four and six in each quantum dot, respectively. Depending on the initial value of the gradient, the nuclear system either runs into the trivial, unpolarized state or into the highly polarized one, if the initial gradient exceeds the critical value; the *blue dotted*, *black dash-dotted* and all other refer to $\Delta_{\text{ext}} = -5\,\mu\text{eV}$, $\Delta_{\text{ext}} = 0$ and $\Delta_{\text{ext}} = 5\,\mu\text{eV}$, respectively. For $\omega_{\text{ext}} \neq 0$, also a homogeneous OH field ω_{OH} builds up which partially compensates ω_{ext}: here, $\omega_{\text{ext}} = 0.1\,\mu\text{eV}$ (*magenta dash-dotted*) and $\omega_{\text{ext}} = -0.1\,\mu\text{eV}$ (*cyan dash-dotted*). Other numerical parameters: $t = 10\,\mu\text{eV}$, $\varepsilon = 30\,\mu\text{eV}$, $\Gamma = 25\,\mu\text{eV}$, $\Gamma_{\pm} = \Gamma_{\text{deph}} = 0.1\,\mu\text{eV}$

3.11.9 Effective Nuclear Master Equation in High-Gradient Regime

This Appendix provides background material for the derivation of the effective nuclear master equation as stated in Eq. (3.55) using projection-operator techniques [46, 70]. We start with

$$\text{Tr}_{\text{el}}\left[\mathcal{P}\mathcal{V}\mathcal{P}\rho\right] = \text{Tr}_{\text{el}}\left[\mathcal{P}\mathcal{L}_{\text{ff}}\mathcal{P}\rho\right] + \text{Tr}_{\text{el}}\left[\mathcal{P}\mathcal{L}_{\text{zz}}\mathcal{P}\rho\right] \tag{3.155}$$

The first term is readily found to be

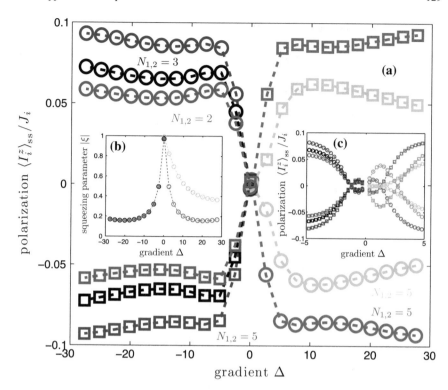

Fig. 3.22 Instability towards nuclear self-polarization: Exact numerical results for small system sizes $J_i = N_i/2$. The exact steady state of the coupled electron-nuclear dynamics is computed as a function of the gradient Δ. The *circles* (*squares*) refer to the polarization in the *left* (*right*) dot, respectively. **a** For $\Delta > \left|\Delta_{\text{OH}}^{\text{crt}}\right|$, we find $\Delta_{\text{OH}} > 0$, whereas for $\Delta < -\left|\Delta_{\text{OH}}^{\text{crt}}\right|$ we get $\Delta_{\text{OH}} < 0$, i.e., outside of the small-gradient regime [see *inset* (**c**)] the nuclear system is seen to be unstable towards the buildup of a OH gradient with opposite polarizations in the two dots. The nuclear polarization depends on the system size J_i and the parameter $|\xi|$; compare *inset* (**b**). **c** The critical value of $\Delta_{\text{OH}}^{\text{crt}} \approx 3\,\mu\text{eV}$ agrees with the semiclassical estimate; it becomes smaller for smaller values of $\Gamma_\pm$. Numerical parameters in μeV: $\varepsilon = 30$, $\Gamma = 10$, $\Gamma_\pm = \Gamma_{\text{deph}} = 0.3$, $\omega_{\text{ext}} = 0$ and $t = 10$ except for the *cyan curve* where $t = 20$ and $\Gamma_\pm = \Gamma_{\text{deph}} = 0.6$ for the *orange curve* in (**c**)

$$\text{Tr}_{\text{el}}\left[\mathcal{P}\mathcal{L}_{\text{ff}}\mathcal{P}\rho\right] = -i\frac{a_{\text{hf}}}{2}\sum_{i,\alpha=\pm}\left\langle S_i^\alpha\right\rangle_{\text{ss}}\left[A_i^{\bar{\alpha}},\sigma\right], \tag{3.156}$$

where $\left\langle\cdot\right\rangle_{\text{ss}} = \text{Tr}_{\text{el}}\left[\cdot\rho_{\text{ss}}^{\text{el}}\right]$ denotes the steady-state expectation value. An analog calculation yields

$$\text{Tr}_{\text{el}}\left[\mathcal{P}\mathcal{L}_{zz}\mathcal{P}\rho\right] = -ia_{\text{hf}}\sum_i\left\langle S_i^z\right\rangle_{\text{ss}}\left[\delta A_i^z,\sigma\right]. \tag{3.157}$$

Using that $\left\langle S_i^\alpha\right\rangle_{\text{ss}} = 0$ and $\left\langle S_i^z\right\rangle_{\text{ss}} = 0$ [the Knight shift seen by the nuclear spins is zero since the electronic quasi steady-state carries no net magnetization], the first two Hamiltonian terms vanish.

The second-order term of interest

$$\mathcal{K}\sigma = \mathrm{Tr}_{\mathrm{el}}\left[\mathcal{P}V\mathcal{Q}\left(-\mathcal{L}_0^{-1}\right)\mathcal{Q}V\mathcal{P}\rho\right] \tag{3.158}$$

can be decomposed as $\mathcal{K}\sigma = \mathcal{K}_{\mathrm{ff}}\sigma + \mathcal{K}_{\mathrm{zz}}\sigma$, where

$$\mathcal{K}_{\mathrm{ff}}\sigma = \mathrm{Tr}_{\mathrm{el}}\left[\mathcal{P}\mathcal{L}_{\mathrm{ff}}\mathcal{Q}\left(-\mathcal{L}_0^{-1}\right)\mathcal{Q}\mathcal{L}_{\mathrm{ff}}\mathcal{P}\rho\right], \tag{3.159}$$

$$\mathcal{K}_{\mathrm{zz}}\sigma = \mathrm{Tr}_{\mathrm{el}}\left[\mathcal{P}\mathcal{L}_{\mathrm{zz}}\mathcal{Q}\left(-\mathcal{L}_0^{-1}\right)\mathcal{Q}\mathcal{L}_{\mathrm{zz}}\mathcal{P}\rho\right]. \tag{3.160}$$

All other second order terms containing combinations of the superoperators $\mathcal{L}_{\mathrm{ff}}$ and $\mathcal{L}_{\mathrm{zz}}$ can be shown to vanish. In the following, we will evaluate the two terms separately.

Hyperfine flip-flop dynamics.—First, we will evaluate $\mathcal{K}_{\mathrm{ff}}$ which can be rewritten as

$$\mathcal{K}_{\mathrm{ff}}\sigma = \underbrace{\int_0^\infty d\tau\,\mathrm{Tr}_{\mathrm{el}}\left[\mathcal{P}\mathcal{L}_{\mathrm{ff}}e^{\mathcal{L}_0\tau}\mathcal{L}_{\mathrm{ff}}\mathcal{P}\rho\right]}_{\text{(a)}}$$

$$\underbrace{-\int_0^\infty d\tau\,\mathrm{Tr}_{\mathrm{el}}\left[\mathcal{P}\mathcal{L}_{\mathrm{ff}}\mathcal{P}\mathcal{L}_{\mathrm{ff}}\mathcal{P}\rho\right].}_{\text{(b)}} \tag{3.161}$$

Here, we used the Laplace transform $-\mathcal{L}_0^{-1} = \int_0^\infty d\tau e^{\mathcal{L}_0\tau}$ and the property $e^{\mathcal{L}_0\tau}\mathcal{P} = \mathcal{P}e^{\mathcal{L}_0\tau} = \mathcal{P}$ [70]. The first term labeled as (a) is given by

$$\text{(a)} = -\int_0^\infty d\tau\,\mathrm{Tr}_{\mathrm{el}}\left(\left[H_{\mathrm{ff}}, e^{\mathcal{L}_0\tau}\left[H_{\mathrm{ff}}, \sigma \otimes \rho_{\mathrm{ss}}^{\mathrm{el}}\right]\right]\right). \tag{3.162}$$

Then, using the relations

$$\mathcal{L}_0\left[|\lambda_k\rangle\langle T_\pm|\right] = -i\left(\delta_k^\pm - i\tilde{\Gamma}_k\right)|\lambda_k\rangle\langle T_\pm|, \tag{3.163}$$

$$\mathcal{L}_0\left[|T_\pm\rangle\langle\lambda_k|\right] = +i\left(\delta_k^\pm + i\tilde{\Gamma}_k\right)|T_\pm\rangle\langle\lambda_k|, \tag{3.164}$$

where (to shorten the notation) $\delta_k^+ = \Delta_k$ and $\delta_k^- = \delta_k$, respectively, we find

$$e^{\mathcal{L}_0\tau}\left(H_{\mathrm{ff}}\sigma\rho_{\mathrm{ss}}^{\mathrm{el}}\right) = \frac{a_{\mathrm{hf}}}{4}\sum_k\left[e^{-i(\delta_k^+ - i\tilde{\Gamma}_k)\tau}|\lambda_k\rangle\langle T_+|L_k\sigma\right.$$

$$\left. + e^{-i(\delta_k^- - i\tilde{\Gamma}_k)\tau}|\lambda_k\rangle\langle T_-|\mathbb{L}_k\sigma\right], \tag{3.165}$$

and along the same lines

$$e^{\mathcal{L}_0\tau}\left(\sigma\rho_{ss}^{el}H_{ff}\right) = \frac{a_{hf}}{4}\sum_k\left[e^{+i(\delta_k^+ +i\tilde{\Gamma}_k)\tau}\,|T_+\rangle\,\langle\lambda_k|\,\sigma L_k^\dagger\right.$$

$$\left. +\, e^{+i(\delta_k^- +i\tilde{\Gamma}_k)\tau}\,|T_-\rangle\,\langle\lambda_k|\,\sigma\mathbb{L}_k^\dagger\right]. \tag{3.166}$$

Plugging Eqs. (3.165) and (3.166) into Eq. (3.162), tracing out the electronic degrees of freedom, performing the integration in τ and separating real and imaginary parts of the complex eigenvalues leads to

$$\boxed{a} = \sum_k\left[\frac{\gamma_k^+}{2}\mathcal{D}[L_k]\sigma + i\frac{\Delta_k^+}{2}\left[L_k^\dagger L_k, \sigma\right]\right. \tag{3.167}$$

$$\left. +\frac{\gamma_k^-}{2}\mathcal{D}[\mathbb{L}_k]\sigma + i\frac{\Delta_k^-}{2}\left[\mathbb{L}_k^\dagger\mathbb{L}_k, \sigma\right]\right]. \tag{3.168}$$

This corresponds to the flip-flop mediated terms given in Eq. (3.55) in the main text. The second term labeled as $\boxed{b}$ can be computed along the lines: due to the additional appearance of the projector $\mathcal{P}$, it contains factors of $\langle S_i^\alpha\rangle_{ss}$ and is therefore found to be zero.

Overhauser fluctuations.—In the next step, we investigate the second-order effect of Overhauser fluctuations with respect to the effective QME for the nuclear dynamics. Our analysis starts out from the second-order expression $\mathcal{K}_{zz}$ which, as above, can be rewritten as

$$\mathcal{K}_{zz}\sigma = \underbrace{\int_0^\infty d\tau\,\mathrm{Tr}_{el}\left[\mathcal{P}\mathcal{L}_{zz}e^{\mathcal{L}_0\tau}\mathcal{L}_{zz}\mathcal{P}\rho\right]}_{\boxed{1}}$$

$$\underbrace{-\int_0^\infty d\tau\,\mathrm{Tr}_{el}\left[\mathcal{P}\mathcal{L}_{zz}\mathcal{P}\mathcal{L}_{zz}\mathcal{P}\rho\right]}_{\boxed{2}}. \tag{3.169}$$

First, we evaluate the terms labeled by $\boxed{1}$ and $\boxed{2}$ separately. We find

$$\boxed{1} = -a_{hf}^2\sum_{i,j}\int_0^\infty d\tau\left[\left\langle S_i^z(\tau)\,S_j^z\right\rangle_{ss}\left[\delta A_i^z, \delta A_j^z\sigma\right]\right.$$

$$\left. -\left\langle S_j^z S_i^z(\tau)\right\rangle_{ss}\left[\delta A_i^z, \sigma\delta A_j^z\right]\right] \tag{3.170}$$

where we used the Quantum Regression theorem yielding the electronic auto-correlation functions

$$\left\langle S_i^z\left(\tau\right)S_j^z\right\rangle_{\rm ss} = {\rm Tr}_{\rm el}\left[S_i^z e^{\mathcal{L}_0\tau}\left(S_j^z\rho_{\rm ss}^{\rm el}\right)\right], \tag{3.171}$$

$$\left\langle S_j^z S_i^z\left(\tau\right)\right\rangle_{\rm ss} = {\rm Tr}_{\rm el}\left[S_i^z e^{\mathcal{L}_0\tau}\left(\rho_{\rm ss}^{\rm el}S_j^z\right)\right]. \tag{3.172}$$

In a similar fashion, the term labeled by ② is found to be

$$② = a_{\rm hf}^2 \sum_{i,j}\int_0^\infty d\tau\,\langle S_i^z\rangle_{\rm ss}\langle S_j^z\rangle_{\rm ss}\left[\delta A_i^z,\left[\delta A_j^z,\sigma\right]\right]. \tag{3.173}$$

Putting together the results for ① and ②, we obtain

$$\mathcal{K}_{zz}\sigma = \sum_{i,j}\Pi_{ij}\left[\delta A_j^z\sigma\delta A_i^z - \delta A_i^z\delta A_j^z\sigma\right]$$
$$+\Upsilon_{ij}\left[\delta A_j^z\sigma\delta A_i^z - \sigma\delta A_i^z\delta A_j^z\right], \tag{3.174}$$

which can be rewritten as

$$\mathcal{K}_{zz}\sigma = \sum_{i,j}\left(\Pi_{ij}+\Upsilon_{ij}\right)\left[\delta A_j^z\sigma\delta A_i^z - \frac{1}{2}\left\{\delta A_i^z\delta A_j^z,\sigma\right\}\right]$$
$$-\frac{i}{2}\left[\frac{1}{i}\left(\Pi_{ij}-\Upsilon_{ij}\right)\delta A_i^z\delta A_j^z,\sigma\right]. \tag{3.175}$$

Here, we have introduced the integrated electronic auto-correlation functions [70]

$$\Pi_{ij} = a_{\rm hf}^2\int_0^\infty d\tau\left(\left\langle S_i^z\left(\tau\right)S_j^z\right\rangle_{\rm ss} - \langle S_i^z\rangle_{\rm ss}\langle S_j^z\rangle_{\rm ss}\right),$$
$$\Upsilon_{ij} = a_{\rm hf}^2\int_0^\infty d\tau\left(\left\langle S_i^z S_j^z\left(\tau\right)\right\rangle_{\rm ss} - \langle S_i^z\rangle_{\rm ss}\langle S_j^z\rangle_{\rm ss}\right).$$

For an explicit calculation, we use the relation

$$S_j^z\rho_{\rm ss}^{\rm el} = \rho_{\rm ss}^{\rm el}S_j^z = \frac{1}{4}\left(|T_+\rangle\langle T_+| - |T_-\rangle\langle T_-|\right), \tag{3.176}$$

and the fact that $|T_+\rangle\langle T_+| - |T_-\rangle\langle T_-|$ is an eigenvector of $\mathcal{L}_0$ with eigenvalue $-5\Gamma_\pm$, which readily yield $\Pi_{ij} = \Upsilon_{ij} = \gamma_{zz}/2$. From this, we immediately obtain the corresponding term appearing in the effective nuclear dynamics as

$$\mathcal{K}_{zz}\sigma = \gamma_{zz}\sum_{i,j}\left[\delta A_j^z\sigma\delta A_i^z - \frac{1}{2}\left\{\delta A_i^z\delta A_j^z,\sigma\right\}\right]. \tag{3.177}$$

3.11.10 Diagonalization of Nuclear Dissipator

The flip-flop mediated terms $\mathcal{K}_{ff}$ in Eq. (3.55) can be recast into the following form

$$\dot{\sigma} = \sum_{i,j} \frac{\gamma_{ij}}{2} \left[A_i \sigma A_j^\dagger - \frac{1}{2} \left\{ A_j^\dagger A_i, \sigma \right\} \right] + i \frac{\Delta_{ij}}{2} \left[A_j^\dagger A_i, \sigma \right], \tag{3.178}$$

where we have introduced the vector $\mathbf{A}$ containing the *local* nuclear jump operators as $\mathbf{A} = \left(A_1^+, A_2^+, A_2^-, A_1^- \right)$. The matrices γ and Δ obey a simple block-structure according to

$$\gamma = \gamma^+ \oplus \gamma^-, \tag{3.179}$$
$$\Delta = \Delta^+ \oplus \Delta^-, \tag{3.180}$$

where the 2-by-2 block entries are given by

$$\gamma^\pm = \begin{pmatrix} \gamma_{11}^\pm & \gamma_{12}^\pm \\ \gamma_{21}^\pm & \gamma_{22}^\pm \end{pmatrix} = \begin{pmatrix} \sum_k \gamma_k^\pm \nu_k^2 & \sum_k \gamma_k^\pm \mu_k \nu_k \\ \sum_k \gamma_k^\pm \mu_k \nu_k & \sum_k \gamma_k^\pm \mu_k^2 \end{pmatrix}, \tag{3.181}$$

and similarly

$$\Delta^\pm = \begin{pmatrix} \Delta_{11}^\pm & \Delta_{12}^\pm \\ \Delta_{21}^\pm & \Delta_{22}^\pm \end{pmatrix}$$
$$= \begin{pmatrix} \sum_k \Delta_k^\pm \nu_k^2 & \sum_k \Delta_k^\pm \mu_k \nu_k \\ \sum_k \Delta_k^\pm \mu_k \nu_k & \sum_k \Delta_k^\pm \mu_k^2 \end{pmatrix}. \tag{3.182}$$

The nuclear dissipator can be brought into diagonal form

$$\tilde{\gamma} = U^\dagger \gamma U = \text{diag} \left(\tilde{\gamma}_1^+, \tilde{\gamma}_2^+, \tilde{\gamma}_1^-, \tilde{\gamma}_2^- \right), \tag{3.183}$$

where

$$\tilde{\gamma}_1^\pm = \frac{1}{2} \left[\gamma_{11}^\pm + \gamma_{22}^\pm + \sqrt{\left(\gamma_{11}^\pm - \gamma_{22}^\pm \right)^2 + 4 \left(\gamma_{12}^\pm \right)^2} \right], \tag{3.184}$$

$$\tilde{\gamma}_2^\pm = \frac{1}{2} \left[\gamma_{11}^\pm + \gamma_{22}^\pm - \sqrt{\left(\gamma_{11}^\pm - \gamma_{22}^\pm \right)^2 + 4 \left(\gamma_{12}^\pm \right)^2} \right], \tag{3.185}$$

and $U = U^+ \oplus U^-$ with

$$U^\pm = \begin{pmatrix} \cos(\theta_\pm/2) & -\sin(\theta_\pm/2) \\ \sin(\theta_\pm/2) & \cos(\theta_\pm/2) \end{pmatrix}. \tag{3.186}$$

Here, we have defined $\theta_\pm$ via the relation $\tan(\theta_\pm) = 2\gamma_{12}^\pm / (\gamma_{11}^\pm - \gamma_{22}^\pm), 0 \le \theta_\pm < \pi$. Introducing a new set of operators $\tilde{\mathbf{A}} = (\tilde{A}_1, \tilde{A}_2, \tilde{B}_1, \tilde{B}_2)$ according to

$$\tilde{A}_k = \sum_j U_{jk} A_j, \qquad (3.187)$$

that is explicitly

$$\tilde{A}_1 = \cos(\theta_+/2) A_1^+ + \sin(\theta_+/2) A_2^+, \qquad (3.188)$$
$$\tilde{A}_2 = -\sin(\theta_+/2) A_1^+ + \cos(\theta_+/2) A_2^+, \qquad (3.189)$$
$$\tilde{B}_1 = \sin(\theta_-/2) A_1^- + \cos(\theta_-/2) A_2^-, \qquad (3.190)$$
$$\tilde{B}_2 = \cos(\theta_-/2) A_1^- - \sin(\theta_-/2) A_2^-, \qquad (3.191)$$

the effective nuclear flip-flop mediated dynamics simplifies to

$$\dot{\sigma} = \sum_l \frac{\tilde{\gamma}_l}{2} \left[\tilde{A}_l \sigma \tilde{A}_l^\dagger - \frac{1}{2} \left\{ \tilde{A}_l^\dagger \tilde{A}_l, \sigma \right\} \right]$$
$$+ i \sum_{k,l} \frac{\tilde{\Delta}_{kl}}{2} \left[\tilde{A}_l^\dagger \tilde{A}_k, \sigma \right], \qquad (3.192)$$

where the matrix $\tilde{\Delta}_{kl} = \sum_{ij} U_{ki}^\dagger \Delta_{ij} U_{jl}$ associated with second-order Stark shifts is in general not diagonal. This gives rise to the Stark term mediated criticality in the nuclear spin dynamics.

In general, the matrices $\gamma^\pm$ have rank $(\gamma^\pm) = 2$, yielding four non-zero decay rates $\tilde{\gamma}_{1,2}^\pm$ and four linear independent Lindblad operators $\tilde{A}_l$; therefore, in general, no pure, nuclear dark state $|\Psi_{dark}\rangle$ fulfilling $\tilde{A}_l |\Psi_{dark}\rangle = 0 \ \forall l$ exists. In contrast, when keeping only the supposedly dominant coupling to the electronic eigenstate $|\lambda_2\rangle$, they simplify to

$$\gamma_{ideal}^\pm = \gamma_2^\pm \begin{pmatrix} \nu_2^2 & \mu_2 \nu_2 \\ \mu_2 \nu_2 & \mu_2^2 \end{pmatrix}, \qquad (3.193)$$

which fulfills rank $(\gamma_{ideal}^\pm) = 1$. Still, also in the non-ideal setting, for realistic experimental parameters we observe a clear hierarchy in the eigenvalues, namely $\tilde{\gamma}_2^\pm / \tilde{\gamma}_1^\pm \lesssim 0.1$.

3.11.11   Nonidealities in Electronic Quasisteady State

In Sect. 3.7 we have analyzed the nuclear spin dynamics in the submanifold of the electronic quasisteady state $\rho_{ss}^{el} = (|T_+\rangle \langle T_+| + |T_-\rangle \langle T_-|)/2$. In this Appendix we

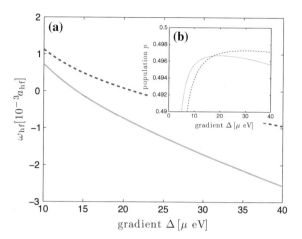

Fig. 3.23 a Knight shift ω_{hf} due to nonzero populations p_k in the electronic quasisteady state for $t = 20\,\mu\mathrm{eV}$ (*solid*) and $t = 30\,\mu\mathrm{eV}$ (*dashed*). **b** In the high gradient regime, for $\Gamma \gg \Gamma_\pm$ the levels $|\lambda_k\rangle$ get depleted efficiently, such that $p_k < 1\,\% \ll p$. Other numerical parameters: $\varepsilon = 30\,\mu\mathrm{eV}$ and $x_\pm = 10^{-3}$

consider (small) deviations from this ideal electronic quasisteady state due to populations of the levels $|\lambda_k\rangle$ ($k = 1, 2, 3$), labeled as p_k. Since all coherences are damped out on electronic timescales, the generalized electronic quasisteady state under consideration is

$$\rho_{\mathrm{ss}}^{\mathrm{el}} = p \left(|T_+\rangle \langle T_+| + |T_-\rangle \langle T_-| \right) + \sum_k p_k |\lambda_k\rangle \langle \lambda_k| . \tag{3.194}$$

Using detailed balance, p_k can be calculated via the equations $p_k \left(\kappa_k^2 + x_\pm \right) = p x_\pm$, where $x_\pm = \Gamma_\pm / \Gamma$ and $p = \left(1 - \sum p_k \right) / 2$ gives the population in $|T_\pm\rangle$, respectively. The electronic levels $|\lambda_k\rangle$ get depleted efficiently for $\Gamma_k \gg \Gamma_\pm$: In contrast to the low-gradient regime where $p_2 \approx 1/3$, in the high-gradient regime, we obtain $p_k < 1\,\% \ll p$ such that the electronic system settles to a quasisteady state very close to the ideal limit where $p = 1/2$; compare Fig. 3.23. In describing the effective nuclear dynamics, nonzero populations p_k lead to additional terms which are second order in ε, but strongly suppressed further as $p_k \ll 1$.

Knight shift.—For nonzero populations p_k, the Knight shift seen by the nuclear spins does not vanish, leading to the following (undesired) additional term for the effective nuclear spin dynamics

$$\dot{\sigma} = -i\omega_{\mathrm{hf}} \left[\delta A_1^z - \delta A_2^z, \sigma \right], \tag{3.195}$$

where

$$\omega_{\mathrm{hf}} = \frac{a_{\mathrm{hf}}}{2} \sum_k p_k \left(\mu_k^2 - \nu_k^2 \right). \tag{3.196}$$

with $a_{\mathrm{hf}} \approx 10^{-4}\,\mu\mathrm{eV}$. As shown in Fig. 3.23, however, $\omega_{\mathrm{hf}} \approx 10^{-7}\,\mu\mathrm{eV}$ is further suppressed by approximately three orders of magnitude; in particular, ω_{hf} is small compared to the dissipative gap of the nuclear dynamics ADR $\approx 2 \times 10^{-5}$ μeV and can thus be neglected.

Hyperfine flip-flop dynamics.—Moreover, nonzero populations p_k lead to additional Lindblad terms of the form $\dot{\sigma} = \cdots + p_k \gamma_k^+ \mathcal{D}[L_k^\dagger]\sigma$. They contain terms which are incommensurate with the ideal two-mode squeezedlike target state. Since $p_k \ll p$, however, they are strongly suppressed compared to the ones absorbed into $\mathcal{L}_{\text{nid}}$ and thus do not lead to any significant changes in our analysis.

References

1. P. Zoller et al., Quantum information processing and communication. Eur. Phys. J. D **36**(2), 203 (2005)
2. M.A. Nielsen, I.L. Chuang, *Quantum Computation and Quantum Information* (Cambridge University Press, Cambridge, 2000)
3. R. Hanson, L.P. Kouwenhoven, J.R. Petta, S. Tarucha, L.M.K. Vandersypen, Spins in few-electron quantum dots. Rev. Mod. Phys. **79**, 1217 (2007)
4. E.A. Chekhovich, M.N. Makhonin, A.I. Tartakovskii, A. Yacoby, H. Bluhm, K.C. Nowack, L.M.K. Vandersypen, Nuclear spin effects in semiconductor quantum dots. Nat Mater **12**, 494 (2013)
5. D.D. Awschalom, N. Samarath, D. Loss, *Semiconductor Spintronics and Quantum Computation* (Springer, Berlin, 2002)
6. D. Loss, D.P. DiVincenzo, Quantum computation with quantum dots. Phys Rev A **57**, 120 (1998)
7. K.C. Nowack, F.H.L. Koppens, Y.V. Nazarov, L.M.K. Vandersypen, Coherent control of a single electron spin with electric fields. Science **318**(5855), 1430 (2007)
8. K.C. Nowack, M. Shafiei, M. Laforest, G.E.D.K. Prawiroatmodjo, L.R. Schreiber, C. Reichl, W. Wegscheider, L.M.K. Vandersypen, Single-shot correlations and two-qubit gate of solid-state spins. Science **333**(6047), 1269 (2011)
9. J.R. Petta, A.C. Johnson, J.M. Taylor, E.A. Laird, A. Yacoby, M.D. Lukin, C.M. Marcus, M.P. Hanson, A.C. Gossard, Coherent manipulation of coupled electron spins in semiconductor quantum dots. Science **309**(5744), 2180 (2005)
10. M.D. Shulman, O.E. Dial, S.P. Harvey, H. Bluhm, V. Umansky, A. Yacoby, Demonstration of entanglement of electrostatically coupled singlet-triplet qubits. Science **336**(6078), 202 (2012)
11. F.H.L. Koppens, C. Buizert, K.J. Tielrooij, I.T. Vink, K.C. Nowack, T. Meunier, L.P. Kouwenhoven, L.M.K. Vandersypen, Driven coherent oscillations of a single electron spin in a quantum dot. Nature **442**, 766 (2006)
12. F.H.L. Koppens, J.A. Folk, J.M. Elzerman, R. Hanson, L.H. Willems van Beveren, I.T. Vink, H.-P. Tranitz, W. Wegscheider, L.P. Kouwenhoven, L.M.K. Vandersypen, Control and detection of singlet-triplet mixing in a random nuclear field. Science **309**, 1346 (2005)
13. J. Schliemann, A. Khaetskii, D. Loss, Electron spin dynamics in quantum dots and related nanostructures due to hyperfine interaction with nuclei. J. Phys.: Condens. Matter **15**(50), R1809 (2003)
14. W.A. Coish, D. Loss, Hyperfine interaction in a quantum dot: non-markovian electron spin dynamics. Phys. Rev. B **70**, 195340 (2004)
15. A.V. Khaetskii, D. Loss, L. Glazman, Electron spin decoherence in quantum dots due to interaction with nuclei. Phys. Rev. Lett. **88**, 186802 (2002)
16. I.A. Merkulov, A.L. Efros, M. Rosen, Electron spin relaxation by nuclei in semiconductor quantum dots. Phys. Rev. B **65**, 205309 (2002)

17. S.I. Erlingsson, Y.V. Nazarov, V.I. Fal'ko, Nucleus-mediated spin-flip transitions in GaAs quantum dots. Phys. Rev. B **64**, 195306 (2001)
18. W.A. Coish, J. Fischer, D. Loss, Exponential decay in a spin bath. Phys. Rev. B **77**, 125329 (2008)
19. Ł. Cywiński, W.M. Witzel, S. Das Sarma, Pure quantum dephasing of a solid-state electron spin qubit in a large nuclear spin bath coupled by long-range hyperfine-mediated interactions. Phys. Rev. B **79**(24), 245314 (2009)
20. I.T. Vink, K.C. Nowack, F.H.L. Koppens, J. Danon, Y.V. Nazarov, L.M.K. Vandersypen, Locking electron spins into magnetic resonance by electron-nuclear feedback. Nat. Phys. **5**, 764 (2009)
21. H. Bluhm, S. Foletti, I. Neder, M. Rudner, D. Mahalu, V. Umansky, A. Yacoby, Dephasing time of GaAs electron-spin qubits coupled to a nuclear bath exceeding 200 µs. Nat. Phys. **7**(2), 109 (2010)
22. S. Foletti, H. Bluhm, D. Mahalu, V. Umansky, A. Yacoby, Universal quantum control of two-electron spin quantum bits using dynamic nuclear polarization. Nat. Phys. **5**, 903 (2009)
23. A.C. Johnson, J.R. Petta, J.M. Taylor, A. Yacoby, M.D. Lukin, C.M. Marcus, M.P. Hanson, A.C. Gossard, Triplet-singlet spin relaxation via nuclei in a double quantum dot. Nature **435**, 925 (2005)
24. K. Ono, S. Tarucha, Nuclear-spin-induced oscillatory current in spin-blockaded quantum dots. Phys. Rev. Lett. **92**, 256803 (2004)
25. K. Ono, D.G. Austing, Y. Tokura, S. Tarucha, Current rectification by Pauli exclusion in a weakly coupled double quantum dot system. Science **297**(5585), 1313 (2002)
26. R. Takahashi, K. Kono, S. Tarucha, K. Ono, Voltage-selective bi-directional polarization and coherent rotation of nuclear spins in quantum dots. Phys. Rev. Lett. **107**, 026602 (2011)
27. J.M. Taylor, C.M. Marcus, M.D. Lukin, Long-lived memory for mesoscopic quantum bits. Phys. Rev. Lett. **90**, 206803 (2003)
28. J.M. Taylor, G. Giedke, H. Christ, B. Paredes, J.I. Cirac, P. Zoller, M.D. Lukin, A. Imamoglu, Quantum information processing using localized ensembles of nuclear spins. arXiv:cond-mat/0407640 (2004)
29. W.M. Witzel, S. Das Sarma, Nuclear spins as quantum memory in semiconductor nanostructures. Phys. Rev. B **76**, 045218 (2007)
30. P. Cappellaro, L. Jiang, J.S. Hodges, M.D. Lukin, Coherence and control of quantum registers based on electronic spin in a nuclear spin bath. Phys. Rev. Lett. **102**, 210502 (2009)
31. Z. Kurucz, M.W. Sørensen, J.M. Taylor, M.D. Lukin, M. Fleischhauer, Qubit protection in nuclear-spin quantum dot memories. Phys. Rev. Lett. **103**, 010502 (2009)
32. M.S. Rudner, L.S. Levitov, Self-polarization and cooling of spins in quantum dots. Phys. Rev. Lett. **99**, 036602 (2007)
33. M.S. Rudner, L.S. Levitov, Electrically-driven reverse overhauser pumping of nuclear spins in quantum dots. Phys. Rev. Lett. **99**, 246602 (2007)
34. M.S. Rudner, L.M.K. Vandersypen, V. Vuletic, L.S. Levitov, Generating entanglement and squeezed states of nuclear spins in quantum dots. Phys. Rev. Lett. **107**, 206806 (2011)
35. F. Verstraete, M.M. Wolf, J.I. Cirac, Quantum computation and quantum-state engineering driven by dissipation. Nat Phys **5**, 633 (2009)
36. S. Diehl, A. Micheli, A. Kantian, B. Kraus, H.P. Büchler, P. Zoller, Quantum states and phases in driven open quantum systems with cold atoms. Nat. Phys. **4**, 878 (2008)
37. R. Sánchez, G. Platero, Dark bell states in tunnel-coupled spin qubits. Phys. Rev. B **87**(8), 081305 (2013)
38. A. Tomadin, S. Diehl, M.D. Lukin, P. Rabl, P. Zoller, Reservoir engineering and dynamical phase transitions in optomechanical arrays. Phys. Rev. **86**(3), 033821 (2012)
39. J.F. Poyatos, J.I. Cirac, P. Zoller, Quantum reservoir engineering with laser cooled trapped ions. Phys. Rev. Lett. **77**(23), 4728 (1996)
40. D.P. DiVincenzo, The physical implementation of quantum computation. Fortschritte der Physik **48**(9–11), 771 (2000)

41. C.A. Muschik, E.S. Polzik, J.I. Cirac, Dissipatively driven entanglement of two macroscopic atomic ensembles. Phys. Rev. A **83**, 052312 (2011)
42. H. Krauter, C.A. Muschik, K. Jensen, W. Wasilewski, J.M. Petersen, J.I. Cirac, E.S. Polzik, Entanglement generated by dissipation and steady state entanglement of two macroscopic objects. Phys. Rev. Lett. **107**, 080503 (2011)
43. Y. Lin, J.P. Gaebler, F. Reiter, T.R. Tan, R. Bowler, A.S. Sørensen, D. Leibfried, D.J. Wineland, Dissipative production of a maximally entangled steady state of two quantum bits. Nature **504**, 415 (2013)
44. J.T. Barreiro, M. Müller, P. Schindler, D. Nigg, T. Monz, M. Chwalla, M. Hennrich, C.F. Roos, P. Zoller, R. Blatt, An open-system quantum simulator with trapped ions. Nat. Phys. **470**(7335), 486 (2011)
45. S. Shankar, M. Hatridge, Z. Leghtas, K.M. Sliwa, A. Narla, U. Vool, S.M. Girvin, L. Frunzio, M. Mirrahimi, M.H. Devoret, Autonomously stabilized entanglement between two supercon-ducting quantum bits. Nature **504**(7480), 419 (2013)
46. M.J.A. Schuetz, E.M. Kessler, J.I. Cirac, G. Giedke, Superradiance-like electron transport through a quantum dot. Phys. Rev. B **86**, 085322 (2012)
47. M.S. Rudner, F.H.L. Koppens, J.A. Folk, L.M.K. Vandersypen, L.S. Levitov, Nuclear spin dynamics in double quantum dots: fixed points, transients, and intermittency. Phys. Rev. B **84**, 075339 (2011)
48. O.N. Jouravlev, Y.V. Nazarov, Electron transport in a double quantum dot governed by a nuclear magnetic field. Phys. Rev. Lett. **96**, 176804 (2006)
49. C. López-Monís, J. Iñarrea, G. Platero, Dynamical nuclear spin polarization induced by elec-tronic current through double quantum dots. New. J. Phys. **13**(5), 053010 (2011)
50. S.E. Economou, E. Barnes, Theory of dynamic nuclear polarization and feedback in quantum dots. Phys. Rev. B **89**, 165301 (2014)
51. J. Danon, I.T. Vink, F.H.L. Koppens, K.C. Nowack, L.M.K. Vandersypen, Y.V. Nazarov, Multiple nuclear polarization states in a double quantum dot. Phys. Rev. Lett. **103**, 046601 (2009)
52. A.M. Lunde, C. López-Monís, I.A. Vasiliadou, L.L. Bonilla, G. Platero, Temperature-dependent dynamical nuclear polarization bistabilities in double quantum dots in the spin-blockade regime. Phys. Rev. B **88**, 035317 (2013)
53. J. Baugh, Y. Kitamura, K. Ono, S. Tarucha, Large nuclear overhauser fields detected in vertically-coupled double quantum dots. Phys. Rev. Lett. **99**, 096804 (2007)
54. J.R. Petta, J.M. Taylor, A.C. Johnson, A. Yacoby, M.D. Lukin, C.M. Marcus, M.P. Hanson, A.C. Gossard, Dynamic nuclear polarization with single electron spins. Phys. Rev. Lett. **100**, 067601 (2008)
55. T. Kobayashi, K. Hitachi, S. Sasaki, K. Muraki, Observation of hysteretic transport due to dynamic nuclear spin polarization in a GaAs lateral double quantum dot. Phys. Rev. Lett. **107**(21), 216802 (2011)
56. C. Barthel, J. Medford, H. Bluhm, A. Yacoby, C.M. Marcus, M.P. Hanson, A.C. Gossard, Relaxation and readout visibility of a singlet-triplet qubit in an overhauser field gradient. Phys. Rev. B **85**, 035306 (2012)
57. M. Gullans, J.J. Krich, J.M. Taylor, H. Bluhm, B.I. Halperin, C.M. Marcus, M. Stopa, A. Yacoby, M.D. Lukin, Dynamic nuclear polarization in double quantum dots. Phys. Rev. Lett. **104**, 226807 (2010)
58. B. Kraus, J.I. Cirac, Discrete entanglement distribution with squeezed light. Phys. Rev. Lett. **92**, 013602 (2004)
59. T. Hayashi, T. Fujisawa, H.-D. Cheong, Y.-H. Jeong, Y. Hirayama, Coherent manipulation of electronic states in a double quantum dot. Phys. Rev. Lett. **91**, 226804 (2003)
60. C. Cohen-Tannoudji, J. Dupont-Roc, G. Grynberg, *Atom-Photon Interactions: Basic Processes and Applications* (Wiley, New York, 1992)
61. G. Petersen, E.A. Hoffmann, D. Schuh, W. Wegscheider, G. Giedke, S. Ludwig, Large nuclear spin polarization in gate-defined quantum dots using a single-domain nanomagnet. Phys. Rev. Lett. **110**, 177602 (2013)

62. G. Giavaras, N. Lambert, F. Nori, Electrical current and coupled electron-nuclear spin dynamics in double quantum dots. Phys. Rev. B **87**(11), 115416 (2013)
63. M.S. Rudner, L.S. Levitov, Self-sustaining dynamical nuclear polarization oscillations in quantum dots. Phys. Rev. Lett. **110**(8), 086601 (2013)
64. F. Qassemi, W.A. Coish, F.K. Wilhelm, Stationary and transient leakage current in the Pauli spin blockade. Phys. Rev. Lett. **102**(17), 176806 (2009)
65. Y. Yamamoto, A. Imamoglu, *Mesoscopic Quantum Optics* (Wiley, New York, 1999)
66. E.M. Kessler, Generalized Schrieffer-Wolff formalism for dissipative systems. Phys. Rev. A **86**, 012126 (2012)
67. M. Pioro-Ladrière, T. Obata, Y. Tokura, Y.S. Shin, T. Kubo, K. Yoshida, T. Taniyama, S. Tarucha, Electrically driven single-electron spin resonance in a slanting Zeeman field. Nat Phys **4**(10), 776 (2008)
68. H. Christ, J.I. Cirac, G. Giedke, Quantum description of nuclear spin cooling in a quantum dot. Phys. Rev. B **75**, 155324 (2007)
69. H. Bluhm, S. Foletti, D. Mahalu, V. Umansky, A. Yacoby, Enhancing the coherence of a spin qubit by operating it as a feedback loop that controls its nuclear spin bath. Phys. Rev. Lett. **105**(21), 216803 (2010)
70. E.M. Kessler, G. Giedke, A. Imamoglu, S.F. Yelin, M.D. Lukin, J.I. Cirac, Dissipative phase transition in a central spin system. Phys. Rev. A **86**(1), 012116 (2012)
71. J. Laurat, G. Keller, J.A. Oliveira-Huguenin, C. Fabre, T. Coudreau, A. Serafini, G. Adesso, F. Illuminati, Entanglement of two-mode Gaussian states: characterization and experimental production and manipulation. J. Opt. B: Quantum Semiclassical Opt. **7**(12), S577 (2005)
72. H. Schwager, J.I. Cirac, G. Giedke, Interfacing nuclear spins in quantum dots to cavity or traveling-wave fields. New J. Phys. **12**, 043026 (2010)
73. G. Giedke, M.M. Wolf, O. Krüger, R.F. Werner, J.I. Cirac, Entanglement of formation for symmetric Gaussian states. Phys. Rev. Lett. **91**, 107901 (2003)
74. M.M. Wolf, G. Giedke, O. Krüger, R.F. Werner, J.I. Cirac, Gaussian entanglement of formation. Phys. Rev. A **69**, 052320 (2004)
75. G. Vidal, R.F. Werner, A computable measure of entanglement. Phys. Rev. A **65**, 032314 (2002)
76. D. Nagy, G. Szirmai, P. Domokos, Critical exponent of a quantum-noise-driven phase transition: The open-system Dicke model. Phys. Rev. A **84**, 043637 (2011)
77. F. Dimer, B. Estienne, A.S. Parkins, H.J. Carmichael, Proposed realization of the Dicke-model quantum phase transition in an optical cavity QED system. Phys. Rev. A **75**, 013804 (2007)
78. S. Diehl, A. Tomadin, A. Micheli, R. Fazio, P. Zoller, Dynamical phase transitions and instabilities in open atomic many-body systems. Phys. Rev. Lett. **105**, 015702 (2010)
79. M. Höning, M. Moos, M. Fleischhauer, Critical exponents of steady-state phase transitions in fermionic lattice models. Phys. Rev. A **86**(1), 013606 (2012)
80. G.A. Álvarez, E.P. Danieli, P.R. Levstein, H.M. Pastawski, Environmentally induced quantum dynamical phase transition in the spin swapping operation. J. Chem. Phys. **124**(19), 194507 (2006)
81. E.P. Danieli, G.A. Álvarez, P.R. Levstein, H.M. Pastawski, Quantum dynamical phase transition in a system with many-body interactions. Solid State Communications **141**(7), 422 (2007)
82. H. Eleuch, I. Rotter, Open quantum systems and Dicke superradiance. Eur. Phys. J. D **68**(3), 74 (2014)
83. I. Lesanovsky, M. van Horssen, M. Gutua, J.P. Garrahan, Characterization of dynamical phase transitions in quantum jump trajectories beyond the properties of the stationary state. Phys. Rev. Lett. **110**, 150401 (2013)
84. H. Schwager, J.I. Cirac, G. Giedke, A quantum interface between light and nuclear spins in quantum dots. Phys. Rev. B **81**, 045309 (2010)
85. D.J. Reilly, J.M. Taylor, E.A. Laird, J.R. Petta, C.M. Marcus, M.P. Hanson, A.C. Gossard, Measurement of temporal correlations of the overhauser field in a double quantum dot. Phys. Rev. Lett. **101**(23), 236803 (2008)

86. D. Paget, Optical detection of NMR in high-purity GaAs: direct study of the relaxation of nuclei close to shallow donors. Phys. Rev. B **25**, 4444 (1982)

87. O.E. Dial, M.D. Shulman, S.P. Harvey, H. Bluhm, V. Umansky, A. Yacoby, Charge noise spectroscopy using coherent exchange oscillations in a singlet-triplet qubit. Phys. Rev. Lett. **110**, 146804 (2013)

88. K. Hammerer, A. Sørensen, E.S. Polzik, Quantum interface between light and atomic ensembles. Rev. Mod. Phys. **82**, 1041 (2010)

89. A. Furusawa, N. Takei, Quantum teleportation for continuous variables and related quantum information processing. Phys. Rep. **443**(3), 97 (2007)

90. W. Wasilewski, K. Jensen, H. Krauter, J.J. Renema, M.V. Balabas, E.S. Polzik, Quantum noise limited and entanglement-assisted magnetometry. Phys. Rev. Lett. **104**(13), 133601 (2010)

91. J. Appel, P.J. Windpassinger, D. Oblak, U.B. Hoff, N. Kjaergaard, E.S. Polzik, Mesoscopic atomic entanglement for precision measurements beyond the standard quantum limit. in Proceedings of the National Academy of Sciences of the United States of America, vol. 106, p. 10960 (2009)

92. M.H. Schleier-Smith, I.D. Leroux, V. Vuletić, States of an ensemble of two-level atoms with reduced quantum uncertainty. Phys. Rev. Lett. **104**(7), 073604 (2010)

93. C. Gross, T. Zibold, E. Nicklas, J. Estève, M.K. Oberthaler, Nonlinear atom interferometer surpasses classical precision limit. Nature **464**(7292), 1165 (2010)

94. H.J. Kimble, The quantum internet. Nature **453**, 1023 (2008)

95. M. Busl et al., Bipolar spin blockade and coherent state superpositions in a triple quantum dot. Nat. Nanotechnol. **8**(4), 261 (2013)

96. F.R. Braakman, P. Barthelemy, C. Reichl, W. Wegscheider, L.M.K. Vandersypen, Long-range coherent coupling in a quantum dot array. Nat. Nanotechnol. **8**, 432 (2013)

97. J.D. Fletcher et al., Clock-controlled emission of single-electron wave packets in a solid-state circuit. Phys. Rev. Lett. **111**(21), 216807 (2013)

98. G. Feve, A. Mahe, J.M. Berroir, T. Kontos, B. Placais, D.C. Glattli, A. Cavanna, B. Etienne, Y. Jin, An on-demand coherent single-electron source. Science **316**(5828), 1169 (2007)

99. E. Bocquillon, V. Freulon, J.M. Berroir, P. Degiovanni, B. Placais, A. Cavanna, Y. Jin, G. Feve, Coherence and indistinguishability of single electrons emitted by independent sources. Science **339**(6123), 1054 (2013)

100. P. Roulleau, F. Portier, P. Roche, A. Cavanna, G. Faini, U. Gennser, D. Mailly, Direct measurement of the coherence length of edge states in the integer quantum hall regime. Phys. Rev. Lett. **100**(12), 126802 (2008)

101. A. Mahé, F.D. Parmentier, E. Bocquillon, J.M. Berroir, D.C. Glattli, T. Kontos, B. Plaçais, G. Fève, A. Cavanna, Y. Jin, Current correlations of an on-demand single-electron emitter. Phys. Rev. B **82**(20), 201309 (2010)

102. S. Hermelin, S. Takada, M. Yamamoto, S. Tarucha, A.D. Wieck, L. Saminadayar, C. Bäuerle, T. Meunier, Electrons surfing on a sound wave as a platform for quantum optics with flying electrons. Nature **477**, 435 (2011)

103. R.P.G. McNeil, M. Kataoka, C.J.B. Ford, C.H.W. Barnes, D. Anderson, G.A.C. Jones, I. Farrer, D.A. Ritchie, On-demand single-electron transfer between distant quantum dots. Nature **477**, 439 (2011)

104. H. Sanada, Y. Kunihashi, H. Gotoh, K. Onomitsu, M. Kohda, J. Nitta, P.V. Santos, T. Sogawa, Manipulation of mobile spin coherence using magnetic-field-free electron spin resonance. Nat. Phys. **9**(5), 280 (2013)

105. M. Yamamoto, S. Takada, C. Bäuerle, K. Watanabe, A.D. Wieck, S. Tarucha, Electrical control of a solid-state flying qubit. Nat. Nanotechnol. **7**(4), 247 (2012)

106. H. Christ, J.I. Cirac, G. Giedke, Entanglement generation via a completely mixed nuclear spin bath. Phys. Rev. B **78**, 125314 (2008)

107. H.J. Carmichael, Analytical and numerical results for the steady state in cooperative resonance fluorescence. J. Phys. B: At. Mol. Phys. **13**(18), 3551 (1980)

108. S. Morrison, A.S. Parkins, Collective spin systems in dispersive optical cavity-QED: quantum phase transitions and entanglement. Phys. Rev. A **77**(4), 043810 (2008)

109. C.-H. Chung, K. Le Hur, M. Vojta, P. Wölfle, Nonequilibrium transport at a dissipative quantum phase transition. Phys. Rev. Lett. **102**, 216803 (2009)
110. A.J. Leggett, S. Chakravarty, A.T. Dorsey, M.P.A. Fisher, A. Garg, W. Zwerger, Dynamics of the dissipative two-state system. Rev. Mod. Phys. **59**, 1 (1987)
111. M.S. Rudner, L.S. Levitov, Phase transitions in dissipative quantum transport and mesoscopic nuclear spin pumping. Phys. Rev. B **82**, 155418 (2010)
112. L. Borda, G. Zarand, D. Goldhaber-Gordon, Dissipative quantum phase transition in a quantum dot. arXiv:cond-mat/0602019 (2006)
113. D. Stepanenko, M. Rudner, B.I. Halperin, D. Loss, Singlet-triplet splitting in double quantum dots due to spin-orbit and hyperfine interactions. Phys. Rev. B **85**, 075416 (2012)
114. A. Rivas, S.F. Huelga, *Open Quantum Systems. An Introduction. SpringerBriefs in Physics* (Springer, Berlin, 2012)
115. A. Uhlmann, The transition probability in the state space of a*-algebra. Rep. Math. Phys. **9**, 273 (1976)
116. L.P. Kouwenhoven, C.M. Marcus, P.L. McEuen, S. Tarucha, R.M. Westervelt, N.S. Wingren, Electron transport in quantum dots, in *Mesoscopic Electron Transport*, ed. by L.L. Sohn, L.P. Kouwenhoven, G. Schön (Proceedings of the NATO Advanced Study Institute. Kluwer Academic Publishers, Dordrecht, 1997)
117. H.-A. Engel, D. Loss, Single-spin dynamics and decoherence in a quantum dot via charge transport. Phys. Rev. B **65**(19), 195321 (2002)
118. P. Recher, E.V. Sukhorukov, D. Loss, Quantum dot as spin filter and spin memory. Phys. Rev. Lett. **85**(9), 1962 (2000)
119. J. Danon, Y.V. Nazarov, Pauli spin blockade in the presence of strong spin-orbit coupling. Phys. Rev. B **80**, 041301 (2009)
120. L.R. Schreiber, F.R. Braakman, T. Meunier, V. Calado, J. Danon, J.M. Taylor, W. Wegscheider, L.M.K. Vandersypen, Coupling artificial molecular spin states by photon-assisted tunnelling. Nature Communications **2**, 556 (2011)

Chapter 4
Universal Quantum Transducers Based on Surface Acoustic Waves

In this chapter we turn our focus towards the question of how to interconnect individual qubits over large distances. To this end, we propose a universal, on-chip quantum transducer based on surface acoustic waves in piezo-active materials. Because of the intrinsic piezoelectric (and/or magnetostrictive) properties of the material, our approach provides a universal platform capable of coherently linking a broad array of qubits, including quantum dots, trapped ions, nitrogen-vacancy centers or superconducting qubits. The quantized modes of surface acoustic waves lie in the gigahertz range, can be strongly confined close to the surface in phononic cavities and guided in acoustic waveguides. We show that this type of surface acoustic excitations can be utilized efficiently as a quantum bus, serving as an on-chip, mechanical cavity-QED equivalent of microwave photons and enabling long-range coupling of a wide range of qubits.

4.1 Introduction

The realization of long-range interactions between remote qubits is arguably one of the greatest challenges towards developing a scalable, solid-state spin based quantum information processor [1]. One approach to address this problem is to interface qubits with a common *quantum bus* which distributes quantum information between distant qubits. The transduction of quantum information between stationary and moving qubits is central to this approach. A particularly efficient implementation of such a quantum bus can be found in the field of circuit QED where spatially separated superconducting qubits interact via microwave photons confined in transmission line cavities [2–4]. In this way, multiple qubits have been coupled successfully over relatively large distances of the order of millimeters [5, 6]. Fueled by the dramatic advances in the fabrication and manipulation of nanomechanical systems [7], an

© Springer International Publishing AG 2017
M.J.A. Schütz, *Quantum Dots for Quantum Information Processing: Controlling and Exploiting the Quantum Dot Environment*, Springer Theses,
DOI 10.1007/978-3-319-48559-1_4

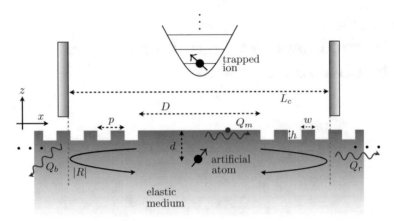

Fig. 4.1 SAW as a universal quantum transducer. Distributed Bragg reflectors made of grooves form an acoustic cavity for surface acoustic waves. The resonant frequency of the cavity is determined by the pitch p, $f_c = v_s/2p$. Reflection occurs effectively at some distance inside the grating; the fictitious mirrors above the surface are not part of the actual experimental setup, but shown for illustrative purposes only. *Red arrows* indicate the relevant decay channels for the cavity mode: leakage through the mirrors, internal losses due to for example surface imperfections, and conversion into bulk modes. Qubits inside and outside of the solid can be coupled to the cavity mode. In more complex structures, the elastic medium can consist of multiple layers on top of some substrate

alternate line of research has pursued the idea of coherent, long-range interactions between individual qubits mediated by mechanical resonators, with resonant *phonons* playing the role of cavity photons [8–13].

In this chapter, we propose a new realization of a quantum transducer and data bus based on surface acoustic waves (SAW). SAWs involve phonon-like excitations bound to the surface of a solid and are widely used in modern electronic devices e.g. as compact microwave filters [14, 15]. Inspired by two recent experiments [16, 17], where the coherent quantum nature of surface acoustic waves (SAWs) has been explored, here we propose and analyze SAW phonon modes in piezo-active materials as a universal mediator for long-range couplings between remote qubits. Our approach involves qubits interacting with a localized SAW phonon mode, defined by a high-Q resonator, which in turn can be coupled weakly to a SAW waveguide serving as a quantum bus; as demonstrated below, the qubits can be encoded in a great variety of spin or charge degrees of freedom. We show that the Hamiltonian for an individual node (for a schematic representation see Fig. 4.1) can take on the generic Jaynes-Cummings form ($\hbar = 1$),

$$H_{\text{node}} = \frac{\omega_q}{2}\sigma^z + \omega_c a^\dagger a + g\left(\sigma^+ a + \sigma^- a^\dagger\right), \qquad (4.1)$$

Table 4.1 Estimates for single-phonon coupling strength g and cooperativity C

	Charge qubit	Spin qubit	Trapped ion	NV-center
Coupling g	(200–450) MHz	(10–22.4) MHz	(1.8–4.0) kHz	(45–101) kHz
Cooperativity C	11–55	21–106	7–36	10–54

We have used A $= (1 - 5)\,\mu$m $\times 40\lambda_c$, $T = 20$ mK [17], (conservative) quality factors of $Q = (1, 1, 3, 1) \times 10^3$ and frequencies of $\omega_c = 2\pi\left(6, 1.5, 2 \times 10^{-3}, 3\right)$ GHz for the four systems listed. For the spin qubit $T_2^\star = 2\,\mu$s [40], and for the trapped ion scenario, $g_{\mathrm{ion}}(C_{\mathrm{ion}})$ is given for $d = 150\,\mu$m due to the prolonged dephasing time further away from the surface (C_{ion} improves with increasing d, even though g_{ion} decreases, up to a point where other dephasing start to dominate). Further details can be found in the main text

where $\vec{\sigma}$ refers to the usual Pauli matrices describing the qubit with transition frequency ω_q and a is the bosonic operator for the localized SAW cavity mode of frequency $\omega_c/2\pi \sim$ GHz.[1] The coupling g between the qubit and the acoustic cavity mode is mediated intrinsically by the piezo-properties of the host material, it is proportional to the electric or magnetic zero-point fluctuations associated with a single SAW phonon and, close to the surface, can reach values of $g \sim 400$ MHz, much larger than the relevant decoherence processes and sufficiently large to allow for quantum effects and coherent coupling in the spin-cavity system as evidenced by cooperativities[2] of $C \sim 10 - 100$ [see Sect. 4.6 and Table 4.1 for definition and applicable values]. For $\omega_q \approx \omega_c$, H_{node} allows for a controlled mapping of the qubit state onto a coherent phonon superposition, which can then be mapped to an itinerant SAW phonon in a waveguide, opening up the possibility to implement on-chip many quantum communication protocols well known in the context of optical quantum networks [11, 18].

4.2 Executive Summary: Reader's Guide

This chapter is organized as follows. In Sect. 4.3, we first summarize the most pertinent characteristics of the proposed sound-based platform for quantum information processing. Thereafter, in Sect. 4.4 we review the most important features of surface acoustic waves, with a focus on the associated zero-point fluctuations. Next, in Sect. 4.5 we discuss the different components making up the SAW-based quantum transducer and the acoustic quantum network it enables: SAW cavities, including a detailed analysis of the achievable quality factor Q, SAW waveguides and a variety

---

[1]Depending on the particular physical implementation of the qubit, the Jaynes-Cummings type Hamiltonian given in Eq. (4.1) may emerge only as an effective description of a potentially driven, two- or multilevel system. However, for e.g. charge or spin qubits embedded in GaAs quantum dots it is a rather straightforward way to describe the coherent dynamics of a single node.

[2]The cooperativity parameter C compares the coherent single phonon coupling strength g with the geometric mean of the relevant incoherent processes, i.e., the qubit's decoherence rate and the cavity's effective linewidth; in direct analogy to cavity QED, it is a key figure of merit in our system as $C > 1$ marks the onset of coherent quantum effects in a coupled spin-oscillator system [8].

of different candidate systems serving as qubits; for the latter see Sect. 4.6. Lastly, as exemplary application, in Sect. 4.7 we show how to transfer quantum states between distant nodes of the network under realistic conditions. Finally, in Sect. 4.8 we draw conclusions and give an outlook on future directions of research.

4.3   Main Results

In this chapter, we propose a novel realization of a quantum transducer and data bus based on surface acoustic waves (SAW) in piezo-active materials (such as conventional GaAs). The most pertinent features of our proposal can be summarized as follows: (1) Our scheme is not specific to any particular qubit realization, but—thanks to the plethora of physical properties associated with SAWs in piezo-active materials (strain, electric and magnetic fields)—provides a common on-chip platform accessible to various different implementations of qubits, comprising both natural (e.g., ions) and artificial candidates such as quantum dots or superconducting qubits. In particular, this opens up the possibility to interconnect dissimilar systems in new electro-acoustic quantum devices. (2) Typical SAW frequencies lie in the gigahertz range, closely matching transition frequencies of artificial atoms and enabling ground state cooling by conventional cryogenic techniques. (3) Our scheme is built upon an established technology [14, 15]: Lithographic fabrication techniques provide almost arbitrary geometries with high precision as evidenced by a large range of SAW devices such as delay lines, bandpass filters or resonators etc. In particular, the essential building blocks needed to interface qubits with SAW phonons have already been fabricated, according to design principles familiar from electromagnetic devices: (i) SAW resonators, the mechanical equivalents of Fabry-Perot cavities, with low-temperature measurements reaching quality factors of $Q \sim 10^5$ even at gigahertz frequencies [19–21], and (ii) acoustic waveguides as analogue to optical fibers [14]. (4) For a given frequency in the gigahertz range, due to the slow speed of sound of approximately $\sim 10^3$ m/s for typical materials, device dimensions are in micrometer range, which is convenient for fabrication and integration with semiconductor components, and about 10^5 times smaller than corresponding electromagnetic resonators. (5) Since SAWs propagate elastically on the surface of a solid within a depth of approximately one wavelength, the mode volume is intrinsically confined in the direction normal to the surface. Further surface confinement then yields large zero-point fluctuations. (6) Yet another inherent advantage of our system is the intrinsic nature of the coupling. In piezoelectric materials, the SAW is accompanied by an electrical potential ϕ which has the same spatial and temporal periodicities as the mechanical displacement and provides an intrinsic qubit-phonon coupling mechanism. For example, recently qubit lifetimes in GaAs singlet-triplet qubits were found to be limited by the piezoelectric electron-phonon coupling [22]. Here, our scheme provides a new paradigm, where coupling to phonons becomes a highly valuable asset for coherent quantum control rather than a liability.

4.4 SAW Properties

Elastic waves in piezoelectric solids are described by

$$\rho \ddot{u}_i - c_{ijkl}\partial_j\partial_l u_k = e_{kij}\partial_j\partial_k\phi, \tag{4.2}$$

$$\varepsilon_{ij}\partial_i\partial_j\phi = e_{ijk}\partial_i\partial_k u_j, \tag{4.3}$$

where the vector $\mathbf{u}\,(\mathbf{x}, t)$ denotes the displacement field ($\mathbf{x}$ is the cartesian coordinate vector), ρ is the mass density and repeated indices are summed over ($i, j = x, y, z$); $\underline{c}, \underline{\varepsilon}$ and $\underline{e}$ refer to the elasticity, permittivity and piezoelectric tensors, respectively [23]; they are largely defined by crystal symmetry [24]. For example, for cubic crystals such as GaAs there is only one non-zero component for the permittivity and the piezoelectric tensor, labeled as ε and e_{14}, respectively [23]. Since elastic disturbances propagate much slower than the speed of light, it is common practice to apply the so-called quasi-static approximation [24] where the electric field is given by $E_i = -\partial_i\phi$. When considering surface waves, Eqs. (4.2) and (4.3) must be supplemented by the mechanical boundary condition that there should be no forces on the free surface (taken to be at $z = 0$ with $\hat{z}$ being the outward normal to the surface), that is $T_{zx} = T_{zy} = T_{zz} = 0$ at $z = 0$ (where $T_{ij} = c_{ijkl}\partial_l u_k + e_{kij}\partial_k\phi$ is the stress tensor), and appropriate electrical boundary conditions [23].

If not stated otherwise, the term SAW refers to the prototypical (piezoelectric) Rayleigh wave solution as theoretically and experimentally studied for example in Refs. [16, 17, 23, 25] and used extensively in different electronic devices [14, 15]. It is non-dispersive, decays exponentially into the medium with a characteristic penetration depth of a wavelength and has a phase velocity $v_s = \omega/k$ that is lower than the bulk velocities in that medium, because the solid behaves less rigidly in the absence of material above the surface [24]. As a result, it cannot phase-match to any bulk-wave [14, 26]. As usual, we consider specific orientations for which the piezoelectric field produced by the SAW is strongest [14, 26], for example a SAW with wavevector along the [110] direction of a (001) GaAs crystal [cf. Refs. [16, 23] and Appendices 4.9.1 and 4.9.2].

SAWs in quantum regime.—In a semiclassical picture, an acoustic phonon associated with a SAW creates a time-dependent strain field, $s_{kl} = (\partial_l u_k + \partial_k u_l)/2$ and a (quasi-static) electrical potential $\phi\,(\mathbf{x}, t)$. Upon quantization, the mechanical displacement becomes an operator that can be expressed in terms of the elementary normal modes as $\hat{\mathbf{u}}\,(\mathbf{x}) = \sum_n [\mathbf{v}_n\,(\mathbf{x})\, a_n + \text{h.c.}]$, where $a_n\,(a_n^\dagger)$ are bosonic annihilation (creation) operators for the vibrational eigenmode n and the set of normal modes $\mathbf{v}_n\,(\mathbf{x})$ derives from the Helmholtz-like equation $\mathcal{W}\mathbf{v}_n\,(\mathbf{x}) = -\rho\omega_n^2\mathbf{v}_n\,(\mathbf{x})$ associated with Eqs. (4.2) and (4.3). The mode normalization is given by $\int d^3\mathbf{x}\rho\mathbf{v}_n^*\,(\mathbf{x})\cdot\mathbf{v}_n\,(\mathbf{x}) = \hbar/2\omega_n$ [22, 27]. An important figure of merit in this context is the amplitude of the mechanical zero-point motion U_0. Along the lines of cavity QED [2], a simple estimate for U_0 can be obtained by equating the *classical* energy of a SAW $\sim \int d^3\mathbf{x}\rho\dot{\mathbf{u}}^2$ with the *quantum* energy of a single phonon, that is $\hbar\omega$. This leads directly to

$$U_0 \approx \sqrt{\hbar/2\rho v_s A}, \tag{4.4}$$

where we used the dispersion relation $\omega = v_s k$ and the intrinsic mode confinement $V \approx A\lambda$ characteristic for SAWs. The quantity U_0 refers to the mechanical amplitude associated with a single SAW phonon close to the surface. It depends only on the material parameters ρ and v_s and follows a generic $\sim A^{-1/2}$ behaviour, where A is the effective mode area on the surface. The estimate given in Eq. (4.4) agrees very well with more detailed calculations presented in Appendix 4.9.3. Several other important quantities which are central for signal transduction between qubits and SAWs follow directly from U_0: The (dimensionless) zero-point strain can be estimated as $s_0 \approx kU_0$. The intrinsic piezoelectric potential associated with a single phonon derives from Eq. (4.3) as $\phi_0 \approx (e_{14}/\varepsilon) U_0$.[3] Lastly, the electric field amplitude due to a single acoustic phonon is $\xi_0 \approx k\phi_0 = (e_{14}/\varepsilon) kU_0$, illustrating the linear relation between electric field and strain characteristic for piezoelectric materials [13]. In summary, we typically find $U_0 \approx 2\,\mathrm{fm}/\sqrt{A\,[\mu\mathrm{m}^2]}$, yielding $U_0 \approx 2\,\mathrm{fm}$ for micron-scale confinement (cf. Appendix 4.9.4). This is comparable to typical zero-point fluctuation amplitudes of localized mechanical oscillators [28]. Moreover, for micron-scale surface confinement and GaAs material parameters, we obtain $\xi_0 \approx 20\,\mathrm{V/m}$ which compares favorably with typical values of $\sim 10^{-3}$ and $\sim 0.2\,\mathrm{V/m}$ encountered in cavity and circuit QED, respectively [2].

For the sake of clarity, we have focused on piezo-*electric* materials so far. However, there are also piezo-*magnetic* materials that exhibit a large magnetostrictive effect. In that case, elastic distortions are coupled to a (quasi-static) magnetic instead of electric field [29, 30]; for details see Appendix 4.9.4. For typical materials such as Terfenol-D the magnetic field associated with a single phonon can be estimated as $B_0 \approx (2-6)\,\mu\mathrm{T}/\sqrt{A\,[\mu\mathrm{m}^2]}$. Finally, we note that composite structures comprising both piezoelectric and piezomagnetic materials can support magneto-electric surface acoustic waves [31, 32].

4.5 SAW Cavities and Waveguides

SAW cavities.—To boost single phonon effects, it is essential to increase U_0. In analogy to cavity QED, this can be achieved by confining the SAW mode in an acoustic resonator. The physics of SAW cavities has been theoretically studied and experimentally verified since the early 1970s [14, 33]. Here, we provide an analysis of a SAW cavity based on an on-chip distributed Bragg reflector in view of potential applications in quantum information science; for details see Appendix 4.9.5. SAW resonators of this type can usually be designed to host a single resonance $f_c = \omega_c/2\pi = v_s/\lambda_c$ ($\lambda_c = 2p$) only and can be viewed as an acoustic Fabry-Perot

---

[3]The numerical values we obtain for ϕ_0 agree very well with material-dependent constants $c_\alpha = |U_\alpha/\phi| \approx (0.1\text{--}1)\,\mathrm{fm/\mu V}$ for $(\alpha = x, z)$ tabulated in Refs. [14, 15].

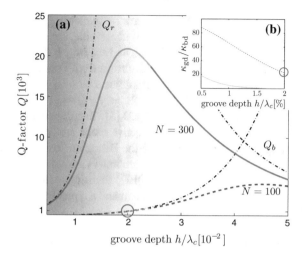

Fig. 4.2 Characterization of a groove-based SAW cavity. **a** Quality factor Q for $N = 100$ (*dashed blue*) and $N = 300$ (*red solid*) grooves as a function of the normalized grove depth h/λ_c. For shallow grooves, Q is limited by leakage losses due to imperfect acoustic mirrors (Q_r-regime, *gray area*), whereas for deep grooves conversion to bulk modes dominates (Q_b-regime); compare asymptotics (*dash-dotted lines*). **b** Ratio of desired to undesired decay rates κ_{gd}/κ_{bd}. The stronger Q is dominated by Q_r, the higher κ_{gd}/κ_{bd}. Here, $w/p = 0.5$, $D = 5.25\lambda_c$, and $f_c = 3$ GHz; typical material parameters for LiNbO$_3$ have been used [cf. Appendix 4.9.5]

resonator with effective reflection centers, sketched by localized mirrors in Fig. 4.1, situated at some effective penetration distance into the grating [14]. Therefore, the total effective cavity size along the mirror axis is $L_c > D$, where D is the physical gap between the gratings. The total cavity line-width $\kappa = \omega_c/Q = \kappa_{gd} + \kappa_{bd}$ can be decomposed into desired (leakage through the mirrors) and undesired (conversion into bulk modes and internal losses due to surface imperfections etc.) losses, labeled as κ_{gd} and κ_{bd}, respectively; for a schematic illustration compare Fig. 4.1. For the total quality factor Q, we can typically identify three distinct regimes [cf. Fig. 4.2]: For very small groove depths $h/\lambda_c \lesssim 2\%$, losses are dominated by coupling to SAW modes outside of the cavity, dubbed as Q_r-regime $(\kappa_{gd} \gg \kappa_{bd})$, whereas for very deep grooves losses due to conversion into bulk-modes become excessive (Q_b-regime, $\kappa_{gd} \ll \kappa_{bd}$). In between, for a sufficiently high number of grooves N, the quality factor Q can ultimately be limited by internal losses (surface cracks etc.), referred to as Q_m-regime $(\kappa_{gd} \ll \kappa_{bd})$. For $N \approx 300$, we find that the onset of the bulk-wave limit occurs for $h/\lambda_c \gtrsim 2.5\%$, in excellent agreement with experimental findings [33, 34]. With regard to applications in quantum information schemes, the Q_r-regime plays a special role in that resonator phonons leaking out through the acoustic mirrors can be processed further by guiding them in acoustic SAW waveguides (see below). To capture this behaviour quantitatively, we analyze κ_{gd}/κ_{bd} [cf. Fig. 4.2]: for $\kappa_{gd}/\kappa_{bd} \gg 1$, leakage through the mirrors is the strongest decay mechanism for the cavity phonon, whereas the undesired decay channels are suppressed. Our analysis shows that, for gigahertz frequencies $f_c \approx 3$ GHz, $N \approx 100$ and $h/\lambda_c \approx 2\%$, a quality factor of $Q \approx 10^3$ is achievable, together with an effective cavity confinement $L_c \approx 40\lambda_c$ (for $D \lesssim 5\lambda_c$) and $\kappa_{\mathrm{gd}}/\kappa_{\mathrm{bd}} \gtrsim 20$ (illustrated by the circle in Fig. 4.2); accordingly, the probability for a cavity phonon to leak through the mirrors (rather than into the bulk for example) is $\kappa_{\mathrm{gd}}/\left(\kappa_{\mathrm{gd}} + \kappa_{\mathrm{bd}}\right) \gtrsim 95\%$. Note that the resulting total cavity linewidth of $\kappa/2\pi = f_c/Q \approx (1-3)$ MHz is similar to the ones typically encountered in circuit QED [5]. To compare this to the effective cavity-qubit coupling, we need to fix the effective mode area of the SAW cavity. In addition to the longitudinal confinement by the Bragg mirror (as discussed above) a transverse confinement length L_{trans} (in direction $\hat{y}$) can be provided, e.g., using waveguiding, etching or (similar to cavity QED) focusing techniques [14, 35, 36]. For transverse confinement $L_{\mathrm{trans}} \approx (1-5)$ μm and a typical resonant cavity wavelength $\lambda_c \approx 1$ μm, the effective mode area is then $A = L_{\mathrm{trans}}L_c \approx (40-200)$ μm^2. In the desired regime $\kappa_{\mathrm{gd}}/\kappa_{\mathrm{bd}} \gg 1$, this is largely limited by the deliberately low reflectivity of a single groove; accordingly, the cavity mode leaks strongly into the grating such that $L_c \gg D$ [cf. Appendix 4.9.5 for details]. While we have focused on this standard Bragg design (due to its experimentally validated frequency selectivity and quality factors), let us shortly mention potential approaches to reduce A and thus increase single-phonon effects even further: (i) The most straightforward strategy (that is still compatible with the Bragg mirror design) is to reduce λ_c as much as possible, down to the maximum frequency $f_c = v_s/\lambda_c$ that can still be made resonant with the (typically highly tunable) qubit's transition frequency $\omega_{\mathrm{q}}/2\pi$; note that fundamental Rayleigh modes with $f_c \approx 6$ GHz have been demonstrated experimentally [37]. (ii) In order to increase the reflectivity of a single groove, one could use deeper grooves. To circumvent the resulting increased losses into the bulk [cf. Fig. 4.2b], free-standing structures (where the effect of bulk phonon modes is reduced) could be employed. (iii) Lastly, alternative cavity designs such as so-called trapped energy resonators make it possible to strongly confine acoustic resonances in the center of plate resonators [38].

SAW waveguides.—Not only can SAWs be confined in cavities, but they can also be guided in acoustic waveguides (WGs) [14, 39]. Two dominant types of design are: (i) Topographic WGs such as ridge-type WGs where the substrate is locally deformed using etching techniques, or (ii) overlay WGs (such as strip- or slot-type WGs) where one or two strips of one material are deposited on the substrate of another to form core and clad regions with different acoustic velocities. If the SAW velocity is slower (higher) in the film than in the substrate, the film acts as a core (cladding) for the guide whereas the unmodified substrate corresponds to the cladding (core). An attenuation coefficient of ~ 0.6 dB/mm has been reported for a 10 μm-wide slot-type WG, defined by Al cladding layers on a GaAs substrate [35, 36]. This shows that SAWs can propagate basically dissipation-free over chip-scale distances exceeding several millimeters. Typically, one-dimensional WG designs have been investigated, but—to expand the design flexibility—one could use multiple acoustic lenses in order guide SAWs around a bend [26].

4.6 Universal Coupling

Versatility.—To complete the analogy with cavity QED, a non-linear element similar to an atom needs to be introduced. In the following, we highlight three different exemplary systems, illustrating the versatility of our SAW-based platform. We focus on quantum dots, trapped ions and NV-centers, but similar considerations naturally apply to other promising quantum information candidates such as superconducting qubits [7, 13, 17, 41], Rydberg atoms [42] or electron spins bound to a phosphorus donor atom in silicon [37]. In all cases considered, a single cavity mode a, with frequency ω_c close to the relevant transition frequency, is retained. We provide estimates for the single-phonon coupling strength and cooperativity [cf. Table 4.1], while more detailed analyses go beyond the scope of this work and are subject to future research.

(i) QD Charge qubit: A natural candidate for our scheme is a charge qubit embedded in a lithographically defined GaAs double quantum dot (DQD) containing a single electron. The DQD can well be described by an effective two-level system, characterized by an energy offset ε and interdot tunneling t_c yielding a level splitting $\Omega = \sqrt{\varepsilon^2 + 4t_c^2}$ [43]. The electron's charge e couples to the piezoelectric potential; the deformation coupling is much smaller than the piezoelectric coupling and can therefore safely be neglected [44]. Since the quantum dot is small compared to the SAW wavelength, we neglect potential effects coming from the structure making up the dots (heterostructure and metallic gates); for a detailed discussion see Appendix 4.9.9. Performing a standard rotating-wave approximation (valid for $\delta, g_{ch} \ll \omega_c$), we find that the system can be described by a Hamiltonian of Jaynes-Cummings form,

$$H_{dot} = \delta S^z + g_{ch} \frac{2t_c}{\Omega} \left(S^+ a + S^- a^\dagger \right), \tag{4.5}$$

where $\delta = \Omega - \omega_c$ specifies the detuning between the qubit and the cavity mode, and $S^\pm = |\pm\rangle \langle \mp|$ (and so on) refer to pseudo-spin operators associated with the eigenstates $|\pm\rangle$ of the DQD Hamiltonian (cf. Appendix 4.9.6). The Hamiltonian H_{dot} describes the coherent exchange of excitations between the qubit and the acoustic cavity mode. The strength of this interaction $g_{ch} = e\phi_0 \mathcal{F}(kd) \sin(kl/2)$ is proportional to the charge e and the piezoelectric potential associated with a single phonon ϕ_0. The decay of the SAW mode into the bulk is captured by the function $\mathcal{F}(kd)$ [d is the distance between the DQD and the surface; see Appendix 4.9.2 for details], while the factor $\sin(kl/2)$ reflects the assumed mode function along the axis connecting the two dots, separated by a distance l. For (typical) values of $l \approx \lambda_c/2 = 250$ nm and $d \approx 50$ nm $\ll \lambda_c$, the geometrical factor $\mathcal{F}(kd) \sin(kl/2)$ then leads to a reduction in coupling strength compared to the bare value $e\phi_0$ (at the surface) by a factor of ~ 2 only. In total, we then obtain $g_{ch} \approx 2$ GHz$/\sqrt{A[\mu m^2]}$. For lateral confinement $L_{trans} \approx (1-5)$ μm, the effective mode area is $A = L_{trans} L_c \approx (20-100)$ μm^2. The resulting charge-resonator coupling strength $g_{ch} \approx (200\text{-}450)$ MHz compares well with values obtained using superconducting qubits coupled to *localized* nano-mechanical resonators made of

piezoelectric material where $g \approx (0.4 - 1.2)\,\text{GHz}$ [7, 13] or superconducting resonators coupled to Cooper pair box qubits ($g/2\pi \approx 6\,\text{MHz}$) [3], transmon qubits
($g/2\pi \approx 100\,\text{MHz}$) [45] and indium arsenide DQD qubits ($g/2\pi \approx 30\,\text{MHz}$) [46].
Note that, in principle, the coupling strength g_{ch} could be further enhanced by additionally depositing a strongly piezoelectric material such as $LiNbO_3$ on the GaAs substrate [16]. Moreover, with a $LiNbO_3$ film on top of the surface, also non-piezoelectric
materials such as Si or Ge could be used to host the quantum dots [47]. The level
splitting $\Omega(t)$ and interdot tunneling $t_c(t)$ can be tuned in-situ via external gate voltages. By controlling δ one can rapidly turn on and off the interaction between the
qubit and the cavity: For an effective interaction time $\tau = \pi/2g_{\text{eff}}$ ($g_{\text{eff}} = 2g_{ch}t_c/\Omega$)
on resonance ($\delta = 0$), an arbitrary state of the qubit is swapped to the absence or
presence of a cavity phonon, i.e., $(\alpha\,|-\rangle + \beta\,|+\rangle)\,|0\rangle \rightarrow |-\rangle\,(\alpha\,|0\rangle - i\beta\,|1\rangle)$, where
$|n\rangle$ labels the Fock states of the cavity mode. Apart from this SWAP operation, fur-
ther quantum control techniques known from cavity QED may be accessible [48].
Note that below we will generalize our results to *spin* qubits embedded in DQDs.

(ii) Trapped ion: The electric field associated with the SAW mode does not only
extend into the solid, but, for a free surface, in general there will also be an electrical
potential decaying exponentially into the vacuum above the surface $\sim \exp[-k\,|z|]$
[23]; cf. Appendix 4.9.2. This allows for coupling to systems situated above the
surface, without any mechanical contact. For example, consider a single ion of charge
q and mass m trapped at a distance d above the surface of a strongly piezoelectric
material such as $LiNbO_3$ or AlN. The electric dipole induced by the ion motion
couples to the electric field of the SAW phonon mode. The dynamics of this system
are described by the Hamiltonian

$$H_{\text{ion}} = \omega_c a^\dagger a + \omega_t b^\dagger b + g_{\text{ion}}\left(ab^\dagger + a^\dagger b\right), \qquad (4.6)$$

where b refers to the annihilation operator of the ion's motional mode and ω_t is
the (axial) trapping frequency. The single phonon coupling strength is given by
$g_{\text{ion}} = qx_0 \times k_c\phi_0 F(k_c d) = q\phi_0 F(k_c d)\,\eta_{\text{LD}}$. Apart from the exponential decay
$F(kd) = \exp[-kd]$, the effective coupling is reduced by the Lamb-Dicke parameter
$\eta_{\text{LD}} = 2\pi x_0/\lambda_c$, with $x_0 = \sqrt{\hbar/2m\omega_t}$, since the the motion of the ion is restricted to
a region small compared with the SAW wavelength λ_c. For $LiNbO_3$, a surface mode
area of $A = (1-5)\,\mu\text{m} \times 40\lambda_c$, the commonly used $^9Be^+$ ion and typical ion trap
parameters with $d \approx 30\,\mu\text{m}$ and $\omega_t/2\pi \approx 2\,\text{MHz}$ [49], we obtain $g_{\text{ion}} \approx (3\text{-}6.7)\,\text{kHz}$.
Here, g_{ion} refers to the coupling between the ion's motion and the cavity. However,
based on H_{ion}, one can in principle generalize the well-known protocols operating
on the ion's spin and motion to operations on the spin and the acoustic phonon mode
[50].

(iii) NV-center: Yet another system well suited for our scheme are NV centers in
diamond. Even though diamond itself is not piezoactive, it has played a key role in the
context of high-frequency SAW devices due to its record-high sound velocity [14]; for
example, high-performance SAW resonators with a quality factor of $Q = 12\,500$ at
$\omega_c \gtrsim 10\,\text{GHz}$ were experimentally demonstrated for AlN/diamond heterostructures
[51, 52]. To make use of the large magnetic coupling coefficient of the NV center

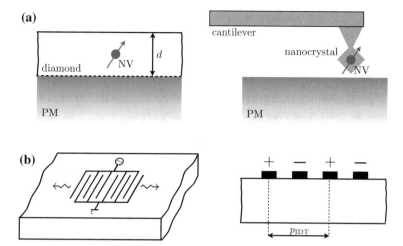

Fig. 4.3 **a** Schematic illustration for coupling to a NV center via a piezo-magnetic (PM) material (see text for details); surface grooves (not shown) can be used to provide SAW phonon confinement. **b** SAWs can be generated electrically based on standard interdigital transducers (IDTs) deposited on the surface. Typically, an IDT consists of two thin-film electrodes on a piezo-electric material, each formed by interdigitated fingers. When an ac voltage is applied to the IDT on resonance (defined by the periodicity of the fingers as $\omega_{IDT}/2\pi = v_s/p_{IDT}$, where v_s is the SAW propagation speed), it launches a SAW across the substrate surface in the two directions perpendicular to IDT fingers [14, 15, 17]

spin $\gamma_{NV} = 2\pi \times 28$ GHz/T, here we consider a hybrid device composed of a thin layer of diamond with a single (negatively charged) NV center with ground-state spin **S** implanted a distance $d \approx 10$ nm away from the interface with a strongly piezo-magnetic material. Equivalently, building upon current quantum sensing approaches [53, 54], one could use a diamond nanocrystal (typically ~ 10 nm in size) in order to get the NV centre extremely close to the surface of the piezo-magnetic material and thus maximize the coupling to the SAW cavity mode; compare Fig. 4.3a for a schematic illustration. In the presence of an external magnetic field $\mathbf{B}_{ext}$,[4] the system is described by

$$H_{NV} = DS_z^2 + \gamma_{NV}\mathbf{B}_{ext} \cdot \mathbf{S} + \omega_c a^\dagger a \qquad (4.7)$$
$$+ g_{NV} \sum_{\alpha=x,y,z} \eta_{NV}^\alpha S^\alpha \left(a + a^\dagger\right),$$

where $D = 2\pi \times 2.88$ GHz is the zero-field splitting, $g_{NV} = \gamma_{NV}B_0$ is the single phonon coupling strength and η_{NV}^α is a dimensionless factor encoding the orientation of the NV spin with respect to the magnetic stray field of the cavity mode.

---

[4]Here, the constant background field generated by the piezo-magnetic material is absorbed into the definition of the external magnetic field $\mathbf{B}_{ext}$. For a typical situation as analyzed in Ref. [30] it is roughly a factor of five smaller than $|\mathbf{B}_{ext}| \approx 0.235$ T.

For $d \ll \lambda_c$, a rough estimate shows that at least one component of η_{NV}^{α} is of order unity [30]. For a NV center close to a Terfenol-D layer of thickness $h \gg \lambda_c$, we find $g_{NV} \approx 400\,\text{kHz}/\sqrt{A\left[\mu\text{m}^2\right]}$. Thus, as compared to direct strain coupling $\lesssim 200\,\text{Hz}/\sqrt{A\left[\mu\text{m}^2\right]}$, the presence of the piezomagnetic layer is found to boost the single phonon coupling strength by three orders of magnitude; this is in agreement with previous theoretical results for a static setting [30].

Decoherence.—In the analysis above, we have ignored the presence of decoherence which in any realistic setting will degrade the effects of coherent qubit-phonon interactions. In this context, the cooperativity parameter, defined as $C = g^2 T_2 / \left[\kappa\left(\bar{n}_{th} + 1\right)\right]$, is a key figure of merit. Here, T_2 refers to the corresponding dephasing time, while $\bar{n}_{th} = (\exp\left[\hbar\omega_c / k_B T\right] - 1)^{-1}$ gives the thermal occupation number of the cavity mode at temperature T. The parameter C compares the coherent single-phonon coupling strength g with the geometric mean of the qubit's decoherence rate $\sim T_2^{-1}$ and the cavity's effective linewidth $\sim \kappa\left(\bar{n}_{th} + 1\right)$; in direct analogy to cavity QED, $C > 1$ marks the onset of coherent quantum effects in a coupled spin-oscillator system, even in the presence of noise; cf. Ref. [8] and Appendix 4.9.8 for a detailed discussion. To estimate C, we take the following parameters for the dephasing time T_2: For system (i) $T_2 \approx 10\,\text{ns}$ has been measured close to the charge degeneracy point $\varepsilon = 0$ [43]. In scenario (ii) motional decoherence rates of $0.5\,\text{Hz}$ have been measured in a cryogenically cooled trap for an ion height of $150\,\mu\text{m}$ and $1\,\text{MHz}$ motional frequency [49]. Since this rate scales as $\sim d^{-4}$ [50, 55], we take $T_2\,[\text{s}] \approx 2(d\,[\mu\text{m}]/150)^4$. Lastly, for the NV-center (iii) $T_2 \approx 0.6\,\text{s}$ has been demonstrated for ensembles of NV spins [56] and we assume an optimistic value of $T_2 = 100\,\text{ms}$, similarly to Ref. [57]. The results are summarized in Table 4.1. We find that $C > 1$ should be experimentally feasible which is sufficient to perform a quantum gate between two spins mediated by a *thermal* mechanical mode [9].

Qubit-qubit coupling.—When placing a pair of qubits into the *same* cavity, the regime of large single spin cooperativity $C \gg 1$ allows for coherent cavity-phonon-mediated interactions and quantum gates between the two spins via the effective interaction Hamiltonian $H_{int} = g_{dr}(S_1^+ S_2^- + \text{h.c.})$, where $g_{dr} = g^2/\delta \ll g$ in the so-called dispersive regime [4]. For the estimates given in Table 4.1, we have restricted ourselves to the Q_r-regime with $Q \approx 10^3$, where leakage through the acoustic mirrors dominates over undesired (non-scalable) phonon losses $\left(\kappa_{gd} \gg \kappa_{bd}\right)$. However, note that small-scale experiments using a *single* cavity only (where there is no need for guiding the SAW phonon into a waveguide for further quantum information processing) can be operated in the Q_m-regime (which is limited only by internal material losses), where the quality factor $Q \approx Q_m \gtrsim 10^5$ is maximized (and thus overall phonon losses minimal).

As a specific example, consider two NV-centers, both coupled with strength $g_{NV} \approx 100\,\text{kHz}$ to the cavity and in resonance with each other, but detuned from the resonator. Since for large detuning δ the cavity is only virtually populated, the cavity decay rate is reduced to $\kappa_{dr} = \left(g^2/\delta^2\right)\kappa \approx 10^{-2}\kappa \approx 1\,\text{kHz}$ (for $f_c = 3\,\text{GHz}$, $Q = 2 \times 10^5$), whereas the spin-spin coupling is $g_{dr} \approx 0.1 g_{NV} \approx 10\,\text{kHz}$. Therefore, $T_2 = 1$ ms is already sufficient to approach the strong-coupling regime where $g_{\mathrm{dr}} \gg \kappa_{\mathrm{dr}}, T_2^{-1}$.

Finally, we note that, in all cases considered above, one could implement a coherent, electrical control by pumping the cavity mode using standard interdigital transducers (IDTs) [14, 15, 17]; compare Fig. 4.3b for a schematic illustration. The effect of the additional Hamiltonian $H_{\mathrm{drive}} = \Xi \cos(\omega_{\mathrm{IDT}} t) \left[a + a^\dagger \right]$ can be accounted for by replacing the cavity state by a coherent state, that is $a \rightarrow \alpha$. For example, in the case of Eq. (4.5), one could then drive Rabi oscillations between the states $|+\rangle$ and $|-\rangle$ with the amplified Rabi frequency $\Omega_R = g\alpha$.

4.7 State Transfer Protocol

The possibility of quantum transduction between SAWs and different realizations of stationary qubits enables a variety of applications including quantum information achitectures that use SAW phonons as a quantum bus to couple dissimilar and/or spatially separated qubits. The most fundamental task in such a quantum network is the implementation of a state transfer protocol between two remote qubits 1 and 2, which achieves the mapping $(\alpha |0\rangle_1 + \beta |1\rangle_1) \otimes |0\rangle_2 \rightarrow |0\rangle_1 \otimes (\alpha |0\rangle_2 + \beta |1\rangle_2)$. In analogy to optical networks, this can be accomplished via coherent emission and reabsorption of a single phonon in a waveguide [11]. As first shown in the context of atomic QED [18], in principle perfect, deterministic state transfer can be implemented by identifying appropriate time-dependent control pulses.

Before we discuss a specific implementation of such a transfer scheme in detail, we provide a general approximate result for the state transfer fidelity $\mathcal{F}$. As demonstrated in detail in Appendix 4.9.8, for small infidelities one can take

$$\mathcal{F} \approx 1 - \varepsilon - \mathcal{C}C^{-1}, \tag{4.8}$$

as a *general* estimate for the state transfer fidelity. Here, individual errors arise from intrinsic phonon losses $\sim \varepsilon = \kappa_{\mathrm{bd}}/\kappa_{\mathrm{gd}}$ and qubit dephasing $\sim C^{-1} \sim T_2^{-1}$, respectively; the numerical coefficient $\mathcal{C} \sim \mathcal{O}(1)$ depends on the specific control pulse and may be optimized for a given set of experimental parameters [58]. This simple, analytical result holds for a Markovian noise model where qubit dephasing is described by a standard pure dephasing term leading to an exponential loss of coherence $\sim \exp(-t/T_2)$ and agrees well with numerical results presented in Ref. [58]. For non-Markovian qubit dephasing an even better scaling with C can be expected [9]. Using experimentally achievable parameters $\varepsilon \approx 5\%$ and $C \approx 30$, we can then estimate $\mathcal{F} \approx 90\%$, showing that fidelities sufficiently high for quantum communication should be feasible for all physical implementations listed in Table 4.1.

In the following, we detail the implementation of a transfer scheme based on *spin* qubits implemented in gate-defined double quantum dots (DQDs).[5] In particular, we consider singlet-triplet-like qubits encoded in lateral QDs, where two electrons are localized in adjacent, tunnel-coupled dots. As compared to the charge qubits discussed above, this system is known to feature superior coherence timescales [59–62]; these are largely limited by the relatively strong hyperfine interaction between the electronic spin and the nuclei in the host environment [62], resulting in a random, slowly evolving magnetic (Overhauser) field for the electronic spin. To mitigate this decoherence mechanism, two common approaches are (i) spin-echo techniques which allow to extend spin-coherence from a time-ensemble-averaged dephasing time $T_2^* \approx 100$ ns to $T_2 \gtrsim 250\,\mu$s [60], and (ii) narrowing of the nuclear field distribution [62, 63]. Recently, real-time adaptive control and estimation methods (that are compatible with arbitrary qubit operations) have allowed to narrow the nuclear spin distribution to values that prolong T_2^* to $T_2^* > 2\,\mu$s [40]. For our purposes, the latter is particularly attractive as it can be done simply before loading and transmitting the quantum information, whereas spin-echo techniques can be employed as well, however at the expense of more complex pulse sequences (see Appendix for details). In order to couple the electric field associated with the SAW cavity mode to the electron *spin* states of such a DQD, the essential idea is to make use of an effective electric dipole moment associated with the exchange-coupled spin states of the DQD [64–67]. As detailed in Appendix 4.9.7, we then find that in the usual singlet-triplet subspace spanned by the two-electron states $\{|{\uparrow}{\Downarrow}\rangle, |{\Downarrow}{\uparrow}\rangle\}$, a single node can well be described by the prototypical Jaynes-Cummings Hamiltonian given in Eq. (4.1). As compared to the direct charge coupling g_{ch}, the single phonon coupling strength g is reduced since the qubit states $|l\rangle$ have a small admixture of the localized singlet $\langle S_{02}|l\rangle$ ($l = 0, 1$) only. Using typical parameters values, we find $g \approx 0.1 g_{\text{ch}} \approx 200\,\text{MHz}/\sqrt{A\left[\mu\text{m}^2\right]}$.[6] In this system, the coupling $g(t)$ can be tuned with great flexibility via both the tunnel-coupling t_c and/or the detuning parameter ε.

The state transfer between two such singlet-triplet qubits connected by a SAW waveguide can be adequately described within the theoretical framework of cascaded quantum systems, as outlined in detail for example in Refs. [11, 18, 68, 69]: The underlying quantum Langevin equations describing the system can be converted into an effective, cascaded Master equation for the system's density matrix ρ. For the relevant case of two qubits, it can be written as $\dot{\rho} = \mathcal{L}_{\text{ideal}}\rho + \mathcal{L}_{\text{noise}}\rho$, where

---

[5]This system is a particulary promising physical realization as it provides a relatively high cooperativity C even in the absence of spin-echo pulses, the spin-resonator Hamiltonian takes on the generic Jaynes-Cummings Hamiltonian Eq. (4.1) in a rather straight-forward way (Appendix 4.9.7) [whereas Eqs. (4.6) and (4.7) require further ingredients such as a MW drive (compare for example Ref. [11]) to effectively obtain Eq. (4.1)], and, outside of the state-transfer time window, it provides better control techniques such as spin-echo pulses than the simple charge-qubit scenario.

[6]Note, however, that more elaborate optimization yields an optimal spot with an even larger coupling $g \approx 0.3 g_{\text{ch}}$ [64].

$$\mathcal{L}_{\text{ideal}}\rho = -i\left[H_S(t) + i\kappa_{\text{gd}}\left(a_1^\dagger a_2 - a_2^\dagger a_1\right), \rho\right]$$
$$+ 2\kappa_{\text{gd}}\mathcal{D}[a_1 + a_2]\rho, \tag{4.9}$$

$$\mathcal{L}_{\text{noise}}\rho = 2\kappa_{\text{bd}}\sum_{i=1,2}\mathcal{D}[a_i]\rho - i\sum_i \delta_i\left[S_i^z, \rho\right]. \tag{4.10}$$

Here, $\mathcal{D}[a]\rho = a\rho a^\dagger - \frac{1}{2}\{a^\dagger a, \rho\}$ is a Lindblad term with jump operator a and $H_S(t) = \sum_i H_i(t)$, with $H_i(t) = g_i(t)[S_i^+ a_i + S_i^- a_i^\dagger]$ describes the coherent Jaynes-Cummings dynamics of the two nodes. The ideal cascaded interaction is captured by $\mathcal{L}_{\text{ideal}}$ which contains the *non-local* coherent environment-mediated coupling transferring excitations from qubit 1 to qubit 2,[7] while $\mathcal{L}_{\text{noise}}$ summarizes undesired decoherence processes: We account for intrinsic phonon losses (bulk-mode conversion etc.) with a rate κ_{bd} and (non-exponential) qubit dephasing. Since the nuclear spins evolve on relatively long time-scales, the electronic spins in quantum dots typically experience non-Markovian noise leading to a non-exponential loss of coherence on a characteristic time-scale $T_2^\star$ given by the width of the nuclear field distribution σ_{nuc} as $T_2^\star = \sqrt{2}/\sigma_{\text{nuc}}$ [40, 62]. Recently, a record-low value of $\sigma_{\text{nuc}}/2\pi = 80\,\text{kHz}$ has been reported [40], yielding an extended time-ensemble-averaged electron dephasing time of $T_2^\star = 2.8\,\mu\text{s}$. In our model, to realistically account for the dephasing induced by the quasi-static, yet unknown Overhauser field, the detuning parameters δ_i are sampled independently from a normal distribution $p(\delta_i)$ with zero mean (since nominal resonance can be achieved via the electronic control parameters) and standard deviation σ_{nuc} [63]; see Appendix 4.9.7 for details. In Appendix 4.9.10 we also provide numerical results for standard Markovian dephasing, showing that non-Markovian noise is beneficial in terms of faithful state transfer.

Under ideal conditions where $\mathcal{L}_{\text{noise}} = 0$, the setup is analogous to the one studied in Ref. [18] and the same time-symmetry arguments can be employed to determine the optimal control pulses $g_i(t)$ for faithful state transfer: if a phonon is emitted by the first node, then, upon reversing the direction of time, one would observe perfect reabsorption. By engineering the emitted phonon wavepacket such that it is invariant under time reversal and using a time-reversed control pulse for the second node $g_2(t) = g_1(-t)$, the absorption process in the second node is a time-reversed copy of the emission in the first and therefore in principle perfect. Based on this reasoning (for details see Ref. [18]), we find the explicit, optimal control pulse shown in Fig. 4.4c.

To account for noise, we simulate the full master equation numerically. The results are displayed in Fig. 4.4a, where for every random pair $\delta = (\delta_1, \delta_2)$ the fidelity of the protocol is defined as the overlap between the target state $|\psi_{\text{tar}}\rangle$ and the actual state after the transfer $\rho(t_f)$, that is $\mathcal{F}_\delta = \langle\psi_{\text{tar}}|\rho(t_f)|\psi_{\text{tar}}\rangle$. The average fidelity $\bar{\mathcal{F}}$ of the protocol is determined by averaging over the classical noise in δ, that

---

[7]In a perfect realization of the transfer scheme the system remains in a dark state throughout the evolution and no field is scattered from the second mirror. Therefore, the assumption of unidirectional propagation is not strictly needed; compare Ref. [18].

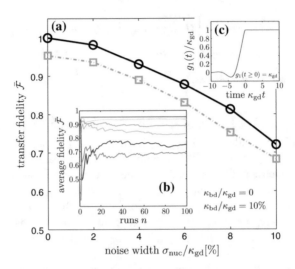

Fig. 4.4 a Average fidelity $\bar{\mathcal{F}}$ of the state transfer protocol for a coherent superposition $|\psi\rangle = (|0\rangle - |1\rangle)/\sqrt{2}$ in the presence of quasi-static (non-Markovian) Overhauser noise, as a function of the root-mean-square fluctuations σ_{nuc} in the detuning parameters δ_i ($i = 1, 2$), for $\kappa_{\mathrm{bd}}/\kappa_{\mathrm{gd}} = 0$ (*solid line, circles*) and $\kappa_{\mathrm{bd}}/\kappa_{\mathrm{gd}} = 10\%$ (*dash-dotted line, squares*). **b** After $n = 100$ runs with random values for δ_i, $\bar{\mathcal{F}}$ approximately reaches convergence. The *curves* refer to $\sigma_{\mathrm{nuc}}/\kappa_{\mathrm{gd}} = (0, 2, \ldots, 10)\%$ (from *top* to *bottom*) for $\kappa_{\mathrm{bd}}/\kappa_{\mathrm{gd}} = 10\%$. **c** Pulse shape $g_1(t)$ for first node

is $\bar{\mathcal{F}} = \int d\delta_1 d\delta_2 p(\delta_1) p(\delta_2) \mathcal{F}_\delta$. Taking an effective mode area $A \approx 100\,\mu\mathrm{m}^2$ as above and $Q \approx 10^3$ to be well within the Q_r-regime where $\kappa_{\mathrm{bd}}/\kappa_{\mathrm{gd}} \approx 5\%$, we have $g \approx \kappa_{\mathrm{gd}} \approx 20\,\mathrm{MHz}$. For two nodes separated by millimeter distances, propagation losses are negligible and $\kappa_{\mathrm{bd}}/\kappa_{\mathrm{gd}} \approx 5\%$ captures well all intrinsic phonon losses during the transfer. We then find that for realistic undesired phonon losses $\kappa_{\mathrm{bd}}/\kappa_{\mathrm{gd}} \approx 5\%$ and $\sigma_{\mathrm{nuc}}/2\pi = 80\,\mathrm{kHz}$ (such that $\sigma_{\mathrm{nuc}}/\kappa_{\mathrm{gd}} \approx 2.5\%$) [40], transfer fidelities close to 95% seem feasible. Notably, this could be improved even further using spin-echo techniques such that $T_2 \approx 10^2 T_2^\star$ [60]. Therefore, state transfer fidelities $\mathcal{F} > 2/3$ as required for quantum communication [70] seem feasible with present technology. Near unit fidelities might be approached from further optimizations of the system's parameters, the cavity design, the control pulses and/or from communication protocols that correct for errors such as phonon losses [71–73]. Once the transfer is complete, the system can be tuned adiabatically into a storage regime which immunizes the qubit against electronic noise and dominant errors from hyperfine interaction with ambient nuclear spins can be mitigated by standard, occasional refocusing of the spins [60, 64]. Alternatively, one could also investigate silicon dots: while this setup requires a more sophisticated hetero-structure including some piezo-electric layer (as studied experimentally in Ref. [37]), it potentially benefits from prolonged dephasing times $T_2^\star > 100\,\mu\mathrm{s}$ [74], since nuclear spins are largely absent in isotopically purified $^{28}\mathrm{Si}$.

4.8 Conclusion and Outlook

In summary, we have proposed and analyzed SAW phonons in piezo-active materials (such as GaAs) as a universal quantum transducer that allows to convert quantum information between stationary and propagating realizations. We have shown that a sound-based quantum information architecture based on SAW cavities and waveguides is very versatile, bears striking similarities to cavity QED and can serve as a scalable mediator of long-range spin-spin interactions between a variety of qubit implementations, allowing for faithful quantum state transfer between remote qubits with existing experimental technology. The proposed combination of techniques and concepts known from quantum optics and quantum information, in conjunction with the technological expertise for SAW devices, is likely to lead to further, rapid theoretical and experimental progress.

Finally, we highlight possible directions of research going beyond our present work: First, since our scheme is not specific to any particular qubit realization, novel hybrid systems could be developed by embedding dissimilar systems such as quantum dots and superconducting qubits into a common SAW architecture. Second, our setup could also be used as a transducer between *different* propagating quantum systems such as phonons and photons. Light can be coupled into the SAW circuit via (for example) NV-centers or self-assembled quantum dots and structures guiding both photons and SAW phonons have already been fabricated experimentally [35, 36]. Finally, the SAW architecture opens up a novel, on-chip test-bed for investigations reminiscent of quantum optics, bringing the highly developed toolbox of quantum optics and cavity-QED to the widely anticipated field of *quantum acoustics* [10, 16, 17, 75]. Potential applications include quantum simulation of many-body dynamics [76], quantum state engineering (yielding for example squeezed states of sound), quantum-enhanced sensing, sound detection, and sound-based material analysis.

4.9 Appendix to Chap. 4

4.9.1 Classical Description of Nonpiezoelectric Surface Acoustic Waves

In this Appendix, we review the general (classical) theoretical framework describing SAW in cubic lattices, such as diamond or GaAs. We derive an analytical solution for propagation in the [110] direction. The latter is of particular interest in piezoelectric systems. The classical description of SAW is explicitly shown here to make our work self-contained, but follows standard references such as Refs. [23].

Wave equation.—The propagation of acoustic waves (bulk and surface waves) in a solid is described by the equation

$$\rho \ddot{u}_i(\mathbf{x}, t) = \frac{\partial T_{ij}}{\partial x_j}, \tag{4.11}$$

where $\mathbf{u}$ denotes the displacement vector with u_i being the displacement along the cartesian coordinate $\hat{x}_i$ ($\hat{x}_1 = \hat{x}$, $\hat{x}_2 = \hat{y}$, $\hat{x}_3 = \hat{z}$), ρ gives the mass density and T is the stress tensor; T_{ij} is the ith component of force per unit area perpendicular to the $\hat{x}_j$-axis. Moreover, $\mathbf{x}$ is the cartesian coordinate vector, where in the following we assume a material with infinite dimensions in $\hat{x}$, $\hat{y}$, and a surface perpendicular to the $\hat{z}$-direction at $z = 0$. The stress tensor obeys a generalized Hooke's law (stress is linearly proportional to strain)

$$T_{ij} = c_{ijkl} u_{kl}, \tag{4.12}$$

where the strain tensor is defined as

$$u_{kl} = \frac{1}{2}\left(\frac{\partial u_k}{\partial x_l} + \frac{\partial u_l}{\partial x_k}\right). \tag{4.13}$$

Using the symmetry $c_{ijkl} = c_{ijlk}$, in terms of displacements we find

$$T_{ij} = c_{ijkl}\frac{\partial u_k}{\partial x_l}, \tag{4.14}$$

such that Eq. (4.11) takes the form of a set of three coupled wave equations

$$\rho \ddot{u}_i(\mathbf{x}, t) - c_{ijkl}\frac{\partial^2 u_k}{\partial x_j \partial x_l} = 0. \tag{4.15}$$

The elasticity tensor $\underline{c}$ obeys the symmetries $c_{ijkl} = c_{jikl} = c_{ijlk} = c_{klij}$ and is largely defined by the crystal symmetry.

Mechanical boundary condition.—The free surface at $z = 0$ is stress free (no external forces are acting upon it), such that the three components of stress across $z = 0$ shall vanish, that is $T_{13} = T_{23} = T_{33} = 0$. This results in the boundary conditions

$$T_{i\hat{z}} = c_{i\hat{z}kl}\frac{\partial u_k}{\partial x_l} = 0 \quad \text{at } z = 0. \tag{4.16}$$

Cubic lattice.—For a cubic lattice (such as GaAs or diamond) the elastic tensor c_{ijkl} has three independent elastic constants, generally denoted by c_{11}, c_{12}, and c_{44}; compare Table 4.2. Taking the three direct two-fold axes as the coordinate axes, the wave equations then read

$$\rho\frac{\partial^2 u_x}{\partial t^2} = c_{11}\frac{\partial^2 u_x}{\partial x^2} + c_{44}\left[\frac{\partial^2 u_x}{\partial y^2} + \frac{\partial^2 u_x}{\partial z^2}\right]$$
$$+ (c_{12} + c_{44})\left[\frac{\partial^2 u_y}{\partial x \partial y} + \frac{\partial^2 u_z}{\partial x \partial z}\right], \tag{4.17}$$

Table 4.2 Material properties [23] for both diamond and GaAs

	c_{11}	c_{12}	c_{44}	$\rho[kg/m^3]$	e_{14}
Diamond	107.9	12.4	57.8	3515	0
GaAs	12.26	5.71	6.00	5307	0.157

The elastic tensor $\underline{c}$ has three independent parameters, given in units of $[10^{10}\,N/m^2]$, while the piezoelectric tensor $\underline{e}$ has a single independent parameter e_{14} for cubic materials (units of C/m^2)

(and cyclic permutations) while the mechanical boundary conditions can be written as

$$T_{13} = c_{44}\left(\frac{\partial u_z}{\partial x} + \frac{\partial u_x}{\partial z}\right) = 0, \tag{4.18}$$

$$T_{23} = c_{44}\left(\frac{\partial u_z}{\partial y} + \frac{\partial u_y}{\partial z}\right) = 0, \tag{4.19}$$

$$T_{33} = c_{11}\frac{\partial u_z}{\partial z} + c_{12}\left(\frac{\partial u_x}{\partial x} + \frac{\partial u_y}{\partial y}\right) = 0, \tag{4.20}$$

at $z = 0$. In the following we seek for solutions which propagate along the surface with a wavevector $\mathbf{k} = k\left(l\hat{x} + m\hat{y}\right)$, where $l = \cos(\theta)$, $m = \sin(\theta)$ and θ is the angle between the $\hat{x}$-axis and $\mathbf{k}$. Following Ref. [77], we make the ansatz

$$\begin{pmatrix} u_x \\ u_y \\ u_z \end{pmatrix} = \begin{pmatrix} U \\ V \\ W \end{pmatrix} e^{-kqz} e^{ik(lx+my-ct)}, \tag{4.21}$$

where the decay constant q describes the exponential decay of the surface wave into the bulk and c is the phase velocity. Plugging this ansatz into the mechanical wave equations can be rewritten as $\mathcal{M}\mathbf{A} = 0$, where

$$\mathcal{M} = \begin{pmatrix} c_{11}l^2 + c_{44}\left(m^2 - q^2\right) - \rho c^2 & lm\,(c_{12} + c_{44}) & lq\,(c_{12} + c_{44}) \\ lm\,(c_{12} + c_{44}) & c_{11}m^2 + c_{44}\left(l^2 - q^2\right) - \rho c^2 & mq\,(c_{12} + c_{44}) \\ lq\,(c_{12} + c_{44}) & mq\,(c_{12} + c_{44}) & c_{11}q^2 - c_{44} + \rho c^2 \end{pmatrix}, \tag{4.22}$$

and $\mathbf{A} = (U, V, iW)$. Nontrivial solutions for this homogeneous set of equations can be found if the determinant of $\mathcal{M}$ vanishes, resulting in the so-called *secular equation* $\det(\mathcal{M}) = 0$. The secular equation is of sixth order in q; as all coefficients in the secular equation are real, there are, in general, three complex-conjugate roots q_1^2, q_2^2, q_3^2, with the phase velocity c and propagation direction θ as parameters. If the medium lies in the half space $z > 0$, the roots with negative real part will lead to a solution which does not converge as $z \to \infty$. Thus, only the roots which lead to vanishing displacements deep in the bulk are kept. Then, the most general solution can be written as a superposition of surface waves with allowed q_r values as

$$\left(u_x, u_y, i u_z\right) = \sum_{r=1,2,3} \left(\xi_r, \eta_r, \zeta_r\right) K_r e^{-k q_r z} e^{i k (l x + m y - c t)}, \tag{4.23}$$

where, for any $q_r = q_r(c, \theta)$, the ratios of the amplitudes can be calculated according to

$$K_r = \frac{U_r}{\xi_r} = \frac{V_r}{\eta_r} = \frac{i W_r}{\zeta_r}, \tag{4.24}$$

where we have introduced the quantities

$$\xi_r = \begin{vmatrix} c_{11} m^2 + c_{44} \left(l^2 - q_r^2\right) - \rho c^2 & m q_r \left(c_{12} + c_{44}\right) \\ m q_r \left(c_{12} + c_{44}\right) & c_{11} q_r^2 - c_{44} + \rho c^2 \end{vmatrix}, \tag{4.25}$$

$$\eta_r = \begin{vmatrix} m q_r \left(c_{12} + c_{44}\right) & l m \left(c_{12} + c_{44}\right) \\ c_{11} q_r^2 - c_{44} + \rho c^2 & l q_r \left(c_{12} + c_{44}\right) \end{vmatrix},$$

and

$$\zeta_r = \begin{vmatrix} l m \left(c_{12} + c_{44}\right) & c_{11} m^2 + c_{44} \left(l^2 - q_r^2\right) - \rho c^2 \\ l q_r \left(c_{12} + c_{44}\right) & m q_r \left(c_{12} + c_{44}\right) \end{vmatrix}. \tag{4.26}$$

Note that for each root q_r and displacement u_i there is an associated amplitude. The phase velocity c, however, is the same for every root q_r, and needs to be determined from the mechanical boundary conditions as described below. Similarly to the acoustic wave equations, the boundary conditions can be rewritten as $\mathcal{B}(K_1, K_2, K_3) = 0$, where the boundary condition matrix $\mathcal{B}$ is

$$\mathcal{B} = \begin{pmatrix} l \zeta_1 - q_1 \xi_1 & l \zeta_2 - q_2 \xi_2 & l \zeta_3 - q_3 \xi_3 \\ m \zeta_1 - q_1 \eta_1 & m \zeta_2 - q_2 \eta_2 & m \zeta_3 - q_3 \eta_3 \\ l \xi_1 + m \eta_1 + a q_1 \zeta_1 & l \xi_2 + m \eta_2 + a q_2 \zeta_2 & l \xi_3 + m \eta_3 + a q_3 \zeta_3 \end{pmatrix}, \tag{4.27}$$

with $a = c_{11}/c_{12}$. Again, nontrivial solutions are found for $\det(\mathcal{B}) = 0$. The requirements $\det(\mathcal{M}) = 0$, $\det(\mathcal{B}) = 0$ together with Eq. (4.24) constitute the formal solution of the problem [77]; $\det(\mathcal{M}) = 0$ and $\det(\mathcal{B}) = 0$ may be seen as determining c^2 and q^2, and Eq. (4.24) then gives the the ratios of the components of the displacement. In the following, we discuss a special case where one can eliminate the q-dependence in $\det(\mathcal{B}) = 0$, leading to an explicit, analytically simple equation for the phase velocity c, which depends only on the material properties.

   Propagation in [110] direction.—The wave equations simplify for propagation in high-symmetry directions. Here, we consider propagation in the [110]-direction, for which $l = m = 1/\sqrt{2}$; we define the diagonal as $\hat{x}' = \left(\hat{x} + \hat{y}\right)/\sqrt{2}$. Subtracting the second row from the first in $\mathcal{M}$, one finds that the common factor $(c_{11} - c_{12})/2 - c_{44} q^2 - \rho c^2$ divides through the first row, which then becomes $(1, -1, 0)$. Thus, $U = V$ and the wave equations can be simplified to $\mathcal{M}_{110}(U, i W) = 0$, where

$$\mathcal{M}_{110} = \begin{pmatrix} c'_{11} - \rho c^2 - c_{44} q^2 & \frac{q}{\sqrt{2}} (c_{12} + c_{44}) \\ \sqrt{2} q (c_{12} + c_{44}) & c_{11} q^2 - c_{44} + \rho c^2 \end{pmatrix}, \tag{4.28}$$

with $c'_{11} = (c_{11} + c_{12} + 2 c_{44})/2$. Then, the secular equation det $(\mathcal{M}_{110}) = 0$ is found to be

$$\left(c'_{11} - \rho c^2 - c_{44} q^2 \right) \left(c_{44} - \rho c^2 - c_{11} q^2 \right)$$
$$+ (c_{12} + c_{44})^2 q^2 = 0, \tag{4.29}$$

yielding the roots q_1^2, q_2^2. We choose the roots commensurate with the convergence condition yielding the general ansatz

$$\begin{pmatrix} u_{x'} \\ i u_z \end{pmatrix} = \sum_{r=1,2} \begin{pmatrix} U'_r \\ i W_r \end{pmatrix} e^{-k q_r z} e^{ik(x'-ct)}. \tag{4.30}$$

with $u_x = u_y = u_{x'}/\sqrt{2}$. The amplitude ratios $\gamma'_r = i W_r / U'_r$ can be obtained from the kernel of $\mathcal{M}$ as

$$\gamma'_r = q_r \frac{c_{12} + c_{44}}{c_{44} - c_{11} \left(X + q_r^2 \right)}, \tag{4.31}$$

where $X = \rho c^2 / c_{11}$. In the coordinate system $\{\hat{x}', \hat{z}\}$, the mechanical boundary conditions read

$$\frac{\partial u_z}{\partial x'} + \frac{\partial u_{x'}}{\partial z} = 0, \quad (z = 0) \tag{4.32}$$

$$c_{12} \frac{\partial u_{x'}}{\partial x'} + c_{11} \frac{\partial u_z}{\partial z} = 0. \quad (z = 0) \tag{4.33}$$

For the ansatz given in Eq. (4.30), they can be reformulated as $\mathcal{B}_{110} \left(U'_1, U'_2 \right) = 0$ with

$$\mathcal{B}_{110} = \begin{pmatrix} \gamma'_1 - q_1 & \gamma'_2 - q_2 \\ 1 + \frac{c_{11}}{c_{12}} q_1 \gamma'_1 & 1 + \frac{c_{11}}{c_{12}} q_2 \gamma'_2 \end{pmatrix}. \tag{4.34}$$

The requirement det $(\mathcal{B}_{110}) = 0$ can be written as

$$q_1 \left[c_{12} + \rho c^2 + c_{11} q_1^2 \right] \left[c_{12} \left(c_{44} - \rho c^2 \right) + c_{11} c_{44} q_2^2 \right] -$$
$$q_2 \left[c_{12} + \rho c^2 + c_{11} q_2^2 \right] \left[c_{12} \left(c_{44} - \rho c^2 \right) + c_{11} c_{44} q_1^2 \right] = 0.$$

From the symmetry of this equation it is clear that one can remove a factor $(q_1 - q_2)$ leading to

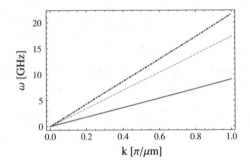

Fig. 4.5 Dispersion relation $\omega_n = v_n k$ of the three $(n = 1, 2, 3)$ Rayleigh-type SAW modes for propagation along $\hat{x}' \parallel [110]$. If not stated otherwise, we refer to the the lowest frequency solution as the SAW mode (*solid line*)

$$c_{12}\left(\frac{c_{12}}{c_{11}} + X\right)\left(\frac{c_{44}}{c_{11}} - X\right) + c_{44}q_1^2 q_2^2 + c_{12} \times$$

$$\left(\frac{c_{44}}{c_{11}} - X\right)\left(q_1^2 + q_2^2 + q_1 q_2\right) - c_{44}q_1 q_2 \left(\frac{c_{12}}{c_{11}} + X\right) = 0.$$

Using simple expressions for $q_1^2 q_2^2$ and $q_1^2 + q_2^2$ obtained from Eq. (4.29), one arrives at the following explicit equation for the wave velocity c [23, 77]

$$\left(1 - \frac{c_{11}}{c_{44}}X\right)\left(\frac{c_{11}c_{11}' - c_{12}^2}{c_{11}^2} - X\right)^2 = X^2\left(\frac{c_{11}'}{c_{11}} - X\right), \tag{4.35}$$

which is cubic in $X = \rho c^2 / c_{11}$. If not stated otherwise, we consider the mode with the lowest sound velocity, referred to as Rayleigh mode; compare Fig. 4.5.

Using the secular equation given in Eq. (4.29) and the mechanical boundary conditions, the ansatz given in Eq. (4.30) can be simplified as follows: The roots compatible with convergence in the bulk are complex conjugate, i.e. $q \equiv q_1 = q_2^*$, and therefore $\gamma \equiv \gamma_1' = \gamma_2^*$. Then, using the first row in the boundary condition matrix [compare Eq. (4.34)], we can deduce

$$U_1' = Ue^{-i\varphi}, \quad U_2' = Ue^{i\varphi}, \tag{4.36}$$

where

$$e^{-2i\varphi} = -\frac{\gamma^* - q^*}{\gamma - q}. \tag{4.37}$$

In summary, we find the following solution [23]

$$u_{x'} = U\left(e^{-qkz-i\varphi} + \text{h.c.}\right)e^{ik(x'-ct)},$$

$$iu_z = U\left(\gamma e^{-qkz-i\varphi} + \text{h.c.}\right)e^{ik(x'-ct)}, \tag{4.38}$$

Fig. 4.6 Depth dependence of the (normalized) vertical displacement u_z/U along $\hat{x}' \| [110]$ for a Rayleigh surface acoustic wave propagating on a (001) GaAs crystal. The acoustic amplitude decays away from the surface into the bulk on a characteristic length scale approximately given by the SAW wavelength $\lambda = 2\pi/k \approx 1\,\mu m$

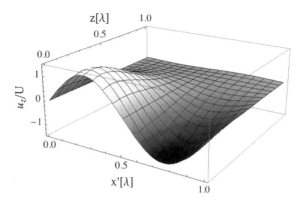

where the material-dependent parameters c, q, γ and φ are determined by Eqs. (4.35), (4.29), (4.31) and (4.35), respectively. For the GaAs parameters given in Table 4.2, we get $c = 2878\,\text{m/s}, q = 0.5 + 0.48i, \gamma = -0.68 + 1.16i$ and $\varphi = 1.05$, respectively. The corresponding (normalized) transversal displacement is displayed in Fig. 4.6.

4.9.2   Surface Acoustic Waves in Piezoelectric Materials

In a piezoelectric material, elastic and electromagnetic waves are coupled. In principle, the field distribution can be found only by solving simultaneously the equations of both Newton and Maxwell. The corresponding solutions are hybrid elasto-electromagnetic waves, i.e., elastic waves with velocity v_s accompanied by electric fields, and electromagnetic waves with velocity $c \approx 10^5 v_s$ accompanied by mechanical strains. For the first type of wave, the magnetic field is negligible, because it is due to an electric field traveling with a velocity v_s much slower than the speed of light c; therefore, one can approximate Maxwell's equations as $\nabla \times \mathbf{E} = -\partial \mathbf{B}/\partial t \approx 0$, giving $\mathbf{E} = -\nabla\phi$. Thus, the propagation of elastic waves in a piezoelectric material can be described within the quasi-static approximation, where the electric field is essentially static compared to electromagnetic fields [24]. The potential ϕ and the associated electric field are not electromagnetic in nature but rather a component of the predominantly mechanical wave propagating with velocity v_s.

General Analysis
Wave equation.—The basic equations that govern the propagation of acoustic waves in a piezoelectrical material connect the mechanical stress T and the electrical displacement $\mathbf{D}$ with the mechanical strain and the electrical field. The coupled constitutive equations are

$$T_{ij} = c_{ijkl} \frac{\partial u_k}{\partial x_l} + e_{kij} \frac{\partial \phi}{\partial x_k},$$

$$D_i = -\varepsilon_{ij} \frac{\partial \phi}{\partial x_j} + e_{ijk} \frac{\partial u_j}{\partial x_k}, \tag{4.39}$$

where e with $\left(e_{ijk} = e_{ikj}\right)$ and ε are the piezoelectric and permittivity tensor, respectively. Here, Hooke's law is extended by the additional stress term due to the piezoelectric effect, while the equation for the displacement D_i includes the polarization produced by the strain. Therefore, Newton's law becomes

$$\rho \ddot{u}_i = c_{ijkl} \frac{\partial^2 u_k}{\partial x_j \partial x_l} + e_{kij} \frac{\partial^2 \phi}{\partial x_j \partial x_k}. \tag{4.40}$$

For an insulating solid, the electric displacement D_i must satisfy Poisson's equation $\partial D_i / \partial x_i = 0$ which yields

$$e_{ijk} \frac{\partial^2 u_j}{\partial x_i \partial x_k} - \varepsilon_{ij} \frac{\partial^2 \phi}{\partial x_i \partial x_j} = 0, \quad z > 0 \tag{4.41}$$

$$\triangle \phi = 0, \quad z > 0. \tag{4.42}$$

Mechanical boundary conditions.—In the presence of piezoelectric coupling the mechanical boundary conditions [compare Eq. (4.16)] generalize to

$$T_{i\hat{z}} = c_{i\hat{z}kl} \frac{\partial u_k}{\partial x_l} + e_{ki\hat{z}} \frac{\partial \phi}{\partial x_k} = 0 \quad \text{at } z = 0. \tag{4.43}$$

Using the symmetries $c_{ijkl} = c_{jikl}$ and $e_{kij} = e_{kji}$ it is easy to check that this is equivalent to Eq. (41) in Ref. [23].

Electric boundary condition.—In addition to the stress-free boundary conditions, piezoelectricity introduces an electric boundary condition: The normal component of the electric displacement needs to be continuous across the surface [36], that is

$$D_z \left(z = 0^+\right) = D_z \left(z = 0^-\right), \tag{4.44}$$

where by definition $D_z = e_{\hat{z}jk} \partial u_j / \partial x_k - \varepsilon_{\hat{z}j} \partial \phi / \partial x_j$. Outside of the medium ($z < 0$), we assume vacuum; thus, $D_z = \varepsilon_0 E_z = -\varepsilon_0 \partial \phi_{\text{out}} / \partial z$, where the electrical potential has to satisfy Poisson's equation $\triangle \phi_{\text{out}} = 0$. The ansatz

$$\phi_{\text{out}} = A_{\text{out}} e^{ik(x'-ct)} e^{\Omega k z} \tag{4.45}$$

gives $\triangle \phi_{\text{out}} = \left(-k^2 + \Omega^2 k^2\right) \phi_{\text{out}} = 0$. Thus, for proper convergence far away from the surface $z \to -\infty$, we take the decay constant $\Omega = 1$; accordingly, the electrical potential decays exponentially into the vacuum above the surface on a typical length scale given by the SAW wavelength $\lambda = 2\pi/k \approx 1\,\mu\text{m}$. Therefore, for the electrical displacement outside of the medium, we find $D_z = -\varepsilon_0 k\phi$. Lastly, the electrical potential has to be continuous across the surface [23], i.e.,

$$\phi\left(z = 0^+\right) = \phi_{\text{out}}\left(z = 0^-\right),$$ (4.46)

which allows us to determine the amplitude A_{out}. In summary, Eq. (4.44) can be rewritten as

$$\left(e_{\hat{z}jk}\partial u_j/\partial x_k - \varepsilon_{\hat{z}j}\partial\phi/\partial x_j + \varepsilon_0 k\phi\right)\big|_{z=0} = 0.$$ (4.47)

Cubic lattice

Cubic lattice.—For a cubic, piezoelectric system there is only one independent nonzero component of the piezoelectric tensor called e_{14} [23, 24]. With this piezo-electric coupling, the wave equations are given by four coupled partial differential equations

$$\rho\frac{\partial^2 u_x}{\partial t^2} = c_{11}\frac{\partial^2 u_x}{\partial x^2} + c_{44}\left[\frac{\partial^2 u_x}{\partial y^2} + \frac{\partial^2 u_x}{\partial z^2}\right]$$ (4.48)

$$+ (c_{12} + c_{44})\left[\frac{\partial^2 u_y}{\partial x\partial y} + \frac{\partial^2 u_z}{\partial x\partial z}\right] + 2e_{14}\frac{\partial^2 \phi}{\partial y\partial z},$$

$$\varepsilon\Delta\phi = 2e_{14}\left[\frac{\partial^2 u_x}{\partial y\partial z} + \frac{\partial^2 u_y}{\partial x\partial z} + \frac{\partial^2 u_z}{\partial x\partial y}\right],$$ (4.49)

and cyclic for u_y and u_z. Here, Δ is the Laplacian and ε is the dielectric constant of the medium. For a cubic lattice, the *mechanical boundary conditions* at $z = 0$ explicitly read

$$T_{13} = c_{44}\left(\frac{\partial u_z}{\partial x} + \frac{\partial u_x}{\partial z}\right) + e_{14}\frac{\partial\phi}{\partial y} = 0,$$ (4.50)

$$T_{23} = c_{44}\left(\frac{\partial u_z}{\partial y} + \frac{\partial u_y}{\partial z}\right) + e_{14}\frac{\partial\phi}{\partial x} = 0,$$ (4.51)

$$T_{33} = c_{11}\frac{\partial u_z}{\partial z} + c_{12}\left(\frac{\partial u_x}{\partial x} + \frac{\partial u_y}{\partial y}\right) = 0,$$ (4.52)

while the *electrical boundary condition* [compare the general relation in Eq. (4.47)] leads to

$$\left[e_{14}\left(\frac{\partial u_x}{\partial y} + \frac{\partial u_y}{\partial x}\right) - \varepsilon\frac{\partial\phi}{\partial z} + \varepsilon_0 k\phi\right]_{z=0} = 0.$$ (4.53)

In general, the wave equations can be formulated into a 4×4 matrix $\mathcal{M}$; the condition $\det\mathcal{M} = 0$ can then used to find the four decay constants. In addition, the mechanical and electrical boundary conditions can be recast to a 4×4 boundary condition matrix $\mathcal{B}$, from which one can deduce the allowed phase velocities of the piezoelectric SAW by solving $\det\mathcal{B} = 0$.

Perturbative treatment.—For materials with weak piezoelectric coupling (such as GaAs), the properties of surface acoustic waves are primarily determined by the elastic constants and density of the medium. Then, within a perturbative treatment of the piezoelectric coupling, one can obtain analytical expressions for the strain and piezoelectric fields. Here, we summarize the results for SAWs propagating along $\hat{x}'||[110]$ of the $\hat{z}||[001]$ surface following Refs. [23, 25]. Since the piezoelectric coupling e_{14} is small, it follows from Eq. (4.49) that ϕ will be order e_{14} smaller than the mechanical displacements u, that is

$$\phi \sim \frac{e_{14}}{\varepsilon}u. \tag{4.54}$$

This results in additional terms in the wave equations that are of order $\sim e_{14}^2/\varepsilon \approx 10^8 \, \text{N/m}^2$. Since the elastic constants are 2 to 3 orders of magnitude bigger than this piezoelectric term, the wave equations Eq. (4.48), and (cyclic versions for u_y, u_z) will be solved by the nonpiezoelectric solution with corrections only at order e_{14}^2. The nonpiezoelectric solution derived in detail in Sect. 4.9.1 can be summarized as

$$u_{x'} = 2U\text{Re}\left[e^{-qkz-i\varphi}\right]e^{ik(x'-vt)},$$
$$u_{y'} = 0,$$
$$u_z = -2iU\text{Re}\left[\gamma e^{-qkz-i\varphi}\right]e^{ik(x'-vt)}, \tag{4.55}$$

where the sound velocity v for the Rayleigh-mode follows from the smallest solution of

$$\left(c_{44} - \rho v^2\right)\left(c_{11}c_{11}' - c_{12}^2 - c_{11}\rho v^2\right)^2 = c_{11}c_{44}\rho^2 v^4\left(c_{11}' - \rho v^2\right), \tag{4.56}$$

with $c_{11}' = c_{44} + (c_{11} + c_{12})/2$. The decay constant q is a solution of

$$\left(c_{11}' - \rho v^2 - c_{44}q^2\right)\left(c_{44} - \rho v^2 - c_{11}q^2\right) + q^2(c_{12} + c_{44})^2 = 0. \tag{4.57}$$

Lastly, the parameters γ, φ can be obtained from

$$\gamma = \frac{(c_{12} + c_{44})q}{c_{44} - c_{11}q^2 - \rho v^2}, \qquad e^{-2i\varphi} = -\frac{\gamma^* - q^*}{\gamma - q}. \tag{4.58}$$

Now, based on the nonpiezoelectric solution given in Eq. (4.55), the potential ϕ is constructed such that both the wave equation in Eq. (4.49) and the electrical boundary condition in Eq. (4.53) are solved. In the $\{\hat{x}', \hat{y}', \hat{z}\}$ coordinate system they read explicitly

$$\varepsilon \Delta \phi = e_{14} \left(2 \frac{\partial^2 u_{x'}}{\partial x' \partial z} + \frac{\partial^2 u_z}{\partial x' \partial x'} \right), \tag{4.59}$$

$$0 = \left. \left(\varepsilon_0 k \phi + e_{14} \frac{\partial u_{x'}}{\partial x'} - \varepsilon \frac{\partial \phi}{\partial z} \right) \right|_{z=0}. \tag{4.60}$$

One can readily check that this is achieved by the form proposed in Ref. [23, 25]

$$\phi = \begin{cases} i \phi_0 \mathcal{F}(kz) e^{ik(x'-vt)}, & z > 0 \\ A_{\text{out}} e^{kz} e^{ik(x'-ct)}, & z < 0, \end{cases} \tag{4.61}$$

where $\phi_0 = (e_{14}/\varepsilon) U$ and $A_{\text{out}} = i \phi_0 \mathcal{F}(0)$. Here, we have introduced the dimensionless function $\mathcal{F}(kz)$ which determines the length scale on which the electrical potential generated by the SAW decays into the bulk. It is given by

$$\mathcal{F}(kz) = 2 |A_1| e^{-\alpha kz} \cos(\beta kz + \varphi + \xi) + A_3 e^{-kz}, \tag{4.62}$$

with $A_1 = |A_1| e^{-i\xi}, q = \alpha + \beta i$, and

$$A_1 = \frac{\gamma - 2q}{q^2 - 1}, \tag{4.63}$$

$$A_3 = -\frac{2}{\varepsilon + \varepsilon_0} \left[\varepsilon \cos \varphi + \varepsilon \text{Re} \left[A_1 q e^{-i\varphi} \right] + \varepsilon_0 \text{Re} \left[A_1 e^{-i\varphi} \right] \right].$$

For $\text{Al}_x \text{Ga}_{1-x} \text{As}$ we obtain the following parameter values [compare Ref. [23]]: $|A_1| \approx 1.59$, $A_3 = -3.1$, $\alpha \approx 0.501$, $\beta \approx 0.472$, $\varphi = 1.06$, and $\xi = -0.33$. The electric potential for this parameter set is shown in Fig. 4.7.

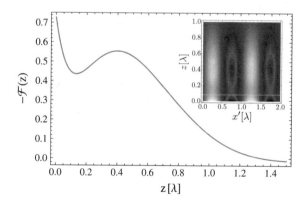

Fig. 4.7 The dimensionless function $\mathcal{F}(z)$ determines the decay of the electrical potential away from the surface into the bulk; the characterisitc length scale is approximately set by the SAW wavelength $\lambda = 2\pi/k \approx 1\,\mu\text{m}$. Inset: Density plot of the (normalized) electric potential $\text{Re}\,[\phi]/\phi_0 = -\mathcal{F}(kz) \sin(kx' - \omega t)$ along $\hat{x}' \| [110]$ for a Rayleigh surface acoustic wave propagating on a (001) GaAs crystal at $t = 0$

4.9.3   Mechanical Zero-Point Fluctuation

In this Appendix we provide more detailed calculations and estimates for the mechanical zero-point motion U_0 of a SAW. We show that they agree very well with the simple estimate given in the main text. Finally, we provide details on the material parameters used to obtain the numerical estimates.

Our first approach follows closely the one presented in Ref. [28]. The analysis starts out from the mechanical displacement operator in the Heisenberg picture

$$\hat{u}(\mathbf{x}, t) = \sum_n \left[\mathbf{v}_n(\mathbf{x}) a_n e^{-i\omega_n t} + \text{h.c.} \right]. \qquad (4.64)$$

To obtain the proper normalization of the displacement profiles, let us assume a single phonon Fock state, that is $|\Psi\rangle = a_n^\dagger |\text{vac}\rangle = |0, \dots, 0, 1_n, 0 \dots\rangle$, where $|\text{vac}\rangle = \prod_n |0\rangle_n$ is the phonon vacuum and compute the expectation value of additional field energy above the vacuum E_{mech}, defined as twice the kinetic energy, since for a mechanical mode half of the energy is kinetic, the other one potential [28]. We find

$$E_{\text{mech}} = 2\omega_n^2 \int d^3\mathbf{r}\rho(\mathbf{r})\, \mathbf{v}_n^*(\mathbf{r}) \cdot \mathbf{v}_n(\mathbf{r}) \qquad (4.65)$$

$$= 2\rho V \omega_n^2 \max\left[|\mathbf{v}_n(\mathbf{r})|^2\right], \qquad (4.66)$$

where the last equality defines the effective mode-volume for mode n. Setting $U_0 = \max[|\mathbf{v}_n(\mathbf{r})|]$, and assuming the phonon energy as $E_{\text{mech}} = \hbar\omega_n$, we arrive at the general result for a phonon mode, $U_0 = \sqrt{\hbar/2\rho V\omega_n}$; this confirms the simple estimate given in the main text.

Explicit example.—Next, we provide a calculation based on the exact analytical results derived in Appendix 4.9.1 and 4.9.2. In what follows, we assume that, in analogy to cavity QED, cavity confinement leads to the quantization $k_n = n\pi/L_c$, where $A = L_c^2$ is the effective quantization area. In a full 3D model, $A = L_x L_y$ where L_y is related to the spread of the transverse mode function as discussed (for example) in Refs. [14]. For simplicity, here we take $L_x = L_y$. Surface wave resonators can routinely be designed to show only one resonance k_0 [14]. Within this single-mode approximation, based on results derived in Appendix 4.9.1 for a SAW traveling wave, we take the quantized mechanical displacement describing a SAW standing wave along the axis $\hat{x}' = (110)$ as

$$\hat{u}(x', z) = U_0 \begin{pmatrix} \chi_0(z)\cos(k_0 x') \\ 0 \\ \zeta_0(z)\sin(k_0 x') \end{pmatrix} \left[a + a^\dagger\right], \qquad (4.67)$$

Here, the functions $\chi_0(z)$ and $\zeta_0(z)$ describe how the SAW decays into the bulk,

Table 4.3 Derived properties for Rayleigh surface waves, for both GaAs and diamond

	Decay const. Ω	$\gamma = |\gamma| e^{-i\Theta}$	φ	$v_s[m/s]$	δ
GaAs	$0.50 + 0.48i$	$-0.68 + 1.16i$	1.05	2878	1.2
Diamond	$0.60 + 0.22i$	$-1.05 + 0.75i$	1.26	11135	0.44

$$\chi_0(z) = 2e^{-\Omega_r k_0 z} \cos(\Omega_i k_0 z + \varphi), \tag{4.68}$$

$$\zeta_0(z) = 2|\gamma| e^{-\Omega_r k_0 z} \cos(\Omega_i k_0 z + \varphi + \theta), \tag{4.69}$$

with material-dependent parameters $\Omega = \Omega_r + i\Omega_i$, $\gamma = |\gamma| \exp[-i\theta]$ and φ; numerical values are presented in Table 4.3. We note that for GaAs we find $\zeta(0)/\chi(0) \approx 1.33$. This is in very good agreement with the numerical values of $c_x = |u_x/\phi| = 0.98$ nm/V and $c_z = |u_z/\phi| = 1.31$ nm/V as given in Ref. [15]. Normalization of the mode-function allows us to determine the parameter U_0. Performing the integration, we find

$$U_0 = \sqrt{\frac{\Omega_r}{\delta}} \sqrt{\frac{\hbar}{2\rho v_s A}}, \tag{4.70}$$

where the parameter δ depends on the material parameters; see Table 4.3. Using typical material parameters, we obtain for GaAs (diamond) $\sqrt{\Omega_r/\delta} = 0.64$ (1.17) and $U_0 \approx 1.2$ (1.36) fm/$\sqrt{A[\mu m^2]}$. This is in very good agreement with the numerical values presented in the main text.

Estimates derived from literature.—In Ref. [23], it is shown that the SAW Rayleigh mode studied in Appendix 4.9.1 has a classical energy density $\mathcal{E}$ (energy per unit surface area) given by

$$\mathcal{E} = kU^2 H, \tag{4.71}$$

where U is the amplitude of the wave, k the wave vector and H a material-dependent factor which is given as $H \approx 28.2 \times 10^{10}$ N/m^2 for GaAs. By equating the classical energy of the SAW given by $\mathcal{E}A$, where A is the quantization area, with its quantum-mechanical analog $N_{ph}\hbar\omega$ (N_{ph} is the number of phonons), we can estimate the single phonon displacement U_0 as

$$U_0 = \sqrt{\frac{\hbar\omega}{kHA}} = \sqrt{\frac{\hbar v_s}{HA}} \approx 1.05 \times 10^{-21} \frac{m^2}{\sqrt{A}}, \tag{4.72}$$

with $U = U_0\sqrt{N_{ph}}$. This estimate is also found to be in very good agreement with a result given in Ref. [36] as

$$U_0 = C\sqrt{\frac{2\hbar}{\rho v_s A}} \approx 1.7 \times 10^{-21} \frac{m^2}{\sqrt{A}}, \tag{4.73}$$

where C is a normalization constant with numerical value $C \approx 0.45$ for GaAs [36]. Therefore, for an effective mode area of $L_c = \sqrt{A} = 1\,\mu\text{m}$ we find a single phonon displacement of $U_0 \approx 1\,\text{fm}$. This confirms the estimates given in the main text.

4.9.4 Zero-Point Estimates

In this Appendix we provide details on piezo-magnetic materials and numerical estimates of the zero-point quantities for several relevant materials. The results are summarized in Table 4.4. The underlying input parameters are given below.

Theoretically, piezo-magnetic materials with a large magneto-strictive effect are typically described in a 1:1 correspondence to Eqs. (4.2) and (4.3), with the appropriate replacements (using standard notation) $\mathbf{E} \to \mathbf{H}$, $\mathbf{D} \to \mathbf{B}$, $\varepsilon_{ij} \to \mu_{ij}$ and $e_{ijk} \to h_{ijk}$ [29]. Coupling between mechanical and magnetic degrees of freedom is described by the piezomagnetic tensor $\underline{h}$ which can reach values as high as $\sim 700\,\text{T/strain}$ [31]; for our estimates we have referred to Terfenol-D, where $h_{15} \approx 167\,\text{T/strain}$. The magnetic field associated with a single phonon can then be estimated as $B_0 \approx h_{15}s_0$, where h_{15} refers to a typical (non-zero) element of $\underline{h}$.

For the piezoelectric materials GaAs, LiNbO$_3$ and Quartz all material parameters have been obtained from Ref. [24]. Phase velocities for typical cut directions have been used, that is (100)[001] GaAs, Y-Z LiNbO$_3$ and ST Quartz. For the piezomagnetic (magnetostrictive) materials CoFe$_2$O$_4$ and Terfenol-D all material parameters have been taken from Ref. [32]. We have used the phase velocities of the bulk shear waves given in there as $v_{\text{sh}} = 3.02 \times 10^3\,\text{m/s}$ and $v_{\text{sh}} = 1.19 \times 10^3\,\text{m/s}$ for CoFe$_2$O$_4$ and Terfenol-D, respectively. This gives an conservative estimate for U_0, since Rayleigh modes have phase velocities that are lower than the ones of bulk modes [14]. For example, in the case of CoFe$_2$O$_4$ in Ref. [31] wave velocities for

Table 4.4 Estimates for zero-point fluctuations (mechanical amplitude U_0, strain s_0, electrical potential ϕ_0, electric field ξ_0 and magentic field B_0) close to the surface ($d \ll \lambda$) for typical piezo-electric and piezo-magnetic (magnetostrictive) materials

	U_0 [fm]	$s_0[10^{-9}]$	$\phi_0\,[\mu\text{V}]$	$\xi_0\,[\text{V/m}]$	$B_0\,[\mu\text{T}]$
GaAs	1.9	11.7	3.1	19.2	–
LiNbO$_3$	1.8	11.3	0.9–25.8	5.8–162.2	–
Quartz	2.75	17.3	2.8–12.0	17.3–75.4	–
Terfenol-D	2.2	13.8	–	–	2.3
CoFe$_2$O$_4$	1.8	11.4	–	–	6.3
Diamond	1.17	7.4	–	–	–

All values must be multiplied by the universal scaling factor $1/\sqrt{A\,[\mu\text{m}^2]}$; thus, they refer to an effective surface mode area of size $A = 1\,\mu\text{m}^2$. Lower (upper) bounds for ϕ_0 and ξ_0 comprise minimum (maximum) non-zero element of $\underline{e}$ with maximum (minimum) non-zero element of $\underline{\varepsilon}$. We have set $k = 2\pi/\mu\text{m}$. Details on cut-directions and material parameters are given in the text

Rayleigh-type surface waves in a piezoelectric-piezomagnetic layered half space are found to be $v_s \approx 2840 \, \text{m/s} < v_{sh}$.

4.9.5 SAW Cavities

In this Appendix, we present a detailed discussion of the theoretical model describing the SAW resonator.

Typically, a SAW cavity is based on an on-chip distributed Bragg reflector formed by a periodic array of either metal electrodes or grooves etched into the surface; see Fig. 4.1. In such a grating, each strip reflects only weakly, but, for many strips $N \gg 1$, the total reflection $|R|$ can approach unity if the pitch p equals half the wavelength, $p = \lambda_c/2$. This Bragg condition defines the center frequency

$$f_c = v_s/2p. \tag{4.74}$$

At $f = f_c$, the total reflection coefficient is given by

$$|R| = \tanh\left(N \, |r_s|\right), \tag{4.75}$$

where N is the number of strips and r_s is the reflection coefficient associated with a single strip [14, 15]. The total reflection coefficient $|R|$ goes to unity in the limit $N \, |r_s| \gg 1$; see Fig. 4.8. Typically, $N \gtrsim 200$ and $|r_s| \approx (1-2)\,\%$ [14]. For $f \approx f_c$, $|r_s|$ increases with the normalized groove depth as $|r_s| = C_1 h/\lambda_c \sin(\pi w/p) + C_2 (h/\lambda_c)^2 \cos(\pi w/p)$, with material-dependent pre-factors [34]. For LiNbO$_3$, $C_1 = 0.67$ and $C_2 = 42$ [34]. As argued in Ref. [34], the first term $\sim C_1$ is due to a impedance mismatch, while the second one $\sim C_2$ is due to the stored energy effect.

Fig. 4.8 Total reflection coefficient $|R|$ as a function of the normalized groove-depth h/λ_c for $N = 100$ (*blue dashed*) and $N = 300$ (*red solid*). Here, $w/p = 0.5$ and material parameters for LiNbO$_3$ have been used (see text)

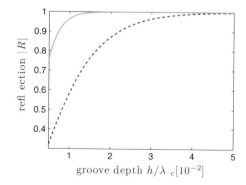

Due to the distributed nature of the mirror, strong reflection occurs over a fractional bandwidth only, given by $\delta f/f_c \approx 2|r_s|/\pi$. In practice, the cavity formed by two reflective gratings can be viewed as an acoustic Fabry-Perot resonator with effective reflection centers, sketched by localized mirrors in Fig. 4.1, situated at some effective penetration distance into the grating, given by $L_p = \tanh[(N-1)|r_s|]\lambda_c/(4|r_s|) \approx \lambda_c/4|r_s|$ [14, 15, 33]. Therefore, the total effective cavity size along the mirror axis is $L_c \approx D + 2L_p$, where D is the physical gap between the gratings; compare Fig. 4.1. For $N \approx 100 - 300$, $h/\lambda_c \approx 2\%$, we then obtain $L_c \approx 38\lambda_c$ and $L_c \approx 42\lambda_c$ for $D = 0.75\lambda_c$ and $D = 5.25\lambda_c$, respectively. In analogy to an optical Fabry-Perot resonator, the mode spacing can then by estimated as $\Delta f/f_c = \lambda_c/2L_c \approx |r_s|$. Since this is larger than $\delta f/f_c$, SAW resonators can be designed to host a single resonance only [14].

The total decay rate of this resonance κ can be decomposed into four relevant contributions [34], $\kappa = \kappa_{bk} + \kappa_d + \kappa_m + \kappa_r$, which includes conversion into bulk modes $\sim \kappa_{bk}$, diffraction losses $\sim \kappa_d$, internal losses due to material imperfections $\sim \kappa_m$, and leakage (radiation) losses due to imperfect mirrors $\sim \kappa_r$. The associated Q-factors are given by $Q_i = \omega_c/\kappa_i$. The desired decay rate is $\kappa_{gd} = \kappa_r$, whereas the undesired one is $\kappa_{bd} = \kappa_{bk} + \kappa_m + \kappa_d$. Here, κ_d is associated with diffraction losses due to spill-over beyond the aperture of the reflector. It can be made negligible by lateral confinement using for example waveguide structures, focusing or etching techniques [14, 35, 36]. Q_m refers to losses due to interaction with thermal phonons, losses due to defects in the material and propagation losses due to contamination [33, 34]. These losses ultimately limit Q: Low temperature experiments on quartz have demonstrated SAW resonators with $Q_m \times f$ [GHz] $> 10^5$ [19, 21]. Another source of losses is due to mode-conversion into bulk-modes. Measurements show that $Q_{bk} = 2\pi N_{\text{eff}}/[C_b(h/\lambda_c)^2]$ with $N_{\text{eff}} = L_c/\lambda_c$ and a material-dependent pre-factor C_b [78]; for LiNbO$_3$ (Quartz), $C_b = 8.7(10)$, respectively [78, 79]. Typically, κ_{bk} is found to be negligible for small groove depths, $h/\lambda_c < 2\%$ [34]. Finally, κ_r arises from leakage through imperfectly reflecting gratings ($|R| < 1$); in direct analogy to optical Fabry-Perot resonators, the associated Q-factor is given by $Q_r = 2\pi N_{\text{eff}}/(1 - |R|^2)$. Assuming negligible diffraction losses (that can be minimized via waveguide-like confinement [14, 39]) and cryostat temperatures, the total Q-factor is then given by

$$Q^{-1} = Q_m^{-1} + Q_{bk}^{-1} + Q_r^{-1}. \tag{4.76}$$

4.9.6   Charge Qubit Coupled to SAW Cavity Mode

We consider a GaAs charge qubit embedded in a tunnel-coupled double quantum dot (DQD) containing a single electron. In the one-electron regime the single-particle orbital level spacing is on the order of ~ 1 meV. Therefore, the system is well described by an effective two-level system: The state of the qubit is set by the position of the electron in the double-well potential, with the logical basis $|L\rangle$, $|R\rangle$ corresponding to the electron localized in the left (right) orbital. The Hamiltonian describing this system reads

$$H_{ch} = \frac{\varepsilon}{2}\sigma^z + t_c\sigma^x, \tag{4.77}$$

with the (orbital) Pauli-operators defined as $\sigma^z = |L\rangle\langle L| - |R\rangle\langle R|$ and $\sigma^x = |L\rangle\langle R| + |R\rangle\langle L|$, respectively. In Eq. (4.5), ε refers to the level detuning between the dots, while t_c gives the tunnel coupling. The level splitting between the eigenstates of H_{ch} is given by $\Omega = \sqrt{\varepsilon^2 + 4t_c^2}$, with a pure tunnel-splitting of $\Omega = 2t_c$ at the charge degeneracy point ($\varepsilon = 0$); typical parameter values are $t_c \sim \mu$eV and $\varepsilon \sim \mu$eV, such that the level splitting $\Omega \sim$ GHz lies in the microwave regime. At the charge degeneracy point, where to first order the qubit is insensitive to charge fluctuations ($d\Omega/d\varepsilon = 0$), the coherence time has been found to be $T_2 \approx 10$ ns [43].

We now consider a charge qubit as described above inside a SAW resonator with a single resonance frequency ω_c close to the qubit's transition frequency, $\omega_c \approx \Omega$, that is the regime of small detuning $\delta = \Omega - \omega_c \approx 0$; note that single-resonance SAW cavities can be realized routinely with today's standard techniques [14]. Within this single-mode approximation, the Hamiltonian describing the SAW cavity simply reads

$$H_{cav} = \omega_c a^\dagger a, \tag{4.78}$$

where $a^\dagger$ (a) creates (annihilates) a phonon inside the cavity. The electrostatic potential associated with this mode is given by $\hat{\phi}(\mathbf{x}) = \phi(\mathbf{x})[a + a^\dagger]$, where the mode-function $\phi(\mathbf{x})$ can be obtained from the corresponding mechanical mode-function $\mathbf{w}(\mathbf{x})$ via the relation $\varepsilon\Delta\phi(\mathbf{x}) = e_{kij}\partial_j\partial_k w_i(\mathbf{x})$; here, Δ is the Laplacian, e_{kij} the piezoelectric tensor and ε the permittivity of the material. The electron's charge e couples to the phonon induced electrical potential $\hat{\phi}$. In second quantization, the piezoelectric interaction reads $H_{int} = e\int d\mathbf{x}\,\hat{\phi}(\mathbf{x})\hat{n}(\mathbf{x})$, where e is the electron's charge, $\hat{n}(\mathbf{x}) = \sum_\sigma \psi_\sigma^\dagger(\mathbf{x})\psi_\sigma(\mathbf{x})$ is the electron number density operator and $\psi_\sigma^\dagger(\mathbf{x})$ creates an electron with spin σ at position $\mathbf{x}$ [22]. Since $\hat{\phi}(\mathbf{x})$ varies on a micron length-scale which is large compared to the spatial extension ~ 40 nm of the electron's wavefunction in a QD [62], the electron density is approximately given by a delta-function at the center of the corresponding dots. For the DQD system under consideration H_{int} is then approximately given by $H_{int} = e\sum_i \hat{\phi}(\mathbf{x}_i)n_i$; here, $\mathbf{x}_i$ refers to the center of the electronic orbital wavefunction $\psi_i(\mathbf{x})$ of dot $i = L, R$. Note that this form of H_{int} becomes exact if the overlap integral vanishes, that is if $\int d\mathbf{x}\,\phi(\mathbf{x})\psi_L^*(\mathbf{x})\psi_R(\mathbf{x}) = 0$ is satisfied. As shown below, for a mode-function $\phi(\mathbf{x})$ of sine-form, this condition maximizes the piezoelectric coupling strength between the electronic DQD system and the phonon mode. For the *charge* qubit system under consideration coupling to the cavity mode is then described by $H_{int} = e(a + a^\dagger)[\phi(x_L)|L\rangle\langle L| + \phi(x_R)|R\rangle\langle R|]$; here, x_i refers to the center of the electronic orbital wavefunction $\varphi_i(\mathbf{x})$ of dot $i = L, R$ and the transverse direction $\hat{y}$ has been integrated out already. To obtain strong coupling between the qubit and the cavity, we assume a mode profile $\phi(x) = \varphi_0\sin(kx)$, with a node tuned between the two dots, such that $\phi(x_L) = \varphi_0\sin(kl/2) = -\phi(x_R)$; here, l gives the distance between the two dots. Note that the single phonon amplitude, defined as $\varphi_0 = \phi_0 \mathcal{F}(kd)$, with $\mathcal{F}(kd) \approx 0$ for $d \gg \lambda$, accounts for the decay of the SAW into the bulk. For $\lambda = (0.5 - 1)\,\mu\text{m}$ and a 2DEG (where the DQD is embedded) situated a distance $d = 50\,\text{nm}$ below the surface, however, the single phonon amplitude is reduced by a factor of ~ 2 only, $\varphi_0 \approx (0.45 - 0.52)\,\phi_0$; see Appendix 4.9.2 for details. Then, the coupling between qubit and cavity reads

$$H_{\text{int}} = g_{\text{ch}}\left(a + a^\dagger\right) \otimes \sigma^z, \tag{4.79}$$

where the single-phonon coupling strength is

$$g_{\text{ch}} = e\phi_0 \mathcal{F}(kd) \sin(kl/2) \approx \frac{1.5\,\mu\text{eV}}{\sqrt{A\left[\mu\text{m}^2\right]}}. \tag{4.80}$$

Here, we have assumed $\lambda \approx 2l$ such that the geometrical factor $\sin(\pi l/\lambda) \approx 1$ [44]. In principle, the coupling strength g_{ch} could be further enhanced by additionally depositing a strongly piezoelectric material such as LiNbO_3 on the GaAs substrate. Moreover, comparison with standard literature shows that the piezoelectric electron-phonon coupling strength can be expressed as $g_{\text{pe}} = \sqrt{P}U_0 \approx e\,(e_{14}/\varepsilon)\,U_0 = e\phi_0$, where $P = (ee_{14}/\varepsilon)^2$ is a material parameter quantifying the piezoelectric coupling strength in zinc-blend structures [80, 81]. Using $P = 5.4 \times 10^{-20}\text{J}^2\text{m}^{-2}$ for GaAs, the single phonon Rabi frequency can be estimated as $g_{\text{pe}} \approx 2.87\,\mu\text{eV}/\sqrt{A[\mu\text{m}^2]}$. This corroborates our estimate for g_{ch}.

In summary, the total system can be described by the Hamiltonian $H = H_{\text{ch}} + H_{\text{cav}} + H_{\text{int}}$,

$$H = \frac{\varepsilon}{2}\sigma^z + t_c\sigma^x + \omega_c a^\dagger a + g_{\text{ch}}\left(a + a^\dagger\right) \otimes \sigma^z. \tag{4.81}$$

This corresponds to the generic Hamiltonian for a qubit-resonator system [28]. It is instructive to rewrite H in the eigenbasis of H_{ch}, given by

$$|+\rangle = \sin\theta\,|L\rangle + \cos\theta\,|R\rangle, \tag{4.82}$$

$$|-\rangle = \cos\theta\,|L\rangle - \sin\theta\,|R\rangle, \tag{4.83}$$

where the mixing angle θ is defined via $\tan\theta = 2t_c/(\varepsilon + \Omega)$. In a rotating wave approximation ($\delta, g_{\text{ch}} \ll \omega_c$), H then reduces to the well-known Hamiltonian of Jaynes-Cummings form

$$H \approx \delta S^z + g_{\text{ch}}\frac{2t_c}{\Omega}\left(S^+ a + S^- a^\dagger\right), \tag{4.84}$$

where $\delta = \Omega - \omega_c$, $S^z = (|+\rangle\langle+| - |-\rangle\langle-|)/2$ and $S^\pm = |\pm\rangle\langle\mp|$.

4.9.7 SAW-Based Cavity QED with Spin Qubits in Double Quantum Dots

In this Appendix, we show in detail how to realize the prototypical Jaynes-Cummings dynamics based on a spin qubit encoded in a double quantum dot (DQD) inside a SAW resonator.

We consider a double quantum dot (in the two-electron regime) coupled to the electrostatic potential generated by a SAW. The system is described by the Hamiltonian

$$H_{DQD} = H_0 + H_{cav} + H_{int},\qquad(4.85)$$

where H_0, H_{cav} and H_{int} describe the DQD, the cavity and the electrostatically mediated coupling between them, respectively. In the following, the different contributions are discussed in detail.

Double Quantum Dot.—The DQD is modeled by the standard Hamiltonian

$$H_0 = H_C + H_t + H_Z.\qquad(4.86)$$

Here, H_C gives the electrostatic energy

$$H_C = \sum_{i,\sigma} \varepsilon_i n_{i\sigma} + U \sum_{i=L,R} n_{i\uparrow} n_{i\downarrow} + U_{LR} n_L n_R,\qquad(4.87)$$

where (due to strong confinement) both the left and right dot are assumed to support a single orbital level with energy ε_i ($i = L, R$) only; U and U_{LR} refer to the on-site and interdot Coulomb repulsion, respectively. As usual, $n_{i\sigma} = d_{i\sigma}^\dagger d_{i\sigma}$ and $n_i = n_{i\uparrow} + n_{i\downarrow}$ refer to the spin-resolved and total electron number operators, respectively, with the fermionic creation (annihilation) operators $d_{i\sigma}^\dagger$ ($d_{i\sigma}$) creating (annihilating) an electron with spin $\sigma = \uparrow, \downarrow$ in the orbital $i = L, R$. We focus on a setting where an applied bias between the two dots approximately compensates the Coulomb energy of two electrons occupying the right dot; that is, $\varepsilon_L \approx \varepsilon_R + U - U_{LR}$. Thus, we restrict ourselves to a region in the stability diagram where only $(1, 1)$ and $(0, 2)$ charge states are of interest. All levels with $(1, 1)$ charge configuration have an electrostatic energy of $E_{(1,1)} = \varepsilon_L + \varepsilon_R + U_{LR}$, while the $(0, 2)$ configuration has $E_{(0,2)} = 2\varepsilon_R + U$. As usual, we introduce the detuning parameter $\varepsilon = \varepsilon_L - \varepsilon_R + U_{LR} - U$. In this regime, the relevant electronic levels are defined as $|T_+\rangle = |\uparrow\uparrow\rangle$, $|T_-\rangle = |\downarrow\downarrow\rangle$, $|T_0\rangle = (|\uparrow\downarrow\rangle + |\downarrow\uparrow\rangle)/\sqrt{2}$, $|S_{11}\rangle = (|\uparrow\downarrow\rangle - |\downarrow\uparrow\rangle)/\sqrt{2}$ and $|S_{02}\rangle = d_{R\uparrow}^\dagger d_{R\downarrow}^\dagger |0\rangle$ with $|\sigma\sigma'\rangle = d_{L\sigma}^\dagger d_{R\sigma'}^\dagger |0\rangle$.

Next, H_t describes coherent, spin-preserving interdot tunneling

$$H_t = t_c \sum_{\sigma} d_{L\sigma}^\dagger d_{R\sigma} + \text{h.c.},\qquad(4.88)$$

where t_c is the interdot tunneling amplitude. Lastly, H_Z accounts for the Zeeman energies,

$$H_Z = g\mu_B \sum_{i=L,R} \mathbf{B}_i \cdot \mathbf{S}_i, \tag{4.89}$$

where g is the electron g-factor and μ_B the Bohr magneton, respectively. In the presence of a micro-/nanomagnet, the two local magnetic fields $\mathbf{B}_i$ are inhomogeneous, $\mathbf{B}_L \neq \mathbf{B}_R$. We can then write $\mathbf{B}_i = \mathbf{B}_0 + \mathbf{B}_m(\mathbf{x}_i)$, where $\mathbf{B}_0$ is the external homogeneous magnetic field, while $\mathbf{B}_m(\mathbf{x}_i)$ is the micromagnet slanting field at the location of dot $\mathbf{x}_i$. In practice, B_0 is a few Tesla, at least larger than the saturation field of the micromagnet $B_0 \gtrsim 0.5\,\mathrm{T}$, while the magnetic gradient $\Delta B = \|\mathbf{B}_m(\mathbf{x}_R) - \mathbf{B}_m(\mathbf{x}_L)\|$ can reach $\Delta B \approx 100\,\mathrm{mT}$, corresponding to an electronic energy scale of $|g\mu_B \Delta B| \approx 2\,\mu\mathrm{eV}$ [82]. Field derivatives realized experimentally are $\partial B_{m,z}/\partial x \approx 1.5\,\mathrm{mT/nm}$. Alternatively, the magnetic gradient can be realized via the Overhauser field, as experimentally demonstrated for example in Ref. [60].

Note that the Fermi contact hyperfine interaction between electron and nuclear spins reads $H_{\mathrm{HF}} = \sum_i \mathbf{h}_i \cdot \mathbf{S}_i$. Here, $\mathbf{h}_i$ is the Overhauser field in QD $i = L, R$. When treating $\mathbf{h}_i$ as a classical (random) variable, H_{HF} is equivalent to H_Z and thus one can absorb $\mathbf{h}_i$ into the definition of the magnetic field $\mathbf{B}_i$ in Eq. (4.89); also see Ref. [82].

To facilitate the discussion, we introduce the magnetic sum field $\mathbf{B} = (\mathbf{B}_L + \mathbf{B}_R)/2$ and the difference field $\Delta\mathbf{B} = (\mathbf{B}_R - \mathbf{B}_L)/2$. While $\mathbf{B}$ conserves the total spin, that is $\left[\mathbf{B}(\mathbf{S}_L + \mathbf{S}_R), (\mathbf{S}_L + \mathbf{S}_R)^2\right] = 0$, the gradient field $\Delta\mathbf{B}$ does not. We set the quantization axis $\hat{z}$ along $\mathbf{B} = B\hat{z}$. For sufficiently large magnetic field B the electronic levels with $S_{\mathrm{tot}}^z = S_L^z + S_R^z \neq 0$ are far detuned and can be neglected for the remainder of the discussion. Therefore, in the following, we restrict ourselves to the $S_{\mathrm{tot}}^z = 0$ subspace. The components $\Delta\mathbf{B}^{x,y}$ give rise to transitions out of the (logical) subspace $S_{\mathrm{tot}}^z = 0$. Since these processes are assumed to be far off-resonance, they are neglected leaving us with the only relevant magnetic gradient $\Delta = g\mu_B \Delta\mathbf{B}^z/2$; compare also Refs. [60, 61]. For a schematic illustration, compare Fig. 4.9.

Fig. 4.9 Illustration of the relevant electronic levels under consideration. The triplet levels with $S_{\mathrm{tot}}^z \neq 1$ can be tuned off-resonance by applying a sufficiently large homogeneous magnetic field

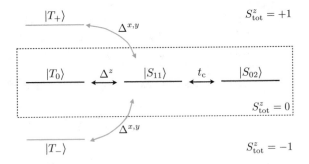

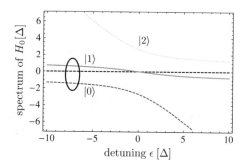

Fig. 4.10 Spectrum of H_0 for $t_c = 5\Delta$. The three eigenstates $|l\rangle$ are displayed in *green dotted* ($l = 2$), *red solid* ($l = 1$) and *blue dashed* ($l = 0$), respectively. The triplets $|T_\pm\rangle$ are assumed to be far detuned by a large external field and not shown. For large negative detuning $\varepsilon > -t_c$, the hybridized levels $\{|0\rangle, |1\rangle\}$ can be used as qubit

In summary, in the regime of interest H_0 simplifies to

$$H_0 = \frac{t_c}{2} \left(|S_{02}\rangle \langle S_{11}| + \text{h.c.} \right) - \varepsilon |S_{02}\rangle \langle S_{02}| \tag{4.90}$$
$$- \Delta \left(|T_0\rangle \langle S_{11}| + \text{h.c.} \right).$$

The eigenstates of H_0 within the relevant $S_{\text{tot}}^z = S_L^z + S_R^z = 0$ subspace can be expressed as

$$|l\rangle = \alpha_l |T_0\rangle + \beta_l |S_{11}\rangle + \kappa_l |S_{02}\rangle, \tag{4.91}$$

with corresponding eigenenergies ε_l ($l = 0, 1, 2$). The spectrum is displayed in Fig. 4.10. For large negative detuning $-\varepsilon \gg t_c$, the level $|2\rangle$ is far detuned, and the electronic subsystem can be simplified to an effective two-level system comprising the levels $\{|0\rangle, |1\rangle\}$, that is

$$H_0 \approx \frac{\omega_0}{2} \left(|1\rangle \langle 1| - |0\rangle \langle 0| \right), \tag{4.92}$$

which can be identified with a 'singlet-triplet'-like logical qubit subspace. Here, $\omega_0 = \varepsilon_1 - \varepsilon_0$ refers to the qubit's transition frequency. Note that the magnetic gradient causes efficient mixing between $|T_0\rangle$ and $|S_{11}\rangle$ for $\Delta \gtrsim |t_c^2/\varepsilon|$. In the regime of interest, the dominant character of the qubit's levels is $|1\rangle \approx |\Downarrow\Uparrow\rangle$, $|0\rangle \approx |\Uparrow\Downarrow\rangle$ (or vice versa) [61] and the transition frequency is approximately $\omega_0 \approx 2\Delta$. For $\Delta \approx 1\,\mu\text{eV}$, the transition frequency $\omega_0 = \varepsilon_1 - \varepsilon_0 \approx 2\,\mu\text{eV} \approx 3\,\text{GHz}$ matches typical SAW frequencies $\sim \text{GHz}$.

Coupling to SAW phonon mode.—Along the lines of Appendix 4.9.6, again we consider a SAW resonator with a single relevant confined phonon mode of frequency ω_c close to the qubit's transition frequency ω_0. For a DQD in the *two*-electron regime, in the basis of Eq. (4.90) coupling to the resoantor mode can be written as [44]

$$H_{\text{int}} = g_0 \left[a + a^\dagger \right] \otimes |S_{02}\rangle \langle S_{02}|, \tag{4.93}$$

where $g_0 = e\varphi_0 \eta_{\text{geo}}$. Here, η_{geo} is a geometrical factor accounting for the DQD's position with respect to the mode-function $\phi(\mathbf{x})$; it is defined according to $\varphi_0 \eta_{\text{geo}} = \phi(\mathbf{x}_R) - \phi(\mathbf{x}_L)$. For example, taking a standing wave pattern along $\hat{x}$ as demonstrated experimentally in Ref. [83], together with a transverse mode function restricting the spread in the $\hat{y}$-direction [35, 39], we obtain $\eta_{\text{geo}} = \sin(2\pi x_R/\lambda) - \sin(2\pi x_L/\lambda)$. It takes on its maximum value η_{opt}, when tuning a node of the standing wave at the center between the two dots, that is $x_R = l/2, x_L = -l/2$; this gives $\eta_{\text{opt}} = 2\sin(\pi l/\lambda)$, where l is the distance between the two dots [44]; as compared to the charge qubit described in Appendix 4.9.6, there is an additional factor of two, since here we consider a DQD in the *two*-electron regime, whereas the charge qubit consists of *one* electron only. For typical parameters ($l = 220\,\text{nm}$, $\lambda \approx 1.4\,\mu\text{m}$) as used in Ref. [44], we get $\eta_{\text{opt}} \approx 0.95$, while $l = 220\,\text{nm}$, $\lambda \approx 0.5\,\mu\text{m}$ leads to the largest possible value of $\eta_{\text{opt}} \approx 2$.

In summary, within the effective electronic two-level subspace $\{|0\rangle, |1\rangle\}$, the system is described by the Hamiltonian

$$H_{\text{DQD}} = \sum_{l=0,1} \left(\varepsilon_l + \kappa_l^2 \hat{V}_{\text{pe}} \right) |l\rangle \langle l| \tag{4.94}$$
$$+ \kappa_0 \kappa_1 \hat{V}_{\text{pe}} \left(|0\rangle \langle 1| + \text{h.c.} \right) + \omega_c a^\dagger a,$$

where $\hat{V}_{\text{pe}} = g_0 \left[a + a^\dagger \right]$. Applying a unitary transformation to a frame rotating at the cavity frequency ω_c, $\tilde{H}_{\text{DQD}} = U H_{\text{DQD}} U^\dagger + i\dot{U} U^\dagger$, with $U = \exp\left[i\omega_c t \left(a^\dagger a + \frac{1}{2} |1\rangle \langle 1| - \frac{1}{2} |0\rangle \langle 0| \right) \right]$, performing a rotating-wave approximation (RWA), and dropping a global energy shift $\tilde{\varepsilon} = (\varepsilon_0 + \varepsilon_1)/2$, we arrive at the effective (time-independent) Hamiltonian of Jaynes-Cummings form

$$\tilde{H}_{\text{DQD}} = \bar{\delta} S^z + g_{\text{QD}} \left[S^+ a + S^- a^\dagger \right], \tag{4.95}$$

where we have introduced the spin operators $S^+ = |1\rangle \langle 0|$ and $S^z = (|1\rangle \langle 1| - |0\rangle \langle 0|)/2$. Moreover, $\bar{\delta} = \omega_0 - \omega_c$ is the detuning between the qubit's transition frequency ω_0 and the cavity frequency ω_c, and the effective single-phonon Rabi frequency is defined as

$$g_{\text{QD}} = \kappa_0 \kappa_1 \eta_{\text{geo}} e\phi_0 \mathcal{F}(kd) \approx 2\kappa_0 \kappa_1 g_{\text{ch}}. \tag{4.96}$$

The coupling between the qubit and the cavity mode is mediated by the piezoelectric potential; therefore, it is proportional to the electron's charge e and the single-phonon electric potential ϕ_0. Due to the prolonged decoherence timescales, here we consider an effective (singlet-triplet like) spin-qubit rather than a charge qubit, such that the coupling g_{QD} is reduced by the (small) admixtures with the localized singlet $\kappa_l = \langle S_{02}|l\rangle$. Increasing $\kappa_0 \kappa_1$ leads to a stronger Rabi frequency g_{QD}, but an increased difference in charge configuration $|\kappa_1^2 - \kappa_0^2|$ makes the qubit more suscep-

Fig. 4.11 The product $\kappa_0\kappa_1$ directly affects the effective single-phonon Rabi frequency $g_{QD}/g_0 = \kappa_0\kappa_1$ [64], while the difference $\left|\kappa_1^2 - \kappa_0^2\right|$ determines the robustness of the qubit against charge noise. Here, $t_c = 5\Delta$

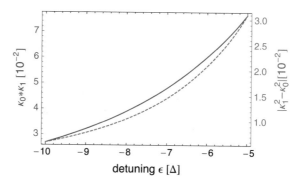

detuning ϵ [Δ]

tible to charge noise. For typical numbers ($t_c \approx 5\,\mu\text{eV}$, $\varepsilon \approx -7\,\mu\text{eV}$, $\Delta \approx 1\,\mu\text{eV}$), we get $\kappa_0\kappa_1 \approx 5 \times 10^{-2}$, $\left|\kappa_1^2 - \kappa_0^2\right| \approx 2 \times 10^{-2}$; see Fig. 4.11. For $l \approx 250\,\text{nm}$, $\lambda \approx 0.5\,\mu\text{m}$, and $d \approx 50\,\text{nm}$ we can then estimate

$$g_{QD}/\hbar \approx \frac{200\,\text{MHz}}{\sqrt{A[\mu\text{m}^2]}}. \qquad (4.97)$$

We take this coupling strength as a conservative estimate, since optimization against the relevant noise sources as done in Ref. [64] yields an optimal point with $\kappa_0\kappa_1 \approx 0.3$ and $\omega_0/2\pi \approx 1.5\,\text{GHz}$. Resonance ($\delta = 0$) yields a SAW wavelength of $\lambda \approx 2\,\mu\text{m}$; accordingly, for a fixed dot-to-dot distance $l = 250\,\text{nm}$, $\eta_{geo} \approx 0.76$ (whereas a larger DQD size $l = 400\,\text{nm}$ as used in Ref. [22] gives $\eta_{geo} \approx 1.18$). For $l = 250\,\text{nm}$, $d = 50\,\text{nm}$, we then obtain $g_{QD}/\hbar \approx 600\,\text{MHz}/\sqrt{A[\mu\text{m}^2]}$, which is a factor of three larger than the estimate quoted above (Table 4.5).

Cooperativity.—In this context, an important figure of merit is the single spin cooperativity [57], $C = g_{QD}^2 T_2/\kappa\,(\bar{n}_{th} + 1)$, where $\kappa = \omega_c/Q$ is the mechanical damping rate and $\bar{n}_{th} = 1/(e^{\hbar\omega_c/k_B T} - 1)$ is the equilibrium phonon occupation number at temperature T; here, since $\hbar\omega_c \gg k_B T$ for cryostatic temperatures, $\bar{n}_{th} \approx 0$. For singlet-triplet qubits in lateral QDs, $T_2^\star \approx 100\,\text{ns}$ [59]; using spin-echo techniques, experimentally this has even been extended to $T_2 = 276\,\mu\text{s}$. Even in the

Table 4.5 Estimates of the single spin cooperativity C for a DQD singlet-triplet qubit with $T_2^\star \approx 100\,\text{ns}$, in a SAW cavity at gigahertz frequencies $\omega_c/2\pi \approx 1.5\,\text{GHz}$ and cryostat temperatures where $\bar{n}_{th} \approx 0$ for both a small, low Q and large, high Q SAW-resonator

	Cavity size $A[\mu\text{m}^2]$	$g_{QD}[\text{MHz}]$	Q	C
Small cavity	1	200	10	4.25
Large cavity	500	9	10^4	8.5

The coupling strength g_{QD} could be further increased by additionally depositing a strongly piezoelectric material such as LiNbO$_3$ on the GaAs substrate and spin-echo (and/or narrowing) techniques allow for dephasing times extended by up to three orders of magnitude [40, 59]

absence of spin-echo pulses, with a far-from-optimistic dephasing time $T_2^\star \approx 100\,\mathrm{ns}$ [40], for a moderately small cavity size $A \approx 100\,\mu\mathrm{m}^2$, a quality factor of $Q = 900$ is sufficient to reach $C \approx g_{\mathrm{QD}}^2 T_2^\star Q/\omega_c \approx 3.8$. Note that $C > 1$ allows to perform a quantum gate between two spins mediated by a thermal mechanical mode [9].

Discussion of approximations.—To arrive at the effective Hamiltonian given in Eq. (4.95), we have made two essential approximations: (i) first, we have neglected the electronic level $|2\rangle$ yielding an effective two-level system (TLS), and (ii) second, we have applied a RWA leading to a major simplification of H_{DQD}; see Eq. (4.95) as compared to Eq. (4.94). In order to corroborate these approximations, we now compare exact numerical simulations of the full system where none of the approximations have been applied to the simplified, approximate description described above. While the dynamics of the former is described by the Master equation

$$\dot{\rho} = -i\,[H_0 + H_{\mathrm{cav}} + H_{\mathrm{int}}, \rho] + \kappa \mathcal{D}\,[a]\,\rho, \qquad (4.98)$$

with H_0, H_{cav} and H_{int} given in Eqs. (4.90), (4.78) and (4.93), respectively, the latter is described by a similar Master equation with the coherent Hamiltonian term replaced by the prototypical Jaynes-Cummings Hamiltonian,

$$\dot{\rho} = -i\,\big[\bar{\delta} S^z + g_{\mathrm{QD}}\left(S^+ a + S^- a^\dagger\right), \rho\big] + \kappa \mathcal{D}\,[a]\,\rho, \qquad (4.99)$$

with ρ referring to the density matrix of the combined system comprising the DQD and the cavity mode. Here, we have also accounted for decay of cavity phonons out of the resonator with a rate κ, described by the Lindblad term $\mathcal{D}\,[a]\,\rho = a\rho a^\dagger - \frac{1}{2}\left\{a^\dagger a, \rho\right\}$. As a figure of merit to validate approximation (i) we determine the population of the level level $|2\rangle$, that is $\mathrm{Tr}\,[\rho\,|2\rangle\langle 2|]$, describing the undesired leakage out of the logical subspace; ideally, this should be zero. Note that leakage into the triplet levels $|T_\pm\rangle$ could be accounted for along the lines, but they can be tuned far off-resonance by another, independent experimental knob, the external homogenous magnetic field. The results are summarized in Fig. 4.12: We find very good agreement between the exact and the approximate model, with a negligibly small error $\mathrm{Tr}\,[\rho\,|2\rangle\langle 2|] \sim \mathcal{O}\left(10^{-5}\right)$. This justifies the approximations made above and shows that (in the regime of interest) the system can simply be described by Eq. (4.99).

Noise Sources for the DQD-based System

Charge noise.—In a DQD device background charge fluctuations and noise in the gate voltages may cause undesired dephasing processes. In a recent experimental study [84], voltage fluctuations in the intedot detuning parameter ε have been identified as the main source of charge noise in a singlet-triplet qubit. Charge noise can be treated by introducing a Gaussian distribution in ε, with a variance σ_ε; typically $\sigma_\varepsilon \approx (1-3)\,\mu\mathrm{eV}$ [82]. The qubit's transition frequency ω_0, however, turns out to be rather insensitive to fluctuations in ε, with a (tunable) sensitivity of approximately $\partial\omega_0/\partial\varepsilon \lesssim 10^{-2}$; see Fig. 4.13. In agreement with experimental results presented in Ref. [84], we find $\partial\omega_0/\partial\varepsilon \sim \omega_0$, indicating ω_0 to be an exponential function of ε.

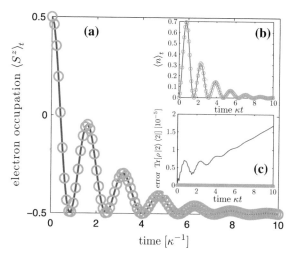

Fig. 4.12 Exact numerical simulations of the full (*blue, solid line*) and approximate (*cyan circles*) Master equations as given in Eqs. (4.98) and (4.99) respectively. Plots are shown for **a** the electronic inversion $\langle S^z \rangle_t$, **b** the cavity occupation $\langle n \rangle_t$ and **c** the error $\mathrm{Tr}\,[\rho \,|2\rangle \langle 2|]$ quantifying the leakage to $|2\rangle$. The latter is found to be negligibly small $\sim 10^{-5}$. We have set $\bar{\delta} = \omega_0 - \omega_c = 0$. Numerical parameters: $t_c = 10\,\mu\mathrm{eV}$, $\varepsilon = -7\,\mu\mathrm{eV}$, $\Delta = 1\,\mu\mathrm{eV}$, $\eta_{\mathrm{geo}} e \varphi_0 = 5.2 \times 10^{-2}\,\mu\mathrm{eV}$ such that $g_{\mathrm{QD}} = 4 \times 10^{-3}\,\mu\mathrm{eV} \approx 6\,\mathrm{MHz}$. The cavity decay rate is $\kappa = g_{\mathrm{QD}}/2$, corresponding to $Q \approx 10^3$

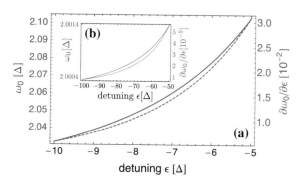

Fig. 4.13 Transition frequency ω_0 (*blue solid*) and its sensitivity against charge noise induced fluctuations in ε for intermediate **a** and large **b** negative detuning; for large negative detuning $\omega_0 \approx 2\Delta$, and the sensitivity $\partial \omega_0 / \partial \varepsilon$ practically vanishes leaving nuclear noise as the dominant dephasing process. By occasional refocusing of the spin states, this regime can be used for long-term storage of quantum information [64]. Other numerical parameters: $t_c = 5\Delta$

At very negative detuning ε, dephasing due to charge noise is practically absent, and $T_2^\star$ will be limited by nuclear noise [84]. Fluctuations in the tunneling amplitude t_c can be treated along the lines: we find ω_0 to be similarly insensitive to noise in t_c, $\partial \omega_0 / \partial t_c \approx 10^{-2}$.

Nuclear noise: Spin echo.—The electronic qubit introduced above has been defined for a fixed set of parameters $(t_c, \varepsilon, \Delta)$; compare Eq. (4.90). Now, let us consider the effect of deviations from this fixed parameters, $H_0 \rightarrow H_0 + \delta H$, where δH can be decomposed as $\delta H = \delta H_{\mathrm{el}} + \delta H_{\mathrm{nuc}}$ with

$$\delta H_{\mathrm{el}} = \frac{\delta t_c}{2} \left(|S_{02}\rangle \langle S_{11}| + \mathrm{h.c.} \right) - \delta\varepsilon \, |S_{02}\rangle \langle S_{02}| , \qquad (4.100)$$

$$\delta H_{\mathrm{nuc}} = -\delta\Delta \left(|T_0\rangle \langle S_{11}| + \mathrm{h.c.} \right), \qquad (4.101)$$

where δt_c and $\delta\varepsilon$ can be tuned electrostatically and basically in-situ. In most practical situations this does not hold for $\delta\Delta$: The primary source of decoherence in this system has been found to come from (slow) fluctuations in the Overhauser field generated by the nuclear spins [40, 60, 61]. In our model, this can directly be identified with a random, slowly time-dependent parameter $\delta\Delta = \delta\Delta\,(t)$. In the relevant subspace $\{|0\rangle\,,|1\rangle\}$, δH_{nuc} is given by $\delta H_{\mathrm{nuc}} = -\delta\Delta \sum_{k,l} \alpha_k \beta_l \left[|k\rangle \langle l| + \mathrm{h.c.} \right]$. Typically, $\delta\Delta \approx 0.1\,\mu\mathrm{eV} \ll \omega_c$ is fulfilled, such that we can apply a RWA yielding $\delta H_{\mathrm{nuc}} \approx -2\delta\Delta \sum_l \alpha_l \beta_l \, |l\rangle \langle l|$; physically, δH_{nuc} is too weak to drive transitions between the electronic levels $|0\rangle$ and $|1\rangle$ which are energetically separated by $\omega_0 \approx \omega_c$. Then, in the spin basis used in Eq. (4.95), we find

$$\delta H_{\mathrm{nuc}} = \delta\,(t)\, S^z, \qquad (4.102)$$

where the gradient noise is given by $\delta\,(t) = 2\delta\Delta\,(t)\,(\alpha_0\beta_0 - \alpha_1\beta_1)$. For $(t_c, \varepsilon, \Delta) \approx (5, -7, 1)$, all in $\mu\mathrm{eV}$, $\alpha_0\beta_0 - \alpha_1\beta_1 \approx 0.9$. Therefore, when also accounting for nuclear noise as described by Eq. (4.102), the full model [compare Eq. (4.95)] reads

$$\tilde{H}_{\mathrm{DQD}} = \left[\bar{\delta} + \delta\,(t) \right] S^z + g_{\mathrm{QD}} \left[S^+ a + S^- a^\dagger \right]. \qquad (4.103)$$

Since the nuclear spins evolve on timescales much longer than all other relevant timescales $\sim \kappa^{-1}, g^{-1}$ [62], the Overhauser noise term can be approximated as quasi-static, that is $\delta\,(t) = \delta$. As experimentally demonstrated in Ref. [60], the slow (nuclear) noise term $\sim \delta\,(t)\,S^z$ can be neutralized by Hahn-echo techniques. Here, the dephasing time of the electron spin qubit was extended by more than three orders of magnitude from $T_2^* \approx (10\text{-}100)\,\mathrm{ns}$ to $T_2 = 276\,\mu\mathrm{s}$.

In the following, assuming nominal resonance $\bar{\delta} = 0$, we detail a sequence of Hahn-echo pulses that cancels the undesired noise term and restores the pure, resonant Jaynes-Cummings dynamics: We consider four short time intervals of length τ, for which the unitary evolution is approximately given by $U_i \approx \mathbb{1} - i H_i \tau$, interspersed by three π-pulses. First, we let the system evolve with $H_1 = \delta S^z + g_{\mathrm{QD}} \left[S^+ a + S^- a^\dagger \right]$, then we apply a π-pulse along $\hat{x}$ ($S^z \rightarrow -S^z$, $S^x \rightarrow S^x, S^y \rightarrow -S^y$) such that $S^\pm \rightarrow S^\mp$ and the system evolves in the second time interval with $H_2 = -\delta S^z + g_{\mathrm{QD}} \left[S^- a + S^+ a^\dagger \right]$. Next, we apply a π-pulse along $\hat{z}$ ($S^z \rightarrow S^z, S^x \rightarrow -S^x, S^y \rightarrow -S^y$) such that $S^\pm \rightarrow -S^\pm$ leading to $H_3 = -\delta S^z - g_{\mathrm{QD}} \left[S^- a + S^+ a^\dagger \right]$. Finally, a π-pulse along $\hat{y}$ ($S^z \rightarrow -S^z, S^x \rightarrow -S^x, S^y \rightarrow S^y$) is applied such that $S^\pm \rightarrow -S^\mp$ giving $H_4 = \delta S^z + g_{\mathrm{QD}} \left[S^+ a + S^- a^\dagger \right]$. In summary, the system evolves over a time interval of 4τ according to $U_{\mathrm{eff}} = U_4 U_3 U_2 U_1 \approx$

$\mathbb{1} - i\tau \sum_i H_i = \mathbb{1} - i H_{\text{eff}} 4\tau$ with the effective Hamiltonian

$$H_{\text{eff}} = \frac{g_{\text{QD}}}{2} \left[S^+ a + S^- a^\dagger \right]. \tag{4.104}$$

Thus, in order to cancel the noise term, the effective single-phonon coupling strength is only lowered by a factor of 1/2.

Different Spin-Resonator Coupling

In Appendix 4.9.7, we have shown how to realize the prototypical Jaynes-Cummings Hamiltonian for SAW phonons interacting with a DQD; see Eq. (4.95). Alternatively, if one does not absorb the gradient Δ into the definition of the qubit basis, one can identify the logical subspace with the electronic states $|T_0\rangle$ and $|S\rangle$, where $|S\rangle$ is one of the two hybridized singlets (while the other one $|S'\rangle$ is far detuned and neglected) [40, 60, 61, 84]. Here, the electronic Hamiltonian reads $H_0 = -J(\varepsilon) |S\rangle \langle S| - \tilde{\Delta} (|T_0\rangle \langle S| + \text{h.c.})$, where $\tilde{\Delta} = \langle S_{11}|S\rangle \Delta$ and $J(\varepsilon)$ describes the exchange interaction. In this regime, the spin-resonator interaction takes on a form that is well known from other (localized) implementations of mechanical resonators [9], namely

$$H_{\text{int}} = g_{\text{qd}} \left(a + a^\dagger \right) \otimes |S\rangle \langle S|, \tag{4.105}$$

which can be viewed as a phonon-state dependent force, leading to a shift of the qubit's transition frequency depending on the position $\hat{x} = \left(a + a^\dagger \right)/\sqrt{2}$. Here, the single-phonon Rabi frequency is $g_{\text{qd}} = \kappa_S^2 \eta_{\text{geo}} e \phi_0 \mathcal{F}(kd)$, with $\kappa_S = \langle S_{02}|S\rangle$. Based on the coupling of Eq. (4.105), one can envisage a variety of experiments known from quantum optics: For example, in the limit of vanishing gradient $\Delta = 0$, the $\hat{x}$ quadrature of the phonon mode could serve as a quantum nondemolition variable, as it is a integral of motion of the coupled system of phonon mode and electronic meter.

4.9.8 Generalized Definition of the Cooperativity Parameter

In this Appendix we provide a generalized discussion of the cooperativity parameter C which in particular accounts for losses of the cavity mode other than leakage through the non-perfect mirrors. Furthermore, we derive a simple, analytical estimate for the state transfer fidelity $\mathcal{F}$ in terms of the parameter C and undesired phonon losses with a rate $\sim \kappa_{\text{bd}}$.

We consider a single qubit $\{|0\rangle, |1\rangle\}$ coupled to a cavity mode. The system is described by the Master equation

$$\dot{\varrho} = \underbrace{\left(\kappa_{\text{gd}} + \kappa_{\text{bd}} \right) \mathcal{D}[a] \varrho}_{\mathcal{L}_0 \varrho} \underbrace{-i [H_{\text{JC}}, \varrho] + \Gamma_{\text{deph}} \mathcal{D}[|1\rangle \langle 1|] \varrho}_{\mathcal{V}\varrho}, \tag{4.106}$$

where $\mathcal{D}[a]\varrho = a\varrho a^\dagger - \frac{1}{2}\{a^\dagger a, \varrho\}$. The first term describes decay of the cavity mode. The corresponding decay rate can be decomposed into desired (leakage through the mirrors) and undesired (bulk mode conversion etc.) contributions, labeled as κ_{gd} and κ_{bd}, respectively. Thus, we write $\kappa = \kappa_{gd} + \kappa_{bd}$. The second term with (on resonance) $H_{JC} = g\left(S^+ a + S^- a^\dagger\right)$ refers to the coherent interaction between qubit and cavity mode, while the last term describes pure dephasing of the qubit with a rate Γ_{deph}.

In the bad cavity limit (where $\kappa \gg g, \Gamma_{deph}$), one can adiabatically eliminate the cavity mode by projecting the system onto the cavity vacuum, $\mathcal{P}\varrho = \text{Tr}_{cav}[\varrho] \otimes \rho_{cav}^{ss} = \rho \otimes |\text{vac}\rangle \langle\text{vac}|$. Standard techniques (perturbation theory up to second order in $\mathcal{V}$, compare Ref. [85]) then yield the effective Master equation for the qubit's density matrix $\rho = \text{Tr}_{cav}[\mathcal{P}\varrho]$ only,

$$\dot{\rho} = \tilde{\kappa}\mathcal{D}\left[S^-\right]\rho + \Gamma_{deph}\mathcal{D}\left[|1\rangle\langle1|\right]\rho, \qquad (4.107)$$

with the effective decay rate $\tilde{\kappa} = 4g^2/\kappa$.

For comparison, the same procedure in standard cavity QED, where $\kappa = \kappa_{gd}$ and $\Gamma_{deph} \times \mathcal{D}[|1\rangle\langle1|]\rho \rightarrow \gamma\mathcal{D}[S^-]\rho$ yields the effective Master equation for the atom only, $\dot{\rho} = \tilde{\kappa}\mathcal{D}\left[S^-\right]\rho + \gamma\mathcal{D}\left[S^-\right]\rho$. Therefore, the atom decays with an effective spontaneous emission rate $\gamma_{tot} = \gamma + \tilde{\kappa} = \left(1 + 4g^2/\kappa\gamma\right)\gamma$, enhanced by the Purcell factor, $\gamma_{tot} = \gamma + \tilde{\kappa} = \left(1 + 4g^2/\kappa\gamma\right)\gamma$. Comparing good $\sim \tilde{\kappa}$ to bad $\sim \gamma$ decay channels, here one defines the cooperativity parameter in a straightforward way as $C_{atom} = g^2/\kappa\gamma$. This is readily read as the cavity-to-free-space scattering ratio, since the effective rate at which an excited atom emits an excitation into the cavity is given by $\tilde{\kappa} \sim g^2/\kappa$. For $C_{atom} > 1$, the atom is then more likely to decay into the cavity mode rather than into another mode outside the cavity. In cavity QED, large cooperativity $C_{atom} \gg 1$ has allowed for a number of key experimental demonstrations such as an enhancement of spontaneous emission [86], photon blockade [87] and vacuum-induced transparency [88].

The Master equation given in Eq. (4.107) describes a two-level system subject to purely dissipative dynamics. The dynamics can be fully described in terms of a set of three simple rate equations for the populations $p_k = \langle k|\rho|k\rangle$ $(k = 0, 1)$ and coherence $\rho_{10} = \langle1|\rho|0\rangle$, summarized as $\vec{p} = (p_1, p_0, \rho_{10})$,

$$\frac{d}{dt}\vec{p} = \begin{pmatrix} -\tilde{\kappa} & 0 & 0 \\ +\tilde{\kappa} & 0 & 0 \\ 0 & 0 & -\gamma_{eff}/2 \end{pmatrix}\vec{p}, \qquad (4.108)$$

where $\gamma_{eff} = \left(\tilde{\kappa} + \Gamma_{deph}\right)$. This allows for a simple analytical solution: For example, for a system initially in the excited state $\rho(t = 0) = |1\rangle\langle1|$ it reads $p_1(t) = \exp\left[-\tilde{\kappa}t\right]$, $p_0 = 1 - \exp\left[-\tilde{\kappa}t\right]$ and $\rho_{10}(t) = 0$.

Here, we aim for a theoretical description that singles out the desired trajectories, where phonon emission through the mirrors happens first, from all others. To do so, we rewrite Eq. (4.107) as

$$\dot{\rho} = -iH\rho + i\rho H^{\dagger} + \mathcal{J}_{\text{gd}}\rho + \mathcal{J}_{\text{bd}}\rho, \tag{4.109}$$

where we have defined an effective (non-hermitian) Hamiltonian H and jump operators according to

$$H = -\frac{i}{2}\gamma_{\text{eff}}\,|1\rangle\,\langle1| = -\frac{i}{2}\left(\tilde{\kappa} + \Gamma_{\text{deph}}\right)|1\rangle\,\langle1|, \tag{4.110}$$

$$\mathcal{J}_{\text{gd}}\rho = \tilde{\kappa}_{\text{gd}}S^{-}\rho S^{+}, \tag{4.111}$$

$$\mathcal{J}_{\text{bd}}\rho = \tilde{\kappa}_{\text{bd}}S^{-}\rho S^{+} + \Gamma_{\text{deph}}\,|1\rangle\,\langle1|\,\rho\,|1\rangle\,\langle1|. \tag{4.112}$$

Here, we have decomposed the effective decay rate $\tilde{\kappa}$ as

$$\tilde{\kappa} = \tilde{\kappa}_{\text{gd}} + \tilde{\kappa}_{\text{bd}} = \frac{4g^{2}}{\kappa^{2}}\kappa_{\text{gd}} + \frac{4g^{2}}{\kappa^{2}}\kappa_{\text{bd}}. \tag{4.113}$$

Formally solving Eq. (4.109) gives

$$\rho(t) = \mathcal{U}(t)\,\rho(0) + \int_{0}^{t}d\tau\,\mathcal{U}(t-\tau)\,\mathcal{J}\rho(\tau), \tag{4.114}$$

with the total jump operator $\mathcal{J}\rho = \mathcal{J}_{\text{gd}}\rho + \mathcal{J}_{\text{bd}}\rho$ and

$$\mathcal{U}(t)\,\rho = e^{-iHt}\rho e^{iH^{\dagger}t} \tag{4.115}$$

The exact solution given in Eq. (4.114) can be iterated, giving an illustrative expansion in terms of the jumps $\mathcal{J}$. It reads

$$\begin{aligned}
\rho(t) = {}& \mathcal{U}(t)\,\rho(0) + \int_{0}^{t}d\tau_{1}\mathcal{U}(t-\tau_{1})\,\mathcal{J}\mathcal{U}(\tau_{1})\,\rho(0) \\
&+ \int_{0}^{t}d\tau_{2}\int_{0}^{\tau_{2}}d\tau_{1}\mathcal{U}(t-\tau_{2})\,\mathcal{J}\mathcal{U}(\tau_{2}-\tau_{1})\times \\
&\mathcal{J}\mathcal{U}(\tau_{1})\,\rho(0) + \dots
\end{aligned}$$

Here, the n-th order term comprises n jumps $\mathcal{J}$ with free evolution $\mathcal{U}$ between the jumps. Now, we can single out the desired events where the first quantum jump is governed by $\mathcal{J}_{\text{gd}}$. This leads to the definition

$$\rho(t) = \mathcal{U}(t)\,\rho(0) + \rho_{\text{gd}}(t) + \rho_{\text{bd}}(t), \tag{4.116}$$

where $\rho_{\text{gd}}(t)$ subsumes all desired trajectories

$$\rho_{gd}(t) = \int_0^t d\tau_1 \mathcal{U}(t - \tau_1) \, \mathcal{J}_{gd}\mathcal{U}(\tau_1) \, \rho(0)$$

$$+ \int_0^t d\tau_2 \int_0^{\tau_2} d\tau_1 \mathcal{U}(t - \tau_2) \, \mathcal{J}\mathcal{U}(\tau_2 - \tau_1) \times$$

$$\mathcal{J}_{gd}\mathcal{U}(\tau_1) \, \rho(0) + \dots \tag{4.117}$$

We focus on a qubit, initially in the excited state, i.e., $\rho(0) = |1\rangle \langle 1|$. Using the relations $\mathcal{U}(\tau_1) \rho(0) = e^{-\gamma_{eff}\tau_1} |1\rangle \langle 1|$, $\mathcal{J}_{gd}\mathcal{U}(\tau_1) \rho(0) = \tilde{\kappa}_{gd} e^{-\gamma_{eff}\tau_1} |0\rangle \langle 0|$ and

$$\mathcal{J}_{gd}\mathcal{U}(\tau_2 - \tau_1) \, \mathcal{J}_{gd}\mathcal{U}(\tau_1) \, \rho(0) = 0, \tag{4.118}$$

$$\mathcal{J}_{bd}\mathcal{U}(\tau_2 - \tau_1) \, \mathcal{J}_{gd}\mathcal{U}(\tau_1) \, \rho(0) = 0, \tag{4.119}$$

the qubit's density matrix evaluates (to all orders in $\mathcal{J}$) to

$$\rho(t) = e^{-\gamma_{eff}t} |1\rangle \langle 1| + \rho_{gd}(t) + \rho_{bd}(t), \tag{4.120}$$

$$\rho_{gd}(t) = \frac{\tilde{\kappa}_{gd}}{\gamma_{eff}} \left(1 - e^{-\gamma_{eff}t}\right) |0\rangle \langle 0|. \tag{4.121}$$

In the long-time limit $t \to \infty$, the system reaches the steady state, $\rho(t \to \infty) = \rho_{gd} + \rho_{bd}$, where $\rho_{gd} = \frac{\tilde{\kappa}_{gd}}{\gamma_{eff}} |0\rangle \langle 0|$. The associated success probability $p_{suc} = \text{Tr}\left[\rho_{gd}\right]$ for faithful decay through the mirrors is then

$$p_{suc} = \frac{\tilde{\kappa}_{gd}}{\tilde{\kappa}_{gd} + \tilde{\kappa}_{bd} + \Gamma_{deph}} = \frac{1}{\frac{\kappa}{\kappa_{gd}} + \frac{1}{4}\frac{\kappa^2 \Gamma_{deph}}{g^2 \kappa_{gd}}}, \tag{4.122}$$

which is a simple branching ratio comparing the strength of the desired decay channel $\sim \tilde{\kappa}_{gd}$ to the undesired ones $\sim \tilde{\kappa}_{bd} + \Gamma_{deph}$. In the limit where $\kappa_{bd} = 0$, i.e., $\kappa = \kappa_{gd}$, the expression for p_{suc} simplifies to

$$p_{suc} = \frac{1}{1 + \frac{1}{4}\frac{1}{C}} \xrightarrow{C \gg 1} 1, \tag{4.123}$$

with the usual definition found in the literature, $C = g^2 / \left(\kappa \Gamma_{deph}\right) = g^2 T_2 / \kappa$; here, $\kappa = \omega_c / Q_{eff} = (\bar{n}_{th} + 1) \omega_c / Q$ is understood to account for thermal occupation of the environment $\bar{n}_{th}$ in terms of a decreased mechanical quality factor (compare for example Refs. [8, 57]).

It is instructive to rewrite the general expression for p_{suc} given in Eq. (4.122) as

$$p_{suc} = \frac{1}{(1 + \varepsilon)\left[1 + \frac{1}{4C}\right]}. \tag{4.124}$$

with $\varepsilon = \kappa_{bd} / \kappa_{gd}$. Based on this definition, it is evident that two conditions need to be satisfied in order to reach $p_{suc} \to 1$ in the regime where $\kappa_{bd} > 0$: (i) a low undesired

Fig. 4.14 Success probability p_{suc} for a qubit excitation to leak through the mirror, as a function of the cooperativity C for $\varepsilon = \kappa_{bd}/\kappa_{gd} = 0$ (*black solid*) and $\varepsilon = 5\%$ (*blue dashed*)

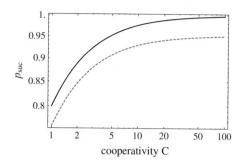

loss rate, $\varepsilon = \kappa_{bd}/\kappa_{gd} \ll 1$, and (ii) high cooperativity, $C \equiv \frac{g^2}{\kappa\Gamma_{deph}} = \frac{g^2 T_2}{\kappa} \gg 1$, with $\kappa = \kappa_{gd} + \kappa_{bd}$. For an illustration, compare Fig. 4.14. This shows, that the usual definition and interpretation of the cooperativity C holds, provided that $\varepsilon \ll 1$ is fulfilled. In order to quantify the cooperativity C for SAW cavity modes both in the $\sim$ MHz ($\bar{n}_{th} \gg 1$) and the $\sim$ GHz ($\bar{n}_{th} \ll 1$) regime, in the main text we take a (conservative) estimate as $C \equiv g^2 T_2 Q / [\omega_c (\bar{n}_{th} + 1)]$. For artificial atoms (quantum dots, superconducting qubits, NV-centers, ...) with resonant frequencies $\sim$ GHz, at cryostatic temperatures this definition reduces to $C \approx g^2 T_2/\kappa$ as discussed above, whereas for a trapped ion with $\omega_t/2\pi \approx \omega_c/2\pi \sim$ MHz it correctly gives $C \approx g^2 T_2 Q / (\omega_c \bar{n}_{th})$ with a decreased effective quality factor $Q_{eff} = Q/\bar{n}_{th}$ [8, 89, 90].

Fidelity estimate.—For small errors, the expression given in Eq. (4.124) can be approximated as $p_{suc} \approx 1 - \varepsilon - 1/(4C)$. Since the absorption process is just the time-reversed copy of the emission process in the state transfer protocol for two nodes, we can estimate the state transfer fidelity as $\mathcal{F} \geq p_{suc} \times p_{suc}$. For small infidelities, we then find

$$\mathcal{F} \gtrsim 1 - 2\varepsilon - \frac{1}{2C}, \tag{4.125}$$

where the individual errors arise from intrinsic phonon losses $\sim \varepsilon$ and qubit dephasing $\sim C^{-1} \sim T_2^{-1}$, respectively. This simple analytical estimate agrees well with numerical results presented in Ref. [58], where (except for noise sources that are irrelevant for our problem) $\mathcal{F} \approx 1 - (2/3)\varepsilon/ (1 + \varepsilon) - \mathcal{C}C^{-1}$ with a numerical coefficient $\mathcal{C} = \mathcal{O}(1)$ depending on the specific pulse sequence. Since $\varepsilon \ll 1$, this relation can be simplified to $\mathcal{F} \approx 1 - (2/3)\varepsilon - \mathcal{C}C^{-1}$. For the state transfer of the coherent superposition $|\psi\rangle = (|0\rangle - |1\rangle)/\sqrt{2}$ as described in detail in Appendix 4.9.10, we have explicitly verified the linear scaling $\sim \varepsilon$ and find numerically $\sim 1/2\varepsilon$ for intrinsic phonon losses [compare also Fig. 4.4 in the main text, where $\mathcal{F} \approx 95\%$ for $\varepsilon \approx 10\%$ and $\sigma_{nuc} = 0$] and take $\sim \varepsilon$ as a simple estimate in Eq. (4.8). Using experimentally achievable parameters $\varepsilon \approx 5\%$ and $C \approx 30$, we can then estimate $\mathcal{F} \approx 90\%$.

4.9.9 Effects Due to the Structure Defining the Quantum Dots

In our analysis of charge and spin qubits defined in quantum dots, we have neglected any potential effects arising due to the structure defining the quantum dots, that is (i) the heterostructure for the 2DEG and (ii) the metallic top gates for confinement of single electrons. In this Appendix, we give several arguments corroborating this approximate treatment, showing that the QD structure does not negatively influence the cavity nor the coupling between qubit and cavity.

Heterostructure.—Following the arguments given in Ref. [23], the 2DEG is taken to be a thin conducting layer a distance d away from the surface of a homogeneous $Al_xGa_{1-x}As$ crystal with typically $x \approx 30\%$. This treatment is approximately correct since the relevant material properties (elastic constants, densities and dielectric constants) of $Al_xGa_{1-x}As$ and GaAs are very similar [23]. The mode-functions and speed of sound are largely defined by the elastic constants [23] which are roughly the same for both $Al_xGa_{1-x}As$ and pure GaAs; for example, the speed of the Rayleigh SAW for $Al_{0.3}Ga_{0.7}As$ is $v_s \approx 3010\,\mathrm{m/s}$ which differs from that of pure GaAs by only 5% [23]. Moreover, the numerical values for the material-dependent parameter H entering the amplitude of the mechanical zero-point motion U_0 according to Eq. (4.72) differ by 2% only [23]; accordingly, the estimate for our key figure of merit U_0 should be rather accurate. Also, the piezo-electric coupling constants are rather similar, with $e_{14} \approx 0.15\,\mathrm{C/m^2}$ for pure GaAs and $e_{14} \approx 0.145\,\mathrm{C/m^2}$ for $Al_{0.3}Ga_{0.7}As$ [22, 23] yielding an accurate estimate for ϕ_0 and ξ_0, respectively. Lastly, the heterostructure is not expected to severely affect the Q-factor, since very high Q-values reaching $Q > 10^4$ have been observed in previous SAW experiments on AlN/diamond heterostructures [51, 52], where the differences in material properties are considerably larger than for the heterostructure making up the 2DEG.

Top gates.—For the following reasons, we have disregarded effects due to the presence of the metallic top gates: (i) In Ref. [25], a closed form analytic solution for the piezoelectric potential $\phi(\mathbf{x}, t)$ accompanying a SAW on the surface of a $GaAs/Al_xGa_{1-x}As$ in the presence of a narrow metal gate has been obtained. In particular, it is shown that $\phi(\mathbf{x}, t)$ is screened *right* below the gate, but remains practically unchanged with respect to the ungated case outside of the edges of the gate. Since the QD electrons are confined to regions outside of the metallic gates, they experience the piezoelectric potential as calculated for the ungated case (see Appendix 4.9.2 for details); therefore, our estimates—where this screening effect has been neglected—remain approximately valid. (ii) In Ref. [44], the coupling of a traveling SAW to electrons confined in a DQD has been experimentally studied. Here, in very good agreement to the experimental results, the potential felt by the QD electrons has been taken as $V_{\mathrm{pe}} \sim \sin(kl)$ (where l is the lithographic distance between the dots) confirming the sine-like mode profile as used in our estimates. Moreover, with $l = 220\,\mathrm{nm}$ and $\lambda = 1.4\,\mu\mathrm{m}$ the parameters used in this experiment perfectly match the ones used in our estimates. Intuitively (in the spirit of the standard electric dipole approximation in quantum optics), since $\lambda \gg l$, the SAW mode cannot resolve the dot structure and thus remains largely unaffected. (iii) In Ref. [16], single phonon SAW pulses have been detected via a single electron transistor (SET) directly deposited on the GaAs substrate with time-resolved measurements clearly identifying the coupling as piezo-electric. Similarly to a QD, the SET is defined by metallic gates. Here, a relation between vertical surface displacement and surface charge induced on the SET is theoretically derived. Using standard tabulated parameter values for GaAs and neglecting any effects due to the presence of the metallic gates, very good agreement with the experimental results is achieved. In particular, based on the results given in Ref. [16], a straightforward estimate gives $U_0 \approx 30$ am for the rather large cavity with $A \approx 10^6 \, \mu\text{m}^2$, whereas Eq. (4.4) yields a smaller, conservative estimate $U_0 \approx 2$ am, due to the averaging over the quantization area A. (iv) The Q-factor of the cavity is not expected to be severely affected by the presence of the metallic gates since metallic Al cladding layers have been used on a GaAs substrate to show basically dissipation-free SAW propagation over millimeter distances [35, 36]. (v) Finally, there is a large body of previous theoretical works on electron-phonon coupling in gate-defined QDs (see for example Refs. [22, 80, 81, 91]) where any effects due to the structure defining the QDs have been neglected as well. As a matter of fact, our description for the electron-phonon coupling emerges directly from these previous treatments in the limit where the continuum of phonon modes is replaced by a single relevant SAW cavity mode (similar to cavity QED, other bulk modes are still present and contribute to the decoherence of the qubit on a timescale T_2). For example, the piezoelectric electron-phonon interaction is given in Ref. [91] as

$$H_{\text{GaAs}} = \beta \sum_{k,\mu} \sqrt{\frac{\hbar}{2\rho V v_\mu k}} \mathcal{M}_k \left[a_{k,\mu} + a^\dagger_{-k,\mu} \right], \qquad (4.126)$$

where $a^\dagger_{k,\mu}$ creates an acoustic phonon with wave vector k and polarization μ and $\mathcal{M}_k$ refers to the matrix element for electron-phonon coupling; for free bulk modes, $\mathcal{M}_k = \sum_{i,j} \sum_\sigma \langle i | \exp [i\mathbf{k}\mathbf{r}] | j \rangle d^\dagger_{i\sigma} d_{j\sigma}$. In agreement with our notation, the coupling constant is $\beta = e e_{14}/\varepsilon$ and the square-root factor can be identified with U_0. Replacing the sum by a single-relevant cavity mode a, we recover the Hamiltonian describing the cavity-qubit coupling with $g \sim \beta U_0 = e\phi_0$; compare Eq. (4.77).

4.9.10   State Transfer Protocol

In this Appendix, we provide further details on the numerical simulation of the state transfer protocol as described by Eqs. (4.9) and (4.10), in the presence of Markovian noise. In line with previous theoretical studies [89], we show that the simple approximate Markovian noise treatment results in a pessimistic estimate for the noise transfer fidelity $\mathcal{F}$.

We first provide results of the full time-dependent numerical simulation of the cascaded Master equation given in Eqs. (4.9) and (4.10), including an exponential

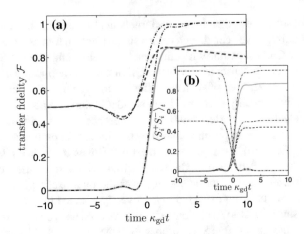

Fig. 4.15 Numerical simulation of state transfer for two different initial states $|1\rangle_1$ (*red solid line*) and $(|0\rangle_1 - |1\rangle_1)/\sqrt{2}$ (*blue dashed line*) in the presence of Markovian noise. *Black (dash-dotted) curves* refer to the ideal, noise-free scenario ($\mathcal{L}_{\text{noise}} = 0$) where perfect transfer is achieved, while colored curves take into account decoherence processes. **a** Transfer fidelity $\mathcal{F}$; **b** Excited-state occupation $\langle S_i^+ S_i^- \rangle_t$ for first and second qubit. Numerical parameters: $g_1 \, (t \geq 0) = \kappa_{\text{gd}}$, $\kappa_{\text{bd}}/\kappa_{\text{gd}} = 5\%$ and $\Gamma_{\text{deph}}/\kappa_{\text{gd}} = 5\%$

loss of coherence for $\Gamma_{\text{deph}} > 0$; see Fig. 4.15. In contrast to the non-Markovian noise model discussed in the main text, the qubits are assumed to be on resonance throughout the evolution, that is $\delta_i = 0$ ($i = 1, 2$), but experience undesired noise as described by

$$\mathcal{L}_{\text{noise}}\rho = 2\kappa_{\text{bd}} \sum_{i=1,2} \mathcal{D}[a_i]\rho + \Gamma_{\text{deph}} \sum_{i=1,2} \mathcal{D}[S_i^z]\rho. \tag{4.127}$$

Here, the second term refers to a standard Markovian pure dephasing term that leads to an exponential loss of coherence $\sim \exp\left(-\Gamma_{\text{deph}}t/2\right)$. As a reference we also show the results for the ideal, noise-free scenario ($\mathcal{L}_{\text{noise}} = 0$), where perfect state transfer is achieved [18]. The results of this type of time-dependent numerical simulations are then summarized in Fig. 4.16. For optimized, but experimentally achievable parameters $T_2^\star \approx 3\,\mu\text{s}$ [40], and accordingly $\Gamma_{\text{deph}}/\kappa_{\text{gd}} \approx 3\%$, we then obtain $\mathcal{F} \approx 0.85$, in the presence of realistic undesired phonon losses $\kappa_{\text{bd}}/\kappa_{\text{gd}} = 5\%$.

This shows that our non-Markovian noise model yields even higher state transfer fidelities than the Markovian noise model. Intuitively, this can be readily understood as follows: The simple Markovian noise model gives a coherence decay $\sim \exp\left(-t/T_2\right)$, whereas our non-Markovian noise model yields $\sim \exp\left(-t^2/T_2^2\right)$. Therefore, for Markovian noise one can estimate the dephasing induced error on the relevant timescale for state transfer $\sim \kappa^{-1}$ as $\sim \kappa^{-1}/T_2 \approx \Gamma_{\text{deph}}/\kappa \ll 1$, whereas non-Markovian noise leads to a considerably smaller error $\sim \left(T_2^{-1}/\kappa\right)^2$.

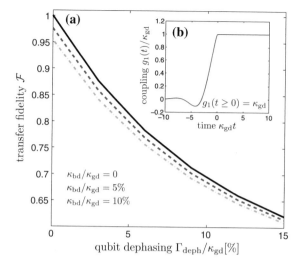

Fig. 4.16 Quantum state-transfer protocol in the presence of Markovian noise. **a** State transfer fidelity $\mathcal{F}$ for a coherent superposition $|\psi\rangle = (|0\rangle - |1\rangle)/\sqrt{2}$ as a function of the qubit's dephasing rate Γ_{deph} for different values of intrinsic phonon losses $\kappa_{\mathrm{bd}}/\kappa_{\mathrm{gd}} = 0$ (*black solid*), $\kappa_{\mathrm{bd}}/\kappa_{\mathrm{gd}} = 5\,\%$ (*blue dashed*) and $\kappa_{\mathrm{bd}}/\kappa_{\mathrm{gd}} = 10\,\%$ (*red dash-dotted*). **b** Pulse shape $g_1(t)$ for first node

References

1. R. Hanson, D.D. Awschalom, Coherent manipulation of single spins in semiconductors. Nature **453**(7198), 1043 (2008)
2. R.J. Schoelkopf, S.M. Girvin, Wiring up quantum systems. Nature **451**(7179), 664 (2008)
3. A. Wallraff, D.I. Schuster, A. Blais, L. Frunzio, R.-S. Huang, J. Majer, S. Kumar, S.M. Girvin, R.J. Schoelkopf, Circuit quantum electrodynamics: coherent coupling of a single photon to a cooper pair box. Nature **431**, 162 (2004)
4. A. Blais, R.-S. Huang, A. Wallraff, S.M. Girvin, R.J. Schoelkopf, Cavity quantum electrodynamics for superconducting electrical circuits: an architecture for quantum computation. Phys. Rev. A **69**, 062320 (2004)
5. M.A. Sillanpää, J.I. Park, R.W. Simmonds, Coherent quantum state storage and transfer between two phase qubits via a resonant cavity. Nature **449**(7161), 438 (2007)
6. J. Majer, Coupling superconducting qubits via a cavity bus. Nature **449**(7161), 443 (2007)
7. A.D. O'Connell, Quantum ground state and single-phonon control of a mechanical resonator. Nature **464**, 697 (2010)
8. S. Kolkowitz, A.C. Bleszynski Jayich, Q.P. Unterreithmeier, S.D. Bennett, P. Rabl, J.G.E. Harris, M.D. Lukin, Coherent sensing of a mechanical resonator with a single-spin qubit. Science **335**(6076), 1603 (2012)
9. P. Rabl, S.J. Kolkowitz, F.H.L. Koppens, J.G.E. Harris, P. Zoller, M.D. Lukin, A quantum spin transducer based on nanoelectromechanical resonator arrays. Nat. Phys. **6**(8), 602 (2010)
10. O.O. Soykal, R. Ruskov, C. Tahan, Sound-based analogue of cavity quantum electrodynamics in silicon. Phys. Rev. Lett. **107**, 235502 (2011)
11. S.J.M. Habraken, K. Stannigel, M.D. Lukin, P. Zoller, P. Rabl, Continuous mode cooling and phonon routers for phononic quantum networks. New J. Phys. **14**(11), 115004 (2012)
12. R. Ruskov, C. Tahan, On-chip cavity quantum phonodynamics with an acceptor qubit in silicon. Phys. Rev. B **88**, 064308 (2013)

13. A.N. Cleland, M.R. Geller, Superconducting qubit storage and entanglement with nanome-chanical resonators. Phys. Rev. Lett. **93**, 070501 (2004)
14. D. Morgan, *Surface Acoustic Wave Filters* (Academic Press, Boston, 2007)
15. S. Datta, *Surface Acoustic Wave Devices* (Prentice-Hall, Upper Saddle River, 1986)
16. M.V. Gustafsson, P.V. Santos, G. Johansson, P. Delsing, Local probing of propagating acoustic waves in a gigahertz echo chamber. Nat. Phys. **8**(4), 338 (2012)
17. M.V. Gustafsson, T. Aref, A.F. Kockum, M.K. Ekström, G. Johansson, P. Delsing, Propagating phonons coupled to an artificial atom. Science **346**(6206), 207 (2014)
18. J.I. Cirac, P. Zoller, H.J. Kimble, H. Mabuchi, Quantum state transfer and entanglement distribution among distant nodes in a quantum network. Phys. Rev. Lett. **78**, 3221 (1997)
19. A. El Habti, F. Bastien, E. Bigler, T. Thorvaldsson, High-frequency surface acoustic wave devices at very low temperature: application to loss mechanisms evaluation. J. Acoust. Soc. Am. **100**(1), 272 (1996)
20. R. Manenti, *Surface Acoustic Wave Resonators for Quantum Information*. Master's thesis, University of Milano (2013)
21. E.B. Magnusson, B.H. Williams, R. Manenti, M.-S. Nam, A. Nersisyan, M.J. Peterer, A. Ardavan, P.J. Leek, Surface acoustic wave devices on bulk ZnO crystals at low temperature. Appl. Phys. Lett. **106**(6), 063509 (2015)
22. V. Kornich, C. Kloeffel, D. Loss, Phonon-mediated decay of singlet-triplet qubits in double quantum dots. Phys. Rev. B **89**, 085410 (2014)
23. S.H. Simon, Coupling of surface acoustic waves to a two-dimensional electron gas. Phys. Rev. B **54**, 13878 (1996)
24. D. Royer E. Dieulesaint. *Elastic Waves in Solids I*. Springer, Berlin (2000)
25. G.R. Aïzin, G. Gumbs, M. Pepper, Screening of the surface-acoustic-wave potential by a metal gate and the quantization of the acoustoelectric current in a narrow channel. Phys. Rev. B **58**, 10589 (1998)
26. R. White, Surface elastic waves. Proceedings of the IEEE **58**(8), 1238 (1970)
27. M.A. Stroscio, Y.M. Sirenko, S. Yu, K.W. Kim, Acoustic phonon quantization in buried waveguides and resonators. J. Phys. Condens. Matter **8**(13), 2143 (1996)
28. M. Aspelmeyer, T.J. Kippenberg, F. Marquardt, *Cavity Optomechanics* (Springer, Berlin, 2014)
29. J.P. Nowacki, *Static and Dynamic Fields in Bodies with Piezoeffects or Polarization Gradient* (Springer, Berlin, 2006)
30. J. Cai, F. Jelezko, M.B. Plenio, Hybrid sensors based on colour centres in diamond and piezoactive layers. Nat. Commun. **5**, 4065 (2014)
31. Y. Pang, J.-X. Liu, Y.-S. Wang, X.-F. Zhao, Propagation of Rayleigh-type surface waves in a transversely isotropic piezoelectric layer on a piezomagnetic half-space. J. Appl. Phys. **103**(7), 074901 (2008)
32. J. xi Liu, D.-N. Fang, W.-Y. Wei, X.-F. Zhao, *Love waves in layered piezoelectric/piezomagnetic structures*. J. Sound Vibr. **315**(12), 146 (2008)
33. D.L.T. Bell, R.C.M. Li, Surface-acoustic-wave resonators. Proc. IEEE **64**(5), 711 (1976)
34. R.C.M. Li, J.A. Alusow, R.C. Williamson, *Experimental exploration of the limits of achievable Q of grooved surface-wave resonators*. 1975 Ultrasonics Symposium, pp. 279–283 (1975)
35. M.M. de Lima, J. Camacho, W. Seidel, H. Kostial, P.V. Santos, Intense acoustic beams for photonic modulation. Proc. SPIE **5450**, 118–125 (2004)
36. M.M. de Lima, P.V. Santos, Modulation of photonic structures by surface acoustic waves. Rep. Prog. Phys. **68**(7), 1639 (2005)
37. S. Büyükköse, B. Vratzov, D. Atac, J. van der Veen, P.V. Santos, W.G. van der Wiel, Ultrahigh-frequency surface acoustic wave transducers on $ZnO/SiO_2/Si$ using nanoimprint lithography. Nanotechnology **23**(31), 315303 (2012)
38. B. Auld, *Acoustic Fields and Waves in Solids II* (Wiley, New York, 1973)
39. A.A. Oliner, *Acoustic Surface Waves, Waveguides for Surface Waves*. (Springer, 1978), pp. 187–223
40. M.D. Shulman, S.P. Harvey, J.M. Nichol, S.D. Bartlett, A.C. Doherty, V. Umansky, A. Yacoby, Suppressing qubit dephasing using real-time Hamiltonian estimation. Nat. Commun. **5**, 5156 (2014)

41. A. Frisk Kockum, P. Delsing, G. Johansson. *Designing frequency-dependent relaxation rates and Lamb shifts for a giant artificial atom.* Phys. Rev. A **90**, 013837 (2014)

42. M. Gao, Y.-X. Liu, X.-B. Wang, Coupling Rydberg atoms to superconducting qubits via nanomechanical resonator. Phys. Rev. A **83**, 022309 (2011)

43. K.D. Petersson, J.R. Petta, H. Lu, A.C. Gossard, Quantum coherence in a one-electron semiconductor charge qubit. Phys. Rev. Lett. **105**, 246804 (2010)

44. W.J.M. Naber, T. Fujisawa, H.W. Liu, W.G. van der Wiel, Surface-acoustic-wave-induced transport in a double quantum dot. Phys. Rev. Lett. **96**, 136807 (2006)

45. D.I. Schuster, et al. Resolving photon number states in a superconducting circuit. Nature **445**(7127), 515 (2007)

46. K.D. Petersson, L.W. McFaul, M.D. Schroer, M. Jung, J.M. Taylor, A.A. Houck, J.R. Petta, Circuit quantum electrodynamics with a spin qubit. Nature **490**(7420), 380 (2012)

47. F.A. Zwanenburg, A.S. Dzurak, A. Morello, M.Y. Simmons, L.C.L. Hollenberg, G. Klimeck, S. Rogge, S.N. Coppersmith, M.A. Eriksson, Silicon quantum electronics. Rev. Mod. Phys. **85**, 961 (2013)

48. J.M. Raimond, M. Brune, S. Haroche, Manipulating quantum entanglement with atoms and photons in a cavity. Rev. Mod. Phys. **73**, 565 (2001)

49. J. Labaziewicz, Y. Ge, P. Antohi, D. Leibrandt, K.R. Brown, I.L. Chuang, Suppression of heating rates in cryogenic surface-electrode ion traps. Phys. Rev. Lett. **100**, 013001 (2008)

50. D. Kielpinski, D. Kafri, M.J. Woolley, G.J. Milburn, J.M. Taylor, Quantum interface between an electrical circuit and a single atom. Phys. Rev. Lett. **108**, 130504 (2012)

51. M. Benetti, D. Cannata, F. Di Pictrantonio, E. Verona, Growth of AlN piezoelectric film on diamond for high-frequency surface acoustic wave devices. IEEE Trans. Ultrason. Ferroelectr. Freq. Control **52**(10), 1806 (2005)

52. J.G. Rodriguez-Madrid, G.F. Iriarte, J. Pedros, O.A. Williams, D. Brink, F. Calle, Super-high-frequency SAW resonators on AlN/diamond. Electron Device Lett. IEEE **33**(4), 495 (2012)

53. L. Rondin, J.-P. Tetienne, T. Hingant, J.-F. Roch, P. Maletinsky, V. Jacques, Magnetometry with nitrogen-vacancy defects in diamond. Rep. Prog. Phys. **77**(5), 056503 (2014)

54. J.M. Taylor, P. Cappellaro, L. Childress, L. Jiang, D. Budker, P.R. Hemmer, A. Yacoby, R. Walsworth, M.D. Lukin, High-sensitivity diamond magnetometer with nanoscale resolution. Nat. Phys. **4**(10), 810 (2008)

55. Q.A. Turchette, et al. Heating of trapped ions from the quantum ground state. Phys. Rev. A **61**, 063418 (2000)

56. N. Bar-Gill, L.M. Pham, A. Jarmola, D. Budker, R.L. Walsworth, Solid-state electronic spin coherence time approaching one second. Nat. Commun. **4**, 1743 (2013)

57. P. Ovartchaiyapong, K.W. Lee, B.A. Myers, A.C.B. Jayich, Dynamic strain-mediated coupling of a single diamond spin to a mechanical resonator. Nat. Commun. **5**, 4429 (2014)

58. K. Stannigel, P. Rabl, A.S. Sørensen, P. Zoller, M.D. Lukin, Optomechanical transducers for long-distance quantum communication. Phys. Rev. Lett. **105**, 220501 (2010)

59. C. Kloeffel, D. Loss, Prospects for spin-based quantum computing in quantum dots. Ann. Rev. Condens. Matter Phys. **4**(1), 51 (2013)

60. H. Bluhm, S. Foletti, I. Neder, M. Rudner, D. Mahalu, V. Umansky, A. Yacoby, Dephasing time of GaAs electron-spin qubits coupled to a nuclear bath exceeding 200 μs. Nat. Phys. **7**(2), 109 (2010)

61. S. Foletti, H. Bluhm, D. Mahalu, V. Umansky, A. Yacoby, Universal quantum control of two-electron spin quantum bits using dynamic nuclear polarization. Nat. Phys. **5**, 903 (2009)

62. E.A. Chekhovich, M.N. Makhonin, A.I. Tartakovskii, A. Yacoby, H. Bluhm, K.C. Nowack, L.M.K. Vandersypen, Nuclear spin effects in semiconductor quantum dots. Nat. Mater. **12**, 494 (2013)

63. I.T. Vink, K.C. Nowack, F.H.L. Koppens, J. Danon, Y.V. Nazarov, L.M.K. Vandersypen, Locking electron spins into magnetic resonance by electron-nuclear feedback. Nat. Phys. **5**, 764 (2009)

64. J.M. Taylor, M.D. Lukin, *Cavity Quantum Electrodynamics with Semiconductor Double-dot Molecules on a Chip.* arXiv:cond-mat/0605144 (2006)

65. M.D. Shulman, O.E. Dial, S.P. Harvey, H. Bluhm, V. Umansky, A. Yacoby, Demonstration of entanglement of electrostatically coupled singlet-triplet qubits. Science **336**(6078), 202 (2012)
66. G. Burkard, A. Imamoglu, Ultra-long-distance interaction between spin qubits. Phys. Rev. B **74**, 041307 (2006)
67. X. Hu, Y.-X. Liu, F. Nori, Strong coupling of a spin qubit to a superconducting stripline cavity. Phys. Rev. B **86**, 035314 (2012)
68. H.J. Carmichael, Quantum trajectory theory for cascaded open systems. Phys. Rev. Lett. **70**, 2273 (1993)
69. C.W. Gardiner, Driving a quantum system with the output field from another driven quantum system. Phys. Rev. Lett. **70**, 2269 (1993)
70. S. Massar, S. Popescu, Optimal extraction of information from finite quantum ensembles. Phys. Rev. Lett. **74**, 1259 (1995)
71. S.J. van Enk, J.I. Cirac, P. Zoller, Ideal quantum communication over noisy channels: a quantum optical implementation. Phys. Rev. Lett. **78**, 4293 (1997)
72. S.J. van Enk, J.I. Cirac, P. Zoller, Photonic channels for quantum communication. Science **279**(5348), 205 (1998)
73. S.J. van Enk, H.J. Kimble, J.I. Cirac, P. Zoller, Quantum communication with dark photons. Phys. Rev. A **59**, 2659 (1999)
74. M. Veldhorst, An addressable quantum dot qubit with fault-tolerant control-fidelity. Nat. Nanotechnol. **9**, 981 (2014)
75. R. Ruskov, C. Tahan, Catching the quantum sound wave. Science **346**(6206), 165 (2014)
76. O.O. Soykal, C. Tahan, Toward engineered quantum many-body phonon systems. Phys. Rev. B **88**, 134511 (2013)
77. R. Stoneley, The propagation of surface elastic waves in a cubic crystal. Proc. R. Soc. Lond. A: Math. Phys. Eng. Sci. **232**(1191), 447 (1955)
78. W.J. Tanski, GHz SAW Resonators, in *1979 Ultrasonics Symposium*, pp. 815–823 (1979)
79. J.P. Parekh, H.S. Tuan, Effect of groove-depth variation on the performance of uniform SAW grooved reflector arrays. Appl. Phys. Lett. **32**(12), 787 (1978)
80. H. Bruus, K. Flensberg, H. Smith, Magnetoconductivity of quantum wires with elastic and inelastic scattering. Phys. Rev. B **48**, 11144 (1993)
81. T. Brandes, B. Kramer, Spontaneous emission of phonons by coupled quantum dots. Phys. Rev. Lett. **83**, 3021 (1999)
82. S. Chesi, Y.-D. Wang, J. Yoneda, T. Otsuka, S. Tarucha, D. Loss, Single-spin manipulation in a double quantum dot in the field of a micromagnet. Phys. Rev. B **90**, 235311 (2014)
83. R.C.M. Li, J.A. Alusow, R.C. Williamson, Surface-wave resonators using grooved reflectors, in 29th Annual Symposium on Frequency. Control **167–176**, (1975)
84. O.E. Dial, M.D. Shulman, S.P. Harvey, H. Bluhm, V. Umansky, A. Yacoby, Charge noise spectroscopy using coherent exchange oscillations in a singlet-triplet qubit. Phys. Rev. Lett. **110**, 146804 (2013)
85. E.M. Kessler, Generalized Schrieffer-Wolff formalism for dissipative systems. Phys. Rev. A **86**, 012126 (2012)
86. E.M. Purcell, Spontaneous emission probabilities at radio frequencies. Phys. Rev. **69**, 681 (1946)
87. K.M. Birnbaum, A. Boca, R. Miller, A.D. Boozer, T.E. Northup, H.J. Kimble, Photon blockade in an optical cavity with one trapped atom. Nature **436**(7047), 87 (2005)
88. H. Tanji-Suzuki, W. Chen, R. Landig, J. Simon, V. Vuletic, Vacuum-induced transparency. Science **333**(6047), 1266 (2011)
89. S.D. Bennett, N.Y. Yao, J. Otterbach, P. Zoller, P. Rabl, M.D. Lukin, Phonon-induced spin-spin interactions in diamond nanostructures: application to spin squeezing. Phys. Rev. Lett. **110**, 156402 (2013)
90. S.D. Bennett, S. Kolkowitz, Q.P. Unterreithmeier, P. Rabl, A.C.B. Jayich, J.G.E. Harris, M.D. Lukin, Measuring mechanical motion with a single spin. New J. Phys. **14**(12), 125004 (2012)
91. V. Srinivasa J. M. Taylor, *Capacitively Coupled Singlet-Triplet Qubits in the Double Charge Resonant Regime*. arXiv:1408.4740 (2014)

Chapter 5
Outlook

Both experimental and theoretical quantum optics has played a pioneering role for understanding the bizarre features of quantum mechanics. This Thesis is based on merging ideas and methods originally rooted in the field quantum optics with modern solid-state semiconductor systems, with a primary focus on gate-defined quantum dots. We have presented specific examples where quantum optical concepts and systems, ranging from superradiance to cavity-QED, find novel counterparts in the solid state, allowing for unique insights in these kinds of systems. This list of examples is unlikely to have come to its end. Conversely, our results motivate both (i) further research in to what extent the proposed analogies between quantum optical and solid-state systems can be extended and generalized to other systems and (ii) the search for novel, emergent relations between these two subfields of modern quantum physics; with the ever improving control of mesoscopic solid-state systems further intriguing insights can be expected.

More specifically, we conclude with potential directions of research going beyond the work presented in this Thesis: In the first part of this Thesis, we have shown that electron transport can serve as a means well-suited to design certain, desired nuclear states. While our analysis has focused on single and double quantum dot systems, involving two nuclear ensembles at most, quantum dot arrays comprising more than two dots [1, 2] may allow to generate more complex nuclear states, such as multipartite entangled states, even between QDs that do not interact directly. Moreover, further theoretical studies might reveal more insights into the role of electron current as a sensor for the state of the ambient nuclei, with potential connections between current fluctuations and collective nuclear dynamics. Lastly, our work might be the starting point for theoretical and experimental studies of dissipative phase transitions in a transport setting, and therefore complement previous studies on dissipative phase transitions in self-assembled quantum dots, which have revealed multiple quantum effects such as first- and second-order phase transitions as well as regions of

M.J.A. Schütz, *Quantum Dots for Quantum Information Processing: Controlling and Exploiting the Quantum Dot Environment*, Springer Theses,
DOI 10.1007/978-3-319-48559-1_5

bistability, squeezing and altered spin pumping dynamics [3]. Dissipative phase transitions have been studied much less than quantum phase transitions and may require the development of new theoretical and numerical tools to find the complete phase diagram associated with the steady states of the system, in particular when accounting for (dipole-dipole) interactions among the nuclei. From an applied point of view, a clear understanding and identification of the different nuclear phases might give access to advanced dynamical nuclear polarization schemes and thus potentially boost electron spin coherence times, or make the nuclear system available as a quantum interface or quantum memory for QIP purposes. In the second part of this Thesis, we have proposed and analyzed a novel quantum bus based on surface acoustic waves (SAWs) in piezo-active media. Building upon this proposal, first of all many technological challenges are still to be addressed, including, for example, improved cavity and waveguide designs, and the realization of hybrid systems. The latter may not only contain different qubit designs such as quantum dots and superconducting qubits, but also combine the established cavity design for microwave photons with the delay-line effects of itinerant SAWs. In a broader context, besides its promise for quantum information processing, the SAW platform gives access to an on-chip testbed for the emergent field of *quantum acoustics*, where the highly developed toolbox of quantum optics is combined with the unique features of the acoustic, on-chip SAW architecture. For example, the realization of a Jaynes-Cummings system using solid-state qubits and SAWs would make available a central tool of quantum optics and the ensuing methods (such as preparation of non-classical states, phonon counting, ...) to the vibrant field of surface physics. While our work has focused on the interaction of SAW phonons with effective two-level systems, giving rise to an acoustic analog of the celebrated Jaynes-Cummings dynamics, multi-level systems may give access to new acoustic versions of quantum optical effects such as electromagnetically induced transparency (EIT). Moreover, SAWs could be used to create regular *acoustic* lattices for charge carriers inside and outside the piezo-electric material, complementing (for example) the study of many-body physics with ultra-cold atoms in *optical* lattices [4]. Lastly, besides the phonon modes associated with SAWs, other approaches to interconnect (solid-state) qubits over large distances definitely deserve further research. Promising candidates for the coherent transport of electron spins over long distances could be QD arrays [1, 2], quantum Hall edge channels [5–9] or semi-classical SAWs forming *moving* quantum dots [10–13]. Here, the nodes of the quantum network would be interconnected with electrons playing the role of photons in more conventional atomic, molecular, and optical (AMO) based approaches [14], providing yet another example of the powerful analogies between quantum optics and nanoscale solid-state systems.

References

1. M. Busl, Bipolar spin blockade and coherent state superpositions in a triple quantum dot. Nat. Nanotechnol. **8**(4), 261 (2013)
2. F.R. Braakman, P. Barthelemy, C. Reichl, W. Wegscheider, L.M.K. Vandersypen, Long-range coherent coupling in a quantum dot array. Nat. Nanotechnol. **8**, 432 (2013)
3. E.M. Kessler, G. Giedke, A. Imamoglu, S.F. Yelin, M.D. Lukin, J.I. Cirac, Dissipative phase transition in a central spin system. Phys. Rev. A **86**(1), 012116 (2012)
4. I. Bloch, J. Dalibard, W. Zwerger, Many-body physics with ultracold gases. Rev. Mod. Phys. **80**, 885 (2008)
5. J.D. Fletcher, Clock-controlled emission of single-electron wave packets in a solid-state circuit. Phys. Rev. Lett. **111**(21), 216807 (2013)
6. G. Feve, A. Mahe, J.M. Berroir, T. Kontos, B. Placais, D.C. Glattli, A. Cavanna, B. Etienne, Y. Jin, An on-demand coherent single-electron source. Science **316**(5828), 1169 (2007)
7. E. Bocquillon, V. Freulon, J.M. Berroir, P. Degiovanni, B. Placais, A. Cavanna, Y. Jin, G. Feve, Coherence and indistinguishability of single electrons emitted by independent sources. Science **339**(6123), 1054 (2013)
8. P. Roulleau, F. Portier, P. Roche, A. Cavanna, G. Faini, U. Gennser, D. Mailly, Direct measurement of the coherence length of edge states in the integer quantum hall regime. Phys. Rev. Lett. **100**(12), 126802 (2008)
9. A. Mahé, F.D. Parmentier, E. Bocquillon, J.M. Berroir, D.C. Glattli, T. Kontos, B. Plaçais, G. Fève, A. Cavanna, Y. Jin, Current correlations of an on-demand single-electron emitter. Phys. Rev. B **82**(20), 201309 (2010)
10. S. Hermelin, S. Takada, M. Yamamoto, S. Tarucha, A.D. Wieck, L. Saminadayar, C. Bäuerle, T. Meunier, Electrons surfing on a sound wave as a platform for quantum optics with flying electrons. Nature **477**, 435 (2011)
11. R.P.G. McNeil, M. Kataoka, C.J.B. Ford, C.H.W. Barnes, D. Anderson, G.A.C. Jones, I. Farrer, D.A. Ritchie, On-demand single-electron transfer between distant quantum dots. Nature **477**, 439 (2011)
12. H. Sanada, Y. Kunihashi, H. Gotoh, K. Onomitsu, M. Kohda, J. Nitta, P.V. Santos, T. Sogawa, Manipulation of mobile spin coherence using magnetic-field-free electron spin resonance. Nat. Phys. **9**(5), 280 (2013)
13. M. Yamamoto, S. Takada, C. Bäuerle, K. Watanabe, A.D. Wieck, S. Tarucha, Electrical control of a solid-state flying qubit. Nat. Nanotechnol. **7**(4), 247 (2012)
14. H.J. Kimble, The quantum internet. Nature **453**, 1023 (2008)